VENTURE INTO SPACE

NASA SP-4301

VENTURE INTO SPACE
EARLY YEARS OF GODDARD SPACE FLIGHT CENTER

ALFRED ROSENTHAL

NASA Center History Series

Scientific and Technical Information Division
OFFICE OF TECHNOLOGY UTILIZATION 1968
NATIONAL AERONAUTICS AND SPACE ADMINISTRATION
Washington, D.C.

Mrs. Robert H. Goddard and James E. Webb, Administrator of the National Aeronautics and Space Administration, unveiled the sculpture of the late rocket pioneer at the dedication of the Goddard Space Flight Center, March 16, 1961.

Foreword

SINCE ITS INCEPTION the Goddard Space Flight Center has magnificently fulfilled its mission to become a symbol of the aims and dedication of my late husband, Robert H. Goddard. As an active division of the National Aeronautics and Space Administration, the Goddard Center has already made many significant contributions to man's knowledge of the upper atmosphere and outer space—the precise goals of my husband's life. Through its televised tracking activities, the name of Goddard has become commonplace in the American home.

Like most scientists, my husband kept a careful and detailed account of his experiments and theories, with occasional summaries and forecasts. It is therefore most appropriate that the Goddard Space Flight Center pause, at intervals, to sum up its activities, evaluate its successes, and plan for even more effective work in the future.

At the dedication of this Center, I remarked that my husband was an extremely happy man, doing what he most wanted to do, with adequate funds in optimum surroundings; and I expressed the hope that many of those who would work at the Goddard Center might be similarly blessed. I feel that this hope is being realized. I also called attention to the opportunities for the "straight thinker and the hard worker," with the wish that the Center would attract such people, and keep them. This, too, has come to pass. With such personnel I have no doubt that this great living memorial will continue to play a vital role in the coming Space Age.

ESTHER C. GODDARD

Preface

THE GODDARD SPACE FLIGHT CENTER is a partnership of many people—scientists, engineers, project managers, and administrators—whose combined efforts are needed to carry on and bring to fruition scientific and technological expeditions into outer space.

While the Goddard Center came into being with the establishment of the National Aeronautics and Space Administration, its antecedents extend much further. Indeed, the Center inherited much scientific and operational competence from groups and individuals who had already achieved professional distinction. Under the guidance of Dr. Harry J. Goett, the Center's first director, 1959–1965, a most competent team came into being. This team successfully developed and launched a wide variety of scientific spacecraft, sent into orbit this Nation's first weather and synchronous communications satellites, and provided the tracking links for America's first man-in-space missions.

The purpose of this preliminary historical report is to describe the Center's historical origins and traditions, as well as the projects and activities which the men and women of Goddard were privileged to make their contribution to the U.S. space program. In doing so, they not only opened a new path of exploration but were carrying on a tradition of scientific and technical curiosity envisioned two generations earlier by a then unknown New England professor—Dr. Robert H. Goddard.

JOHN F. CLARK
Director, Goddard Space Flight Center

Contents

		PAGE
INTRODUCTION		1

Part I—Origins of the Goddard Space Flight Center

1	FROM ROBERT H. GODDARD TO THE INTERNATIONAL GEOPHYSICAL YEAR	5
2	FROM PROJECT VANGUARD TO THE GODDARD SPACE FLIGHT CENTER	13
3	ESTABLISHMENT OF THE GODDARD SPACE FLIGHT CENTER	27

Part II—Goddard Space Flight Center Goes to Work

4	THE EARLY YEARS	39
5	ORGANIZING FOR SPACE SCIENCE	43
6	TRACKING, DATA ACQUISITION, AND DATA REDUCTION	65
7	GODDARD-MANAGED SATELLITES AND SPACE PROBES	79
8	BOOSTERS AND SOUNDING ROCKETS	121
9	GODDARD LOOKS TO THE FUTURE	131
FOOTNOTES		136

Appendixes

A	INTRODUCTION TO THE UNITED STATES SPACE SCIENCES PROGRAM	147
B	GODDARD SPACE FLIGHT CENTER SATELLITE AND SPACE PROBE PROJECTS	155
C	NASA SOUNDING ROCKET FLIGHTS	181
D	A CHRONOLOGY OF EVENTS RELATED TO THE GODDARD SPACE FLIGHT CENTER	203
E	REPORTS OF PROCUREMENT ACTIONS, 1960–1963	255
F	ORGANIZATION CHARTS	261
G	SCIENTIFIC EXPLORATION OF SPACE AND ITS CHALLENGE TO EDUCATION	267
H	EXHIBITS	279
I	ROBERT H. GODDARD CONTRIBUTIONS AND MEMORABILIA	329
J	SELECTED BIBLIOGRAPHY	333
INDEX		341

LIST OF ILLUSTRATIONS

	PAGE
Part I	
Dedication of the Goddard Space Flight Center, March 16, 1961	v
Dr. Goddard at work on his rocket, October 1935	3
Dr. Goddard and colleagues holding the rocket used in the flight of April 19, 1932	10
Dr. Goddard and colleagues after the successful test of May 19, 1937	10
Dr. Goddard at work, October 1935	11
Project Vanguard staff	18
Technicians mate *Vanguard I* satellite to rocket	21
Vanguard I, March 17, 1958	23
Vanguard III, September 18, 1959	24
Center dedication ceremony, March 16, 1961	32
Dignitaries and guests at Center's dedication	33
Joseph Anthony Atchison at work on bust of Goddard	34
Mrs. Goddard at dedication ceremonies	34
Part II	37
The wooded site selected for the Space Center	40
First computers are moved to the Goddard Center	41
Orbiting Astronomical Observatory project management chart	46
Goddard Space Flight Center project assignments	48
Employee orientation	49
Aerial view of Goddard Space Flight Center, June 1962	50
Location plan of Center, June 1962	52
Location plan of Center, 1963 estimates	53
The site for Building 1, June 1959	54
Building 1 under construction, October 1959	54
Building 1	55
Aerial view of Buildings 1, 2, 3	56
Building 8 under construction, July 1962	57
Staff of the Institute for Space Studies	60
Welding satellite circuitry	61
Space environment simulator under construction at Goddard	62
A Wallops Island antenna that receives video signals from Tiros satellites	66
Fairbanks, Alaska, tracking station	67
The two tracking ships of the Mercury network	69
Mercury network map	70–71
Goldstone antenna	72
Goddard monitoring the MA–6 flight	74

CONTENTS

LIST OF ILLUSTRATIONS (Cont'd)

	PAGE
Communications area	74
During the MA–6 mission	75
Goddard computer room	76
Explorer VI, launched August 7, 1959	83
Pioneer V in final checkout	86
Echo inflation test sequence	90
Explorer X, an interplanetary probe, launched March 25, 1961	96
Explorer XII, an energetic particles satellite, launched August 15, 1961	100
Orbiting Solar Observatory I, launched March 7, 1962	104
Ariel I, launched April 26, 1962	106
Tiros V is mated to launch vehicle	108
Tiros V liftoff, June 19, 1962	108
Telstar I, launched July 10, 1962	109
Tiros VI photograph of Cape Blanc, April 23, 1963	111
Canadian scientists check *Alouette I*, launched September 29, 1962	112
Explorer XIV	113
Relay I, December 13, 1962	115
Syncom I	116
Explorer XVII readied for launch	117
Delta on launch pad; Scout launches *Explorer XVI*	124
Sounding rocket at Fort Churchill	126
Launch of Japanese electron temperature experiment	126
Nike-Cajun; Javelin; Nike-Apache	128
Sounding rocket used in first joint flight by the United States and Japan	128
Splash crater on the moon	132
Goddard's top management staff, 1962	133
Goddard's top management staff, 1965	134
Appendixes	145
Explorer I, January 31, 1958	207
Building 1 under construction	210
Building 2 under construction	210
Vanguard vehicle in a gantry	212
Architect's drawing of Buildings 7 and 10	213
Antenna at Goldstone, California	214
Conference on Origins of the Solar System, New York City	216
Italy and the Mediterranean as seen from a Tiros satellite	217
Rosman tracking facility	220

LIST OF ILLUSTRATIONS (Cont'd)

	PAGE
The digital solar aspect sensor	221
Dr. Kunio Hirao and Toshio Muraoka at the launch of the first U.S.-Japanese sounding rocket	223
The Nile delta as seen from a Tiros satellite	225
A picture transmitted by *Telstar I*	225
Aerobee sounding rocket fired from Wallops Island, Va.	229
Comsat picture of unveiling "Mona Lisa" at National Gallery of Art, Washington, D.C.	230
Fifth Anniversary of Space Tracking ceremonies	232
Delta Day ceremony, March 1, 1963	235
Pictures of Italian Premier Fanfani's Chicago trip were transmitted to Europe via *Relay I*	237
Relay comsat transmits ceremony awarding U.S. citizenship to Sir Winston Churchill	239
Tiros VII photograph of U.S. eastern seaboard, June 23, 1963	242
Specially constructed instrument that photographed solar eclipse, July 20, 1963	243
Nigerian Governor General Nnamdi Azikiwe, via *Syncom II* satellite	244
The Rosman, N.C., tracking facility	248
Artist's conception of Relay mission received in Japan	250
Space Communications Laboratory, Ibaraki, Japan	252
Antenna at the Space Communications Laboratory, Ibaraki, Japan	252
Tiros VIII, December 26, 1963	253
Procurement actions, January 1, 1960, through June 30, 1960	255
Procurement actions, Fiscal Year 1961	256
Procurement actions, Fiscal Year 1962	257
Procurement actions, Fiscal Year 1964	259
Organization chart, July 1959	262
Organization chart, March 1960	263
Organization chart, January 1961	264
Organization chart, November 1962	265
Goddard Center missions	268
Divisions of space in the earth-sun region	269
Scope of Tiros photographs	270
Tiros coverage	270–271
Temperature distribution derived from *Tiros VII*	271
Infrared data from *Nimbus I*	272
The upper atmosphere	273

LIST OF ILLUSTRATIONS (Cont'd)

	PAGE
Earth-sun relationships	274
The earth and "empty" space	274
Orbit of *Orbiting Solar Observatory I*	275
Deflection of cosmic and solar particles by earth's magnetic field	276
Effect of solar flare on the earth's magnetic field	276

Introduction

THIS HISTORICAL REPORT represents a preliminary record of the efforts of this NASA Center, from its antecedents through 1963. Any cutoff date for such a report must be necessarily arbitrary; 1963 has been selected as terminal date because that year saw the culmination of many of the early efforts: the organization achieved the form its planners had envisioned; many of the physical facilities were completed; and, perhaps most important, scientific findings produced by "first generation" satellites began to be returned to curious scientists. As a consequence of the new scientific knowledge and technological advances, the years beyond 1963 would feature more advanced missions, utilizing "second generation" spacecraft with more sophisticated instrumentation. Weather and communications satellites developed during the early years had by 1963 demonstrated such utility as to make operational systems a reality. The Goddard-operated manned space flight tracking network contributed to the successful completion of Project Mercury, the United States' first man-in-space program. In brief, for Goddard Space Flight Center the year 1963 could be considered the end of Act One of the Space Age, and the curtain raiser of Act Two.

This volume is a mosaic of what was considered to be reliable information, gleaned from many sources. Historical documents, a GSFC chronology, and a bibliography are included as appendixes. In its preparation, valuable assistance and advice was received from numerous officials at the National Aeronautics and Space Administration Headquarters—particularly Dr. Eugene M. Emme, NASA Historian; his deputy, Dr. Frank W. Anderson, Jr.; and Mr. Thomas E. Jenkins, NASA Management Reports Director, formerly of the Goddard Space Flight Center—and from the entire Center staff. Without the considerate support from virtually every element of the agency, the preparation of this document would have been an almost impossible task. Comments, suggestions, or corrections of fact are sincerely invited so as to assist our historical efforts.

ALFRED ROSENTHAL
Historian, Goddard Space Flight Center

PART ONE

ORIGINS OF THE GODDARD SPACE FLIGHT CENTER

*No single thing abides;
    but all things flow.
Fragment to fragment clings
    —the things thus grow
Until we know them
    and name them.*
        —HALLOCK

From Robert H. Goddard to the International Geophysical Year

1

THE HISTORICAL TRADITIONS of the Goddard Space Flight Center have antecedents in a period long before America's awakening to the Space Age. Named in honor of Dr. Robert H. Goddard, the Center continues in the scientific tradition of this New England scientist, who has been recognized not only as the "Father of American Rocketry" but also as one of the pioneers in the theory of space exploration.[1]

Dr. Goddard was one of those rare combinations occasionally appearing in the history of science and technology; he was a theoretical scientist and a practical and exacting engineer, but he was also a dreamer who was considerably ahead of his own time. His particular dream was the scientific conquest of the upper atmosphere and ultimately of the void of space through the use of rocket propulsion. To the fulfillment of this dream Dr. Goddard devoted his talents, his energies, and his life. He had the drive and single-mindedness found among those who today are probing the innermost secrets of the space environment.

The Life of Dr. Goddard

Robert Hutchings Goddard was born in Worcester, Massachusetts, on October 5, 1882. Childhood illnesses prevented the young boy from expressing his energy in the usual boyish activities. As a consequence, he developed his imagination and read voraciously. He was greatly influenced by H. G. Wells' *War of the Worlds*.[2] Jules Verne also stimulated his imagination. Dr. Goddard is known to have read Verne's *From the Earth to the Moon* annually and ultimately to have rewritten it to include a rocket launch instead of a cannon-powered flight.[3]

A germ of his later work might be seen in his effort in the spring of 1898 to construct a hydrogen-filled balloon made of thin aluminum. Although it was too heavy to fly, Goddard was not discouraged; he turned his attention to such other things as "how birds fly" (which also interested two brothers named Wright, in Ohio) and the marvels of electricity. This was the situation when, on the afternoon of October 19, 1899, he found himself high in a cherry tree, assigned the duty of clipping its dead limbs. It was, he said, "one of those quiet, colorful afternoons of sheer beauty which we have in October in New England . . . as I looked toward the fields to the east, I imagined how wonderful it would be to make some device which had even the *possibility* of ascending to Mars, and how it would look on a small scale if sent up from the meadow at my feet . . . I was a different boy when I descended the ladder. Life now had a purpose for me."[4]

The young Robert Goddard began to construct models of his own design and devoted himself, in high school, to the study of physics and mathematics. By the time he graduated, he had, he said, "a set of models which would not work and a number of suggestions which, from the physics I had learned, I knew were erroneous." He gathered up all his models and his carefully catalogued notes and burned them.

"But the dream would not go down," wrote Dr. Goddard in later years, "and inside of two months I once again caught myself making notes of further suggestions, for even though I reasoned with myself that the thing was impossible, there was something inside me which simply would not stop working." As early as 1909, Goddard conceived the multiple-stage rocket, the general theory of hydrogen-and-oxygen rocket propulsion, and the use of a plane-like structure for rocket guidance.[5]

By this time, Goddard had graduated from Worcester Polytechnic Institute with a B.S., holding high honors in physics and mathematics. All through college the young scientist had been absorbed by his obsession with propulsion. Near the end of his senior year, he filled the basement of the Worcester Polytechnic Institute physics building with smoke, the result of a static test of a small rocket.[6]

The years between graduation from Worcester Polytechnic Institute and 1919 were full and arduous ones for Goddard. He stayed on at Worcester as an instructor in physics while doing graduate work at Clark University, where he received his A.M. in 1910 and his Ph.D. in physics in 1911. He then spent a year at Clark as an honorary fellow in physics, where he worked on various rocket methods. In 1912 he went to Princeton as a research fellow and worked on electrical theory (the subject of his Ph.D. dissertation) during the day and on rocket propulsion theory during the evenings. Through 1913–1914 illness prevented him from teaching but did not hinder his speculations on rocketry. In fact, during his convalescence Dr. Goddard laid the foundations for two patents, received in July 1914, which developed his idea of a multistage rocket and liquid propellants.[7]

In the autumn of 1914, Dr. Goddard returned to Clark University as an instructor and began his basic work on rocketry, which led to the now-famous 1919 Smithsonian publication. It was during this period that he proved his theory, by static laboratory test, that a rocket would perform in a vacuum and was therefore capable of operating in space. By the middle of 1916, he had reached the limit of what he could accomplish on his own funds. He wrote up his experiments, entitled the manuscript "A Method of Reaching Extreme Altitudes," and sent it to three organizations that he thought might be interested enough to aid him financially. The only encouraging reply he received was from the Smithsonian Institution, which was looking for a very-high-altitude device to extend meteorological research and asked for further details. This was in December 1916. After receiving the Smithsonian's commendation for his work, Dr. Goddard requested $5,000 to continue his experiments. The next letter from Washington granted the $5,000 and enclosed an advance of $1,000.[8]

Thus began, in January 1917, the years of experimentation which Dr. Goddard continued unceasingly for the remainder of his life. When the United States entered World War I in 1917, he volunteered to direct his rocket research toward ends which might prove useful to the military. This work led to the development of a solid-fuel prototype that became the World War II "bazooka"; however, the war ended five days after he demonstrated it. During World War II, Dr. Goddard's invention was to be "dusted off" by his colleague, Dr. Clarence N. Hickman, and was perfected to provide the American soldier with the first effective hand weapon against the tank.[9]

After returning to Clark in 1919, Dr. Goddard persuaded the Smithsonian to publish his revised "Method of Reaching Extreme Altitudes"—a rather dry and factual report on his experiments during the preceding several years which had been designed to show the Smithsonian where its money was going. The Smithsonian agreed, provided the cost of publication came from the $5,000 grant to Goddard. The study was released on January 11, 1920, as Smithsonian Miscellaneous Publication Number 2540. In this publication, Dr. Goddard mildly referred to the space potential of rocket thrust and suggested the possibility that someday a rocket such as the one he was designing might be used to hit the moon. The newspapers picked up the story, some portrayed him as a crackpot, others as an amateur who did not know that reactive thrust would not work in a vacuum. This adverse publicity had a deep and lasting effect on Dr. Goddard. From that day forward, he rarely spoke of anything which might pertain to space flight, and he avoided publicity for himself and his work. In private, however, he continued to speculate on the possibilities of the rocket in space, on possible manned and unmanned missions, on methods of navigating in space, and on the potentialities of a solar-powered engine. All this he kept locked away in a cabinet in a folder marked "Formulae for Silvering Mirrors." [10]

In the early 1920s Dr. Goddard began his pioneering experiments with liquid-fuel propulsion. He had considered the idea of a hydrogen and oxygen fuel supply as early as 1909. After several experiments he discovered that liquid oxygen and gasoline made the most practical fuel, and the first static test of liquid-fuel propulsion was made on November 1, 1923. After overcoming numerous problems with the apparatus, the fuel pumps in particular, he was ready to try again. On March 16, 1926, Professor Goddard, his wife Esther, and two assistants drove to "Aunt Effie" Ward's farm near Auburn, Massachusetts, and prepared his rocket for launch. Ignition was accomplished by a blowtorch attached to a pole. The rocket rose from the ground, traveled a distance of 184 feet, reached an average speed of 60 miles per hour, and stayed in the air for 2½ seconds. It was the first liquid-fuel rocket flight in the world, an event comparable to Kitty Hawk in its significance. Not a word reached a newspaper.[11]

Dr. Goddard continued his experiments at Auburn. In 1928, after another test, he reported to the Smithsonian that he had demonstrated the rocket's potential for study of the ultraviolet; upper air composition, electrical conditions, and movement; and mapmaking by several simultaneous ground observations of high-altitude light flashes. Then on July 17, 1929, he launched a scientific payload of a barometer, a thermometer, and a camera. On that date his rocket was mistaken for an airplane crashing in flames, which caused the State fire marshal to forbid Goddard to conduct any more launches in Massachusetts. The Smithsonian succeeded in persuading the Army to allow him to launch his rockets on Federal property at Camp Devens, Massachusetts; but as a precaution against fires he could experiment only after a rain or a snowfall. The "moon-rocket man" publicity from this affair brought Goddard an unexpected windfall. Up to this time he had been relying on the steady but relatively small grants from the Smithsonian for the necessary financial assistance. Charles A. Lindbergh, who was at the height of his popularity, read the unfavorable press accounts of Goddard's work. Lindbergh was interested in the use of rockets to provide emergency thrust for airplanes. He visited Dr. Goddard and was impressed with his work. Lindbergh took Goddard's cause to Daniel and Harry Guggenheim. Subsequently the Daniel and Florence Guggenheim Foundation began to supply him with money for his experiments. Between 1929 and 1941, Dr. Goddard received over $150,000 from this source.[12]

The first Guggenheim grant enabled Dr. Goddard to leave Massachusetts for a more suitable testing ground. He moved his work to Roswell, New Mexico, which was to be his headquarters for the remaining fifteen years of his life. All through the 1930s Dr. Goddard and his small staff worked at improving his rockets and their components. While at Roswell, he devised and patented a gyroscopic control for rockets and an ingenious system for cooling the combustion chamber, called "curtain cooling," in which the fuel of the rocket acted as the cooling agent. "There was never so much invention with so little manpower," remarked one of Dr. Goddard's mechanics.[13]

When war broke out in Europe in 1939, Dr. Goddard visited the U.S. War Department and tried to interest the military in his work, but nothing tangible came of it. After the United States became involved in 1941, the Navy and the Army Air Corps asked him to work for them, not to develop his rocket as an offensive or defensive missile but merely to develop a jet-assisted takeoff (JATO) device for helping aircraft take off from short runways or aircraft carriers. Dr. Goddard's repeated efforts to convince the American military of the potential of the rocket were to no avail. So it happened that JATO and the revived 1918 "bazooka" were the major contributions which this genius was allowed to make to the American war effort.[14]

The Germans, however, had not neglected their rocket technicians as had the Americans. By September 1944, German V-2 ballistic rockets began to fall on Britain. The Allies were startled at the great lead of German rocket technology. When details of the V-2 reached Annapolis, where Dr. Goddard was working in the Navy's research laboratories, he noted the similarity between the German missile and his own liquid-fuel rocket. Although the 5½-ton V-2 was much larger than anything that Dr. Goddard (or anyone in the U.S.) had ever constructed, the two rockets were almost identical in basic design. Out of this similarity arose a controversy over the extent to which the Germans may have worked from Goddard's patent designs.[15]

Finally illness took its toll. In Baltimore, in 1945, Dr. Goddard was operated on for throat cancer. His lungs, already weakened from an earlier attack of tuberculosis, gave out and the American rocket pioneer died on August 10, 1945. His passing went practically unnoticed except among his faithful small group of family and friends.[16]

The Goddard Legacy

Robert Hutchings Goddard's rocket research was perhaps as fundamental to the opening of the Space Age as was the Wright Brothers' research to the Air Age. Yet his work attracted little serious attention during his lifetime and he did not encourage it. When the United States began to prepare for the conquest of space in the 1950s, American rocket scientists began to recognize the enormity of the early debt which their science owed to the New England professor. They discovered that it was virtually impossible to construct a rocket or launch a satellite without acknowledging the work of Dr. Goddard. This great legacy was covered by more than 200 patents, many of which were issued after his death.

Belated honors have begun to pour upon the name of Robert Goddard in recent years. On September 16, 1959, the Congress of the United States authorized the issuance of a gold medal in his honor. The Smithsonian

Dr. Goddard and colleagues holding the rocket used in the successful experimental flight of April 19, 1932. They are, from left to right, L. Mansur, A. Kisk, C. Mansur, Dr. R. H. Goddard, and N. L. Jungquist.

Dr. Robert H. Goddard and colleagues at Roswell, New Mexico, after the successful test of May 19, 1937. Dr. Goddard is holding the cap and the pilot parachute.

Dr. Goddard at work on his rocket in his shop at Roswell, New Mexico, October 1935.

Institution, long familiar with his work, awarded him its coveted Langley Medal, in honor of his discoveries in rocketry, on June 28, 1960. Subsequently Clark University, Worcester, Massachusetts, which was made the depository of his papers, established the Robert H. Goddard Memorial Library. In 1964 a commemorative airmail postage stamp was issued in his honor.

On May 1, 1959, the National Aeronautics and Space Administration named its new Space Flight Center at Greenbelt, Maryland, the Goddard Space Flight Center.[17] It is hardly a coincidence or accident that his name was chosen to inspire the work being done by this team of scientists and engineers engaged in the scientific exploration of space. It is perhaps one of the most fitting of the many belated honors which have come to the name of Goddard, because it established a Center which is realizing the dream of space exploration the young Goddard had conceived at the turn of the century.

The essence of his philosophy, as he expressed it in his high school oration in 1904, serves well as the motto of the Goddard Space Flight Center:

> It is difficult to say what is impossible, for the dream of yesterday
> is the hope of today and the reality of tomorrow.

From Project Vanguard to the Goddard Space Flight Center
2

AFTER WORLD WAR II, interest in rocket technology gradually developed in the United States.[18] This was generated mainly by the impact of the German V-2 rocket upon American military and scientific circles during and after the war. At the end of the war, a number of the German rocket experts and almost 100 V-2 rockets were brought to the United States. Few realized their full potential. Scientists saw the rocket as a new tool of high-altitude research, while military considerations aroused the interest of the Army, the Air Force, and the Navy.

Early Rocket Development

In January 1946, the U.S. Army announced that a firing program for the V-2 rockets would begin later that year at White Sands, New Mexico. Government agencies and several universities were invited to consider using the V-2s for high-altitude (sounding rocket) research and experimentation.[19] The first V-2 to be used in the sounding rocket research program was launched in June 1946, and in the next 6 years over 60 were launched. As the result of the V-2 program in the United States, valuable knowledge was gained in two areas. First, the rockets enabled soundings to be made to an altitude of about 100 miles, and measurements of high-energy particle radiation, found at high altitudes but absorbed at lower levels, were made. Second, a great deal was learned about rocket technology and men were trained so that similar-size American rockets could be built as the supply of V-2s became depleted.[20]

Several organizations, partially staffed with personnel who had engaged in United States V-2 research, began to develop rockets. The first of these rockets was the Aerobee, designed by the Applied Physics Laboratory (APL) of The Johns Hopkins University. It was capable of carrying a small payload to an altitude of about 80 miles.[21] In 1947 the Naval Research Laboratory (NRL), in Washington, D.C., proposed the construction of a rocket which would replace the V-2 in the American sounding rocket program. This rocket, at first called Neptune but later identified as Viking, was smaller than the V-2, but more powerful.[22] It could lift a larger payload to a height of about 150 miles with a high order of stability. In the 6 years between 1949 and 1955, 12 of these rockets were launched, carrying payloads as high as 158 miles. None attained the hoped-for altitude, but new altitude records were established and valuable scientific information was gained.[23] Other groups benefiting from the experience of the American V-2 program were the Army, which began work on its Redstone missile after 1950, and the U.S. Air Force, which began work on the Atlas ICBM in 1954.[24] In addition, the Jet Propulsion Laboratory (JPL) of the California Institute of Technology developed the WAC Corporal research rocket.[25]

Early U.S. Satellite Proposals

Although the main emphasis in these years was on the development of an improved rocket-powered vehicle, a germinal program was initiated in earth satellites. The U.S. Navy's Bureau of Aeronautics was one of the first Government organizations to initiate a satellite study program. In October 1945, a committee of the Bureau of Aeronautics recommended that an earth satellite development program be undertaken for scientific purposes. The Aerojet Corporation and the California Institute of Technology were given the task of determining whether such a project was technically feasible using the single-stage rocket vehicle which the Navy had proposed.[26]

In March 1946, the Navy took its proposal to the Army Air Force, suggesting a joint satellite program to aid funding. Although the first effort at such a program appeared promising, the Air Force informed the Navy that it could not cooperate. In the meantime, the Air Force had established Project RAND (later to become the RAND Corporation) to begin a satellite feasibility study. RAND drew up a proposal for the Air Force entitled "Preliminary Design of an Experimental World-Circling Spaceship." The RAND proposal ruled out the satellite as a military weapon because no rocket could be constructed which could lift the heavy A-bomb into orbit and no explosive force short of an atomic one would inflict enough damage to warrant the expense of putting it into orbit. The problem was not one of capability (it was assumed that the U.S. could launch a 500-pound satellite by 1951) but rather one of devising a useful function for the satellite to perform once it was in orbit. Because a satellite was not a potential weapon, there were no funds available for its development.[27]

The RAND-Air Force proposal, like its Navy counterpart, urged the early adoption of a satellite program for scientific purposes. They argued for its capabilities in the fields of meteorology, communications, and astronomy. In October 1946, RAND issued an additional study entitled "The Time Factor in The Satellite Program," in which they emphasized the psychological and political factors which could result from the first satellite launch. Even this dramatic prognosis was insufficient to overcome the factor that the satellite was not a potential weapon.

When the Department of Defense (DOD) was created in 1947, none of the three military services was authorized to continue development of a rocket with satellite capability. The Air Force discontinued its satellite studies in mid-1947, but did resume them in 1949. By that time, the Navy had discontinued its studies because of lack of funds. Early in 1948, DOD reviewed the existing satellite proposals but again concluded that "neither the Navy nor the USAF has as yet established either a military or a scientific utility commensurate with the presently expected cost of a satellite vehicle." The work was so neglected at DOD that, in November 1954, the Secretary of Defense remarked publicly that he knew of no American satellite program.[28]

The RAND Corporation proved to be prophetic in its prediction of the great psychological-propaganda impact of the first satellite launching. They had emphasized, as early as 1946, that a satellite would be an "instrument of political strategy." When the Soviet Union launched *Sputnik I* in October 1957, it had exactly the impact that RAND had said it would have —only in reverse. It was the United States which did the soul searching and suffered a drop in world opinion. It was only after this 1957 propaganda defeat that the U.S. Government fully understood the wider significance of these early satellite studies.[29] Yet, at the time of the first Sputniks, there actually was a satellite program in the United States.

The International Geophysical Year

While missile development and satellite proposals were progressing within the military services, an important boost was given to the scientific use of rocket technology. By 1951, the American Rocket Society (ARS) had grown to a point where its voice could be heard. In the winter of that year, Commander Robert Truax, who had been championing rocket propulsion in the Navy, strongly and bluntly told the members at their annual meeting that they were too complacent in their attitude toward space flight, that time was catching up with them, and that definite action was called for. As a consequence of this meeting, the American Rocket Society formed an Ad Hoc Committee on Space Flight.[30]

In 1954, this ARS committee proposed that the Government sponsor the development of a small scientific satellite and use available military hardware to launch it. This proposal was informally submitted to Dr. Alan T. Waterman of the National Science Foundation. The satellite idea was alive in many forms in many scientific circles. The International Scientific Committee of the National Academy of Sciences, in making plans for the International Geophysical Year (IGY),[31] recommended that the launch of small scientific satellites be considered by individual groups preparing their own programs for the IGY. The United States National Committee for the IGY, formed by the National Academy of Sciences, also studied the possibility of having an earth satellite launched as part of the U.S. contribution to IGY. Interest in satellite projects had also been revived among the military; the Army and the Navy had proposed in early 1955 a joint program (Project Orbiter) to launch an elementary, uninstrumented satellite in 2 or 3 years.[32]

It was in 1954 that the International Geophysical Year (1957–1958) was proposed. Its American spokesmen were among those scientists who had participated in the V–2 sounding rocket program. That summer, the International Scientific Radio Union and the International Union of Geodesy and Geophysics adopted resolutions calling for the launch of an artificial earth satellite during the forthcoming IGY. Both the United States and the Soviet Union picked up this proposal. On July 29, 1955, the White House announced that the United States would launch "small, unmanned, earth-circling satellites as a part of the U.S. participation in IGY." The next day the Soviet Union made a similar announcement.[33]

The Vanguard Project

The White House announcement of the proposed satellite launchings was the product of coordinated efforts within the National Academy of Sciences (NAS), the National Science Foundation (NSF), and the Department of Defense. The announcement stated that NAS would determine the experiments to be orbited, NSF would supply the necessary funds, and DOD would launch the satellite. A Committee on Special Capabilities was established in DOD to determine the means for launching the U.S. satellite. This Committee, chaired by Dr. Homer J. Stewart, had three proposals from which to select.[34] One proposal was based on the as yet incomplete Atlas missile, one on the Army's Redstone (Project Orbiter), and one on the Naval Research Laboratory's Viking. The Navy proposal was based on sounding rocket research experience of the Naval Research Laboratory (NRL) and the Martin Company, builders of the Viking. In essence it would use the Viking as a first stage, the Aerobee as a second stage, and an as yet undetermined rocket as a third stage.[35] After lengthy deliberation, a majority of the Stewart Committee recommended the NRL satellite proposal in August 1955. The recommendation was accepted and endorsed by the Policy Committee of DOD. The U.S. IGY satellite program under Navy management and DOD monitoring was established and designated "Project Vanguard."

> **PROJECT VANGUARD**
>
> *Objectives:*
>
> - To develop and procure a satellite-launching vehicle.
> - To place at least one satellite in orbit during IGY.
> - To accomplish one scientific experiment with the satellite.
> - To track the satellite's flight to demonstrate that it had actually attained orbit.
>
> *Criteria:*
>
> - The first stage was to be based on the Viking rocket, which had been developed by the Navy to replace the dwindling supply of captured V-2s.
> - The second stage was to be an improved Aerobee rocket.
> - The third stage was to be a solid-fuel rocket weighing about 500 pounds, necessitating a real advance in the existing solid-fuel rocket technology.
> - On top of this vehicle would be placed a nose cone weighing 20 pounds, including the IGY scientific experiment to be orbited.

On September 9, 1955, Project Vanguard was officially authorized when the Department of Defense notified the Secretary of the Navy to proceed with the project. Project Vanguard was to be accomplished without a specific appropriation from Congress. All funds came from the emergency fund of the Secretary of Defense. Only after *Sputnik I* had been launched and the Vanguard project had reached its final stages of completion did Congress authorize the Secretary of Defense to make available additional funds for Vanguard by reprograming the Defense budget. Two years, 6 months, and 8 days after the Department of Defense authorized the project the first successful Vanguard satellite was launched (March 17, 1958).[36]

At NRL, a special task force, headed by Dr. John P. Hagen, was assembled to handle the Vanguard program.[37] In a letter to the Navy Department, this group clarified its definition of what Project Vanguard really would be: "a complete system for space exploration." They had a difficult task before them. In addition to the development of a new satellite launching rocket, they had to place a reliable scientific experiment into earth orbit

Project Vanguard staff members meet with Dr. John P. Hagen, Director of Project Vanguard, at the U.S. Naval Research Laboratory, Washington, D.C. Left to right: Dr. J. W. Siry, Head of the Theory and Analysis Staff; D. G. Mazur, Manager of the Vanguard Operations Group at Cape Canaveral, Fla.; J. M. Bridger, Head of the Vehicle Branch; Cdr. W. E. Berg, Navy Program Office; Dr. Hagen; Dr. J. P. Walsh, Deputy Project Director; M. W. Rosen, Technical Director; J. T. Mengel, Head of the Tracking and Guidance Branch; and Dr. H. E. Newell, Jr., Science Program Coordinator. L. Winkler, Engineering Consultant, was not present when this picture was taken.

and not only prove that it was in orbit but gather data from the satellite via telemetry. This had never been done before.

Dr. Hagen's small NRL team had mountains of problems to overcome. One difficulty might be used for illustration. At the Martin Company, which NRL had selected to build the Vanguard missile, the original NRL-Viking engineering team had been broken up. Unknown to the Navy, the Martin Company had received a prime contract from the Air Force to develop the second-generation ICBM, the Titan. Some of the leading Viking engineers already had been put on this project. This was, Dr. Hagen noted, "a shock, as we had cleared our intentions with the DOD before letting our letter of intent." The Navy stuck with the Martin Company, but "things could have been much easier for the Vanguard group if the original Viking team of Martin had remained intact." [38]

While the NRL was busy preparing the launch vehicle, the National Academy of Sciences established a technical panel, under its IGY committee, to select the experiments to be launched. The Vanguard group, through the liaison of Dr. Homer E. Newell, insisted on only one requirement for each experiment: that it must have a very high reliability of performance and must be tested thoroughly to prove this reliability. [39]

The National Academy of Sciences requested the Vanguard group to make the satellite spherical in shape; in fact, a 30-inch sphere was requested. This caused some concern, as it originally had been planned that Vanguard would orbit merely a simple nose cone. The Vanguard group agreed that they could change their design to launch a 20-inch sphere, but this would require a complete redesign of the second stage, which would have to have a large diameter. Therefore, in the fall of 1955, a redesign of the Vanguard vehicle was undertaken to fulfill the new requirements.[40] Since Vanguard was scientific in purpose, there was no alternative.

By March 1956, the redesign of the Vanguard rocket was completed and a full schedule of six test vehicles and seven satellite-launching vehicles was prepared. As prime contractor for the launch vehicle, the Martin Company was constructing the first stage; Aerojet Corporation had received a subcontract for the second stage; and the Grand Central Rocket Company and the Allegany Ballistics Laboratory were each building separate third stages based on different designs. Two major problems remained to be solved: choosing a launch site, and constructing the necessary satellite tracking system.[41]

With the rocket thrust then attainable, it was virtually mandatory that the satellite be launched eastward in order to gain, rather than lose, the earth's rotational velocity of some 1,300 feet per second (about 1,000 mph). An eastward launching could be made only from the east coast, lest the spent rocket stages fall on inhabited areas. This ruled out the otherwise most natural launch site, White Sands, New Mexico, where the Viking launch facilities were available. The best available site was Cape Canaveral, Florida, which then was being expanded to accommodate the testing of large, liquid-fuel ballistic missiles. The only other serious "competitor" was Roosevelt Roads, in Puerto Rico. Cape Canaveral was selected for many reasons, the main one being financial. However, a number of problems arose from this selection. When the Navy requested that the Army Ballistic Missile Agency (ABMA) share its launch facilities at the Cape with the Vanguard operation, the Army refused on the grounds that it would interfere with the Redstone program and thus be detrimental to the Nation's ballistic missile program. It was then decided that Vanguard would construct its own checkout hangar, blockhouse, and launch pad at Cape Canaveral. This was an 18-month program but was still within the limited time remaining to complete a satellite launching during IGY. Additional down-range facilities also had to be constructed. Unlike the ballistic missile of that day, Vanguard was multistage, requiring command and control points as far away as a thousand miles from the launch site to inject a satellite into orbit. Even a gantry (service tower) was unavailable; the Vanguard group had to disassemble the Viking gantry at White Sands, transport it to the Cape, and reassemble it there. By the time of the first launch, the Vanguard group had constructed the Nation's first complete satellite launch facility, almost from the ground up.[42]

Tracking facilities proved to be a problem. Two types of tracking were necessary—electronic and optical. The electronic tracking system had to have a series of ground stations equipped with radio transmitting and receiving equipment, timing facilities, and data-acquisition (telemetry) equipment. These facilities had to be constructed before the launch could take place and had to be located in various parts of the world to provide the degree of orbital coverage considered necessary. A contract was awarded to the Bendix Corporation to construct this system, which later became known as Minitrack (for *Minimum weight tracking,* because it required only the simplest and lightest transmitter in the satellite). A system of optical tracking stations was established and managed by the Smithsonian Astrophysical Observatory. A communications network centered at the Naval Research Laboratory, Washington, D.C., tied the 13 Minitrack and 12 SAO stations together.[43]

Twelve Vikings had been built and fired in the normal course of NRL's upper atmosphere research. Viking 13 was at White Sands awaiting preparation for launch when the Navy was given the Vanguard mission. Viking 14 was modified by adding an ejectable sphere and a solid-fuel second stage to test its ignition and separation at altitude. This vehicle was designated as Test Vehicle 1 (TV-1). It was then decided to use Viking 13 to check out the new launch facilities at Cape Canaveral. To make this the first of the Vanguard series and to avoid having to renumber all the vehicle designations, Viking 13 was placed ahead of the rest of the planned series and designated Test Vehicle 0 (TV-0).[44]

On December 8, 1956, TV-0 was successfully launched at Cape Canaveral. It reached an altitude of 126 miles and dropped into the ocean 183 miles away.[45] TV-1 was launched on May 1, 1957; this was a redesigned rocket and the only one of its kind flown. The first stage was the Viking 14; the second stage (which was actually the Vanguard third stage) ignited, separated successfully, and flew 450 miles farther, carrying a heavy instrumented nose cone. A milestone had been reached; a solid-fuel upper stage had been ignited in flight, and the feasibility of the Vanguard rocket had been proven.[46]

In July 1957, an important change was made in the Vanguard program; NRL directed that the Vanguard team replace the instrument test packages previously flown on its test vehicles with small (6-inch) satellite spheres. The 6-inch spheres had orginally been developed to give an extra margin of reliability over the heavier 20-inch sphere when used on the launch vehicles. The decision to use the 6-inch sphere on the test vehicles was an indication that emphasis was being shifted from the testing of the vehicles to the earlier launching of satellites. This alteration was not made on TV-2, the first of the true Vanguard vehicles. This rocket was already on the launch stand going through prelaunch checkout when, on October 4, it was announced that the Soviet Union had launched an earth satellite at 7:30 in the evening from the Tyuratam Range in Kazakhstan, U.S.S.R.[47]

After Sputnik

The launch of *Sputnik I* caused a great deal of turmoil in the United States. Great pressure was exerted on the Vanguard team to get an American satellite into orbit.[48] The launch of TV–2 on October 23, 1957, seemed anticlimactic, since the launch was not designed to place a satellite in orbit but simply to test the vehicle. The vehicle consisted of the first of the new Viking first stages and dummy second and third stages, with some of the control system of the last two stages operational. *Sputnik II* was launched on November 3, 1957. On December 6, an attempt was made to launch TV–3; this was the first test of the complete live three-stage vehicle and control system and was the first Vanguard rocket with potential orbital capabilities. The first-stage engine lost thrust after two seconds, and the

Technicians mate the *Vanguard I* satellite to its slender booster rocket in preparation for its successful flight on March 17, 1958.

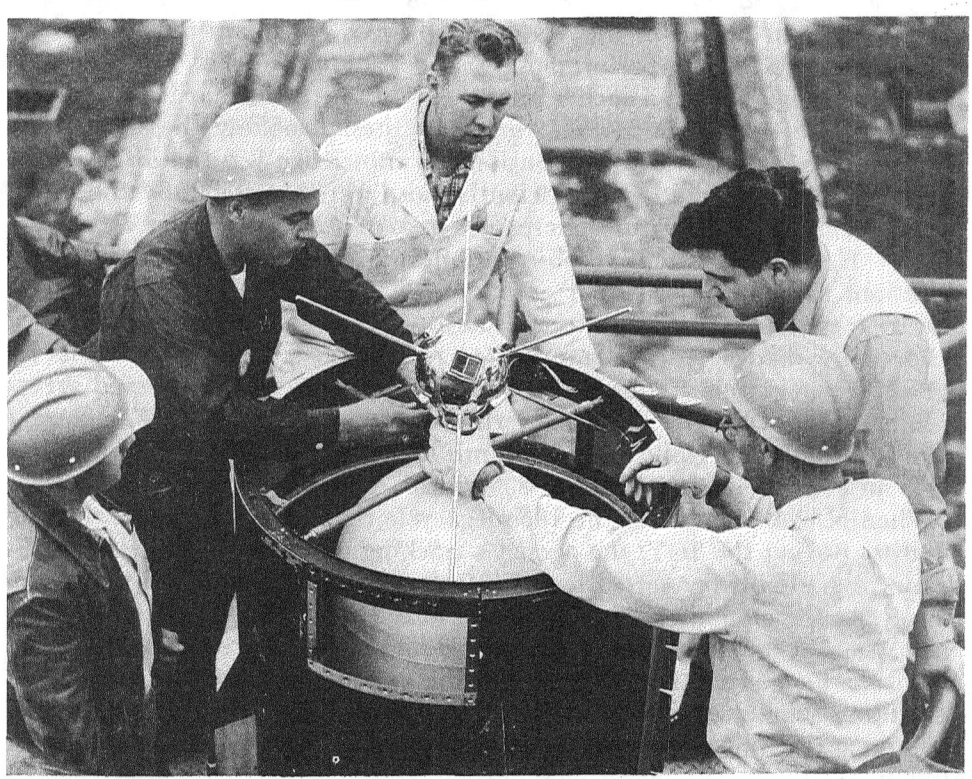

vehicle burned up on the launching pad. Because of the Russian first with *Sputniks I* and *II* and because a White House statement that the next Vanguard launch would place an American satellite in orbit was wrongly construed to apply to this test launch, news of the Vanguard failure reverberated around the world.[49]

The unfortunate turn of events in the early Vanguard test launches, which reflected the troubles inevitable in development of a new three-stage booster, plus the "space race" pressure generated by the Sputniks, led to a relaxation of the ban on use of military missiles for the IGY satellite project. The Army Ballistics Missile Agency was authorized in November to attempt a satellite launching with its proven Redstone missile. As a result, the Army and the Jet Propulsion Laboratory were able to launch the first U.S. satellite, *Explorer I*, on January 31, 1958. The *Explorer I* and its IGY experiment of James A. Van Allen boosted the prestige of the U.S. space program. But this event, as well as the breakup of the TV–3 backup Vanguard on February 5, brought more unkind comments in public about the Vanguard satellite program.[50]

On March 17, 1958, TV–4 successfully launched into orbit *Vanguard I*, a 6-inch sphere weighing 4 pounds. Although this was far from the final objective of a 20-pound instrumented satellite, it did justify the confidence which had been placed in the Vanguard project. Primary purpose of the launch was a test of the performance of the Vanguard rocket, but the small sphere it carried achieved such a remarkably stable orbit that it proved one of the Nation's most important early satellites. Probably the most noteworthy of its many major contributions to knowledge was the discovery of the "pear shape" of the Earth. Scientists also were able to study and measure the density of the atmosphere in a region some 465 miles above the Earth. It provided extensive observation and measurements of air density variations associated with solar activity and the first quantitative data on how solar radiation pressure affects a satellite's orbit.

For more than 6 years it transmitted radio signals from space on its assigned 108-megacycle frequency, powered only by six quartz-covered arrays of solar cells. Officially known internationally as 1958 Beta 2, *Vanguard I* is still circling the globe every 134 minutes and has an apogee of about 2,400 miles and a perigee of about 400 miles.

When NASA phased out the 108-megacycle radio band used for scientific satellites during the IGY, the agency's tracking and data acquisition facilities were gradually converted to the internationally allocated 136-megacycle band. At the close of 1964, the station near Quito, Ecuador, was the only NASA station still monitoring on the 108-megacycle frequency, and the signals from *Vanguard I* had degraded to the extent that Quito was unable to detect any signals, even at optimum conditions (when the satellite was in sunlight at the time of its perigee).

The successful launch of *Vanguard I* confirmed the merit of the rocket

The liftoff of *Vanguard I* on March 17, 1958.

design; it also demonstrated that the Vanguard group had become a well-integrated professional and technical team. Three other Vanguard rockets were launched before the Vanguard team was transferred from NRL to the newly created civilian space agency, NASA. The first satellite launch vehicle (SLV–1), launched May 27, 1958, was successful except for a premature second-stage burnout; in the second (SLV–2), launched June 26, 1958, the second stage cut off prematurely; the third (SLV–3), launched September 26, 1958, reached an altitude of 265 miles.[51]

Vanguard and NASA

The launch of *Sputnik I* in the fall of 1957 was a real jolt to the complacency of the American people. In true American tradition, a great clamor went up as to why the Soviet Union was ahead of the United States, who was to blame for the situation, and what was to be done about it. The people engaged in existing satellite programs had a difficult time explaining that their best efforts had been slowed by limitations over which they had no control. The end results of the new "space consciousness" were beneficial, since there developed a general realization that the American effort had to be greatly expanded and financially supported.

About 6 months after the launch of *Sputnik I*, the President's Science Advisory Committee and the President's Advisory Committee on Governmental Organization recommended the establishment of a civilian agency to direct nonmilitary space activity. President Eisenhower delivered a message to Congress on April 2, 1958, which stated that "aeronautical and space science activities sponsored by the United States should be conducted under the direction of a civilian agency except for those projects primarily associated with military requirements." As a result of this message and with a clear public demand for such action, Public Law 85–568, the National Aeronautics and Space Act, was enacted and signed by the President on July 29, 1958. This law established the National Aeronautics and Space Administration and gave the new agency the responsibility for conducting the scientific exploration of space for peaceful purposes. The law also gave the

Vanguard III, launched September 18, 1959.

President the authority to transfer to NASA "any function of any other department or agency of the United States, or of any officer or organizational entity thereof, which relate primarily to the functions, powers, and duties of . . . NASA." NASA opened its doors on October 1, 1958. Project Vanguard was transferred to NASA, with other DOD space projects.[52]

The Vanguard project was continued under the direction of NASA. *Vanguard II* (SLV-4), launched on February 17, 1959, was the first full-scale (21-pound) Vanguard payload to achieve orbit. It was also the first satellite designed to observe and record the cloud cover of the earth and was a forerunner of the Television Infrared Observation Weather Satellites (Tiros). *Vanguard III* (SLV-7), launched on September 18, 1959, was a 20-inch sphere weighing about 50 pounds.[53]

Vanguard Helped Shape the Future

As it happened, Vanguard did not put the first U.S. satellite into orbit. Nonetheless its contributions to the U.S. space effort were great indeed. Vanguard research became the basis for later launch vehicles, particularly the remarkably reliable Delta. Vanguard pioneered the use of advanced state-of-the-art techniques, including the first utilization of solar cells, which have since become commonplace components of American satellites. The scientific experiments which were flown on the Vanguard satellites increased the amount of scientific knowledge of space and opened the way for more sophisticated experiments.[54]

Perhaps the most significant achievement of Project Vanguard was to bring together a group of dedicated and talented scientists and engineers who came to understand the complexities and challenges of the space sciences program. This team was assimilated into the National Aeronautics and Space Administration, where it became the human core of the Goddard Space Flight Center and served as the foundation for the distinguished space sciences programs which were to emerge.[55]

Establishment of the Goddard Space Flight Center
3

ON SEPTEMBER 25, 1958, Administrator T. Keith Glennan announced the activation of the National Aeronautics and Space Administration (NASA), effective October 1, 1958.[56] Approximately 8,000 people and five laboratories of the 43-year-old National Advisory Committee for Aeronautics (NACA) were to be assimilated into the new agency. NACA's facilities would then become NASA's facilities, including Wallops Station (Wallops Island, Va.), and four research centers: Langley Research Center (Hampton, Va.), Lewis Research Center (Cleveland, Ohio), Ames Research Center (Moffett Field, Calif.), and the Flight Research Center (Edwards, Calif.).

On October 1, 1958, an executive order of the President effected the transfer to the National Aeronautics and Space Administration of the responsibilities involving several space research projects, including the Navy's Vanguard project.[57]

By this executive order, about 150 Project Vanguard personnel were transferred from the U.S. Naval Research Laboratory to the National Aeronautics and Space Administration.[58] The transfer became effective on November 30, 1958, and this group became known as the NASA-Vanguard Division. In December 1958, this group was transferred from the Naval Research Laboratory to the Space Science Division of NASA. During December 1958 and January 1959, 15 people from the Naval Research Laboratory were transferred to the Theoretical Division of NASA. Early in 1959, these elements, with others, were designated by NASA Headquarters to serve as the nucleus of a new Space Projects Center. Its staff was temporarily housed at the Naval Research Laboratory, Washington, D.C., and at the Colemont Building, Silver Spring, Maryland.

This assemblage was composed of some of the most experienced men engaged in space research. It included upper atmosphere scientific research teams and scientists and engineers from all three military services, the Project Mercury (manned satellite) team culled from the experienced staff of the former NACA laboratories, and the Navy's Project Vanguard staff. These groups gave immediate, mature capabilities in many vital areas of space flight research and development, since each was a "going concern" when transferred. From these and other groups, the initial team of senior personnel, around which was built the organization of the new Space Center, was assembled.

NASA Deputy Administrator Dr. Hugh L. Dryden appears to have been a key figure in selection of the Beltsville site. When the need for the new Space Center became apparent, he remembered the availability of surplus Government land near the Beltsville Agricultural Research Center. Believing that most of the Project Vanguard staff lived in Maryland, he had encouraged consideration of the Beltsville site. "Later, I learned that I may have been mistaken, since many of the Vanguard people actually lived in Virginia," Dr. Dryden recalled.

The New Beltsville Space Center

On August 1, 1958, Senator J. Glenn Beall of Maryland announced in a press release that the new "outer space agency" (NASA) would establish a laboratory and plant at Greenbelt, Maryland. This was the first time public notice was drawn to what was to become Goddard Space Flight Center.[59]

Planning of the new Center continued through the rest of 1958 and by the end of the year events were ripening. On January 15, 1959, by action of the NASA Administrator, four divisions (Construction Division, Space Sciences Division, Theoretical Division, and the Vanguard Division) of NASA were designated as the new Beltsville Space Center.[60]

On January 22, 1959, a NASA General Notice announced the establishment of the Beltsville Space Center to be operated under the direction of the Director of Space Flight Development in NASA Headquarters, Dr. Abe Silverstein.[61]

In a meeting held on February 12, 1959, for the purpose of surveying the organization and functions of the Beltsville Space Center, it was generally agreed that the Center probably would perform five major interrelated space science functions on behalf of NASA:[62]

- Project management
- Research
- Development and fabrication
- Advanced planning
- Operations

At the meeting it was agreed that the Beltsville Space Center should conduct an active space science program, launch six or seven vehicles for communications and meteorological satellites, and carry out research with geodetic satellites as well as fulfill other Vanguard Division follow-on programs. In addition to the scientific satellites and the meteorological and communications programs, the Beltsville center was to assume administrative responsibility for the early phases of the Mercury project—the first U.S. man-in-space program. Vehicles under consideration in these activities, in addition to the Center-managed Delta vehicle, were the Vega, Centaur, Thor-Vanguard (which became Thor-Delta), Juno V, and the Nova. Another extremely important function of the Beltsville center would be the global tracking operation which included tracking, data acquisition, and data reduction for both NASA's manned and scientific space missions.

Beltsville Becomes Goddard

On May 1, 1959, Dr. T. Keith Glennan, NASA Administrator, in a public release, formally announced that the Beltsville Space Center would be redesignated the Goddard Space Flight Center "in commemoration of Dr. Robert H. Goddard, American pioneer in rocket research." The Center would be under the overall guidance of Dr. Abe Silverstein, then Director of Space Flight Development at NASA Headquarters.

The organization of Goddard Space Flight Center (GSFC) was to include a director, not yet appointed; three major research and development groups, each headed by an assistant director; and business administration and technical services departments.

In the announcement, Dr. John W. Townsend, Jr., Chief of NASA's Space Sciences Division and previously Chief of the Rocket Sonde Branch of the Naval Research Laboratory, was named Assistant Director for Space Science and Satellite Applications. John T. Mengel, who was responsible for the development of the Project Vanguard Minitrack satellite tracking system, was named Assistant Director for Tracking and Data Systems. Dr. Robert R. Gilruth, who would become Director of Project Mercury and who had been Chief of the Pilotless Aircraft Research Division, Langley Research Center, was named Assistant Director for Manned Satellites. The three Assistant Directors temporarily reported to Dr. Silverstein. The announcement also stated that the Office of Business Administration would be headed by Dr. Michael J. Vaccaro, transferring from the NASA Lewis Research Center, Cleveland, Ohio, where he had served as Director of Organization and Personnel. This was the first formal announcement of the Goddard organization, mission, and appointment of key personnel.[63]

Two other key appointments followed a few months later. In May 1959 Leopold Winkler, who had transferred to NASA with the Vanguard program, was appointed Chief, Technical Services. And in September 1959, Dr. Harry J. Goett was named Director of Goddard Space Flight Center. Goett came from Ames Research Center, where he had been Chief of the Full Scale and Flight Research Division.

> **GODDARD SPACE FLIGHT CENTER RESPONSIBILITIES, 1959**
> - Conducting advanced planning and theoretical studies
> - Conducting necessary supporting research
> - Developing payloads for approved programs
> - Supervising GSFC flight operations
> - Supervising tracking, data acquisition, communications, and computing operations
> - Interpreting results of flight programs
> - Furnishing technical management of projects
> - Exercising procurement and contract administration authority
> - Providing support of space program activities of other organizations
> - Reporting status of approved programs
> - Providing administrative and management support

With the new space agency, NASA, specifically responsible for activities in space devoted to peaceful purposes, the question arose as to which space programs initiated by the Department of Defense under its Advanced Research Projects Agency should be continued by NASA.[64] Spacecraft and meteorological satellites had been developed by the Army Signal Corps' Research and Development Laboratory, Fort Monmouth, New Jersey; vehicle development had progressed under the Army's Ballistic Missile Agency. Upon transfer of the meteorological program to NASA (April 1959), the mission was assigned to the Goddard Space Flight Center. The space communications program also had been a military project. One phase of it—indeed, the earliest phase—had been the passive balloon technique, with experiments conducted by the Army Signal Corps at Fort Monmouth and at NACA's Langley Laboratory. The other phase was the experimental hardware for active repeater communications satellites. With creation of NASA and the establishment of Goddard, both projects were assigned to the new Center.

Having acquired programs and people from other agencies, Goddard immediately needed money to operate. Some money had been inherited along with the programs and people. The executive order transferring Project Vanguard to NASA also transferred remaining project funds totaling approximately $6 million, plus about $300,000 earmarked for special equipment ordered earlier by the Navy's Vanguard staff. Also available to the newly established Center were certain funds which had been appropriated to NACA. These resources were not enough to meet the new Center's needs. Since the Fiscal Year 1959 Independent Offices Appropriations Bill already had cleared the House of Representatives, the Bureau of the Budget authorized the budget request to be included in the 1959 Independent Offices Appropriations Bill as a supplemental item while the bill was being considered by the Senate. The item was subsequently considered by the House-Senate Conference Committee without further referral to the House.

The Fiscal Year 1960 budget without the manned space flight program totaled somewhat less than $100 million (the manned flight program was about $140 million, giving the Center a total budget of about $240 million). The program mushroomed in Fiscal Year 1961 to about $160 million, plus an additional $140 million designated for the manned space missions. The Center's scientific and technical programs for Fiscal Year 1962 came to about $250 million; for Fiscal Year 1963 it was about $354 million.

During the early period of its development, contractual operations, which became a vital and integral part of Goddard's business operations, were handled for the Center by NASA Headquarters.

Meanwhile on April 24, 1959, construction of the new space laboratory began on a site located on a 550-acre tract formerly part of the U.S. Department of Agriculture's Agricultural Research Center at Beltsville, Maryland. By September 1960, Building 1 was fully occupied and other buildings were well underway. Although much of the occupancy was on a temporary basis and the personnel complement was widely scattered from Anacostia, D.C., to Silver Spring, Maryland, and points between, the Goddard Space Flight Center had become a physical reality.

The Dedication

On February 8, 1961, Dr. Harry J. Goett, the Director, announced dedication ceremonies to be held on March 16, 1961. A committee with Dr. Michael J. Vaccaro as chairman and Robert C. Baumann as co-chairman planned the ceremonies.

The dedication included opening remarks by Dr. Goett and a welcoming address by James E. Webb, NASA Administrator. The event also marked the presentation of a Congressional Medal awarded posthumously to Dr. Robert H. Goddard, which was accepted by his widow, Mrs. Esther C. Goddard. In presenting the Medal, Representative Overton Brooks said:[65] "From the Congress of which I am Chairman of the House Committee on Space and Aeronautics, we present this medal, but truly it comes not from the Congress of the United States but from the heart ... of the American people as a whole." Senator Robert S. Kerr, Chairman of the Committee on Aeronautical and Space Sciences of the Senate, was unable to be present but sent the following message to Mrs. Goddard:

Mrs. Goddard, I am more than honored to have the opportunity of joining my good friend, Overton Brooks, in presenting to you this Congressional Medal in recognition of the creative achievements of your late husband. It was just 35 years ago today that he launched the world's first successful liquid fuel rocket and it is most appropriate that we make this presentation on this auspicious anniversary. It is only through the genius of a man like Dr. Goddard, who was not afraid to work for what he believed in, that we shall maintain the spirit and vitality that has made our country great. This medal, authorized by Congress on behalf of all the people, is but a small token from a grateful nation.

Dr. Hugh L. Dryden, Deputy Administrator of NASA, introduced Dr. Detlev W. Bronk, President, National Academy of Sciences, who made the dedication address. Dr. Bronk said in part:

There are two quotations I would like to repeat. The one appropriate to the mission of this institution, the other with regard to the man we honor. The first is from Louis Pasteur, speaking at a time when his beloved country was not doing well. "Oh, my country," said he, "You who so long held the sceptre of thought, why did you neglect your noblest creations? Take interest, I beseech you, in those sacred institu-

Dedication ceremony, March 16, 1961.

Dignitaries and guests attending the Center's dedication.

tions which we designate under the expressive name of laboratories. Demand that they be multiplied and adorned for they are the temples of wealth and of the future. There it is that humanity grows, becomes stronger and better . . . it learns to read in the work of nature symbols of progress with universal harmony." And from Pliny the Younger, "It is a noble employment to save from oblivion those who deserve to be remembered."

A bronze bust of Dr. Goddard was unveiled by his wife, assisted by Dr. Abe Silverstein, NASA Director of Space Flight Programs. The bust was created by the Washington sculptor Joseph Anthony Atchison, noted for his creative work in the Shrine of the Immaculate Conception in Washington, the World Flight Memorial for the Smithsonian Institution, and the Second Inaugural Medal of President Franklin D. Roosevelt.

Responding to the recognition paid her late husband, Mrs. Goddard remarked: "I hope that this bust and the man it represents will serve as an inspiration not only to the brilliant and dedicated people who are now at work at this tremendous Space Flight Center but to all who may work here in years to come. My husband would be deeply proud and happy for this very great tribute."

Joseph Anthony Atchison at work on the bust of Dr. Robert H. Goddard.

Mrs. Robert H. Goddard participating in the dedication of the NASA Center named in honor of her late husband.

Tours of the Center were conducted for invited guests, and "open house" was held for employees and their families. Included was a Control Room demonstration with simulation of prelaunch and countdown procedures, followed by a simulated satellite injection into orbit.

Lectures reviewed the Center's operation of global satellite networks, including Minitrack and Project Mercury. The cooperative role of the Center in the international exploration of space was explained. Guests saw an animated miniature tracking station and a scale model of the forthcoming United States-United Kingdom spacecraft, *Ariel I*, the first international satellite to be flown under Goddard project direction. There were also displays of spacecraft instrumentation and Goddard's family of sounding rockets, including an Aerobee 150A with a new attitude control system.

Other models on display included the Tiros weather satellites; *Explorer X*, the magnetometer spacecraft; *Explorer VIII*, the Direct Measurement Satellite; and *Vanguard I*. There was also a demonstration of a micrometeoroid detector, and vacuum, vibration, and spin-balancing equipment used to simulate space environmental conditions was shown.

Under authorization for construction at the time of dedication were eight buildings, representing a $27 million investment. They would provide the necessary facilities for 2,000 scientific, technical, and administrative personnel. The 550-acre tract once devoted to agricultural research was rapidly assuming a new role—the peaceful exploration of space.

PART TWO

GODDARD SPACE FLIGHT CENTER GOES TO WORK

Nature to be commanded must be obeyed.
—Francis Bacon

The Early Years
4

THE OPERATIONAL CONCEPTS which have been developed and applied at the Goddard Space Flight Center go beyond the traditional men, money, and machines concept of management. Although the early years of Goddard were deeply concerned with men (or manpower), money (in terms of budgetary activity, procurement activity, and operational costs), and machines (in terms of buildings and support equipment resources), other factors had far-reaching effects. These included such elements as Goddard's approach to project organization, its plan for uniting different disciplines into a group serving a common purpose, its program for disseminating scientific information, and many others.

Dr. T. Keith Glennan, then NASA Administrator, said: "We are not an operating organization in the ordinary sense of that term. We do not expect to operate meteorological or communications systems. Our product is knowledge—new and fundamental knowledge—the techniques, processes, and systems by means of which we acquire that knowledge. The rocket-powered launch vehicles we design and buy are not an end in themselves—they are cargo-carrying trucks of space, discarded when their fuel is exhausted." [66]

While the direct relation between some particular element or effort of Goddard and the acquisition of space knowledge sometimes appeared tenuous, the fact remained that the primary reason for the Center's existence was to acquire new knowledge. To do so required the coordinated effort of many scientists, engineers, technicians, and support personnel—often located in remote areas throughout the world—as well as buildings, equipment, and facilities.

The Center's growth may be measured by several factors: (1) the rapidity with which it expanded its work force from a few people formerly with the Vanguard project at NRL to a staff of some 3,000 with widely varied skills and backgrounds; (2) the growth of its physical plant from a wooded area near Greenbelt, Maryland, to a complex of modern space science laboratories, testing facilities, and worldwide tracking, data acquisition, and reduction facilities; and (3) the growth in financial responsibility from the

The wooded site selected for the Space Center.

approximately $6,300,000 transferred by the Navy to an R&D budget of about $354.03 million for Fiscal Year 1963.

A Nucleus Goes to Work

The first employees of the activity later designated as the Goddard Space Flight Center were some 150 individuals of the Vanguard group, transferred from the Navy to the newly created National Aeronautics and Space Administration. By mutual agreement between DOD and NASA it was decided that this cadre would remain physically at the Naval Research Laboratory "until suitable space is available at the projected NASA Space Projects Center in Beltsville."

In December 1958 another 46 employees were transferred to the Beltsville Center from NRL's Space Sciences group. By the end of 1958 the new Center had a total of 216 employees. By June 1959 the Center had grown to 391 people in the Washington area. In 1959 recruitment activities were stepped up significantly, and by the year's end there were 579 employees. By June 30, 1960, through transfer and recruitment the personnel complement had grown to 707 people.

First computers are moved to the Goddard Space Flight Center.

As previously indicated, NASA's manned space flight program was an integral part of the early Goddard mission. For this mission, the Center had the talent and technical capabilities from the early Vanguard days, including the worldwide Minitrack network. Increasing emphasis on scientific, meteorological, and communications satellite projects, together with recognition that the manned space flight program demanded an independent organization, led to the Space Task Group (STG) at Langley being separated from its organizational assignment to Goddard as of January 3, 1961. Goddard retained its responsibilities in connection with the Project Mercury tracking network. As a result of this transfer, 667 people left the Goddard roster to form the nucleus of what later became NASA's Manned Spacecraft Center, at Houston, Texas.[67]

Organizing for Space Science
5

> The secret of good administration . . . lies not in the administrator's vast and exact knowledge, but in his skill in navigating areas of ignorance. . . . It is the daily experience of an administrator that he make decisions in areas outside his expertise on what a scholar would consider to be insufficient evidence.[68]

EVEN AS THE INVESTIGATION AND EXPLORATION of space became a national goal, the effective direction and administration of the space program became an urgent necessity for the newly created agency. Here was a national effort which was to be conducted under the closest scrutiny of the public, the Congress, and the scientific community. It was a program involving vast human and financial resources which had to be given sound, and in many ways, novel, direction and guidance.

The national commitment to space did not come as a smooth, steady, acceleration of effort, but instead as a series of challenges and responses. We have seen in earlier chapters the experiments of one New England professor, how such efforts were multiplied many times during World War II, received postwar government and scientific endorsement and support in the IGY, blossomed into a national space program with civil and military components in the wake of *Sputnik I*, and leapfrogged into the exclusive bracket of top Federal program expenditures following the shock of the world's first manned space flight, made by the Soviet Union's Cosmonaut Gagarin in April 1961. It was then that President John F. Kennedy and his administration rallied the Nation; he said on May 25, 1961: "Now it is time to take longer strides—time for a great new American enterprise —time for this Nation to take a clearly leading role in space achievement which in many ways may hold the key to our future on earth."[69] Later President Kennedy predicted that this major expansion of the space program would be considered "as one of the most important decisions that will be made during my incumbency."[70] The goal was not only to land a man on the moon within the decade, but also to gain American competence and preeminence in all space activities.

Responsible for three distinct phases of the U.S. space program—scientific investigation of cislunar space, applications satellites (weather and communications), and space tracking of manned and scientific satellites (tracking, data acquisition, and data reduction)—Goddard Space Flight Center's missions were vital to the U.S. position in space. Within four years after its establishment, the Goddard Space Flight Center had an annual research and development budget of some $354.03 million—about one million dollars per day. The accompanying charts graphically illustrate the rapid growth in expenditures at Goddard.

With the establishment of the Goddard Center from the Vanguard project and the Upper Atmosphere group of the Naval Research Laboratory, varied capabilities, practices, and management concepts were brought together. The early organization did not fit neatly into simple categories

Funding at Goddard Space Flight Center, 1959–1963
[All amounts are in millions of dollars]

	Major GSFC missions						
Year	Sounding rockets	Satellites			Tracking and data acquisition	Delta launch vehicle	Total
		Scientific	Meteor- ological	Communi- cations			
1959	3.56	21.31	0.99	3.57	3.10	12.93	45.46
1960	9.68	20.24	7.93	3.05	16.19	12.48	69.57
1961	8.25	35.14	17.50	31.15	29.65	9.58	131.27
1962	7.29	67.64	26.97	21.42	45.13	0.70	169.15
1963	9.51	89.87	42.43	31.30	86.59	0.70	261.40

	Salaries and plant support				Advanced research and technology
Year	Salaries and expenses	Construction and equipment (on site and tracking)	Plant support	Total	
1959	2.02	3.95	0.14	6.11	0
1960	11.40	17.74	3.56	32.70	0
1961	16.31	14.63	4.97	35.91	0
1962	26.68	32.46	11.47	70.61	2.78
1963	38.83	35.41	13.81	88.05	4.58

of programs, disciplines, or functions. It contained most of the needed "across-the-board" capability with many specialized skills. This in-house competence, further increased by those who joined the staff later, was one of the Center's greatest assets. It provided the basic capability to assure intelligent control of its programs and to conduct enough in-house research and development of a significant and challenging nature to ensure the professional excellence of its scientific and technical staff.

The complexity of the Center's missions made for intricate patterns of communications and decision-making, calling for new skills in management and in conduct of organizational relations. Since some 90 percent of the Center's research and development funds were expended with private industry, nonprofit and educational institutions, and other Government agencies, the need for effective management techniques became increasingly important. Also needed were effective communications with the scientific community; new scientific knowledge was the fundamental objective of Goddard's space program. Suggestions and proposals for scientific experiments were evaluated by subcommittees of the NASA Headquarters Space Sciences Steering Committee for: (a) scientific merit; (b) the capabilities of the proposer and his institution. The experimenter chosen by the Committee could elect to build the hardware himself or subcontract to industry.

Where management responsibility for a major space project was assigned to the Goddard Center, project groups were created and became the backbone of the Center's management structure. Headed by a project manager, each project group included support elements from the Center's Office of Administration (for fiscal, procurement, scheduling, and other administrative details) and representatives for test and evaluation (reliability) and tracking (data acquisition and data reduction). It was the project manager's responsibility to ensure that Goddard's resources, both internal and contractual, were effectively used to serve the needs of a particular project.

The Center management considered the following relationships essential to effective project operations:

- Project needs must be communicated to line supervisors.
- Project manager must have rapid and direct access to top management to report how adequately requirements are being served.
- Top management must be able to step in to resolve such problems as arise from conflict between the needs of the various projects, between project demands and the more general discipline activity.
- Engineers and scientists in the project groups must keep in constant touch with the contractors and major subcontractors to follow the progress of project elements.
- A project support staff must assist the project group by performing such functions as procurement, financial management, PERT analysis, progress reporting.

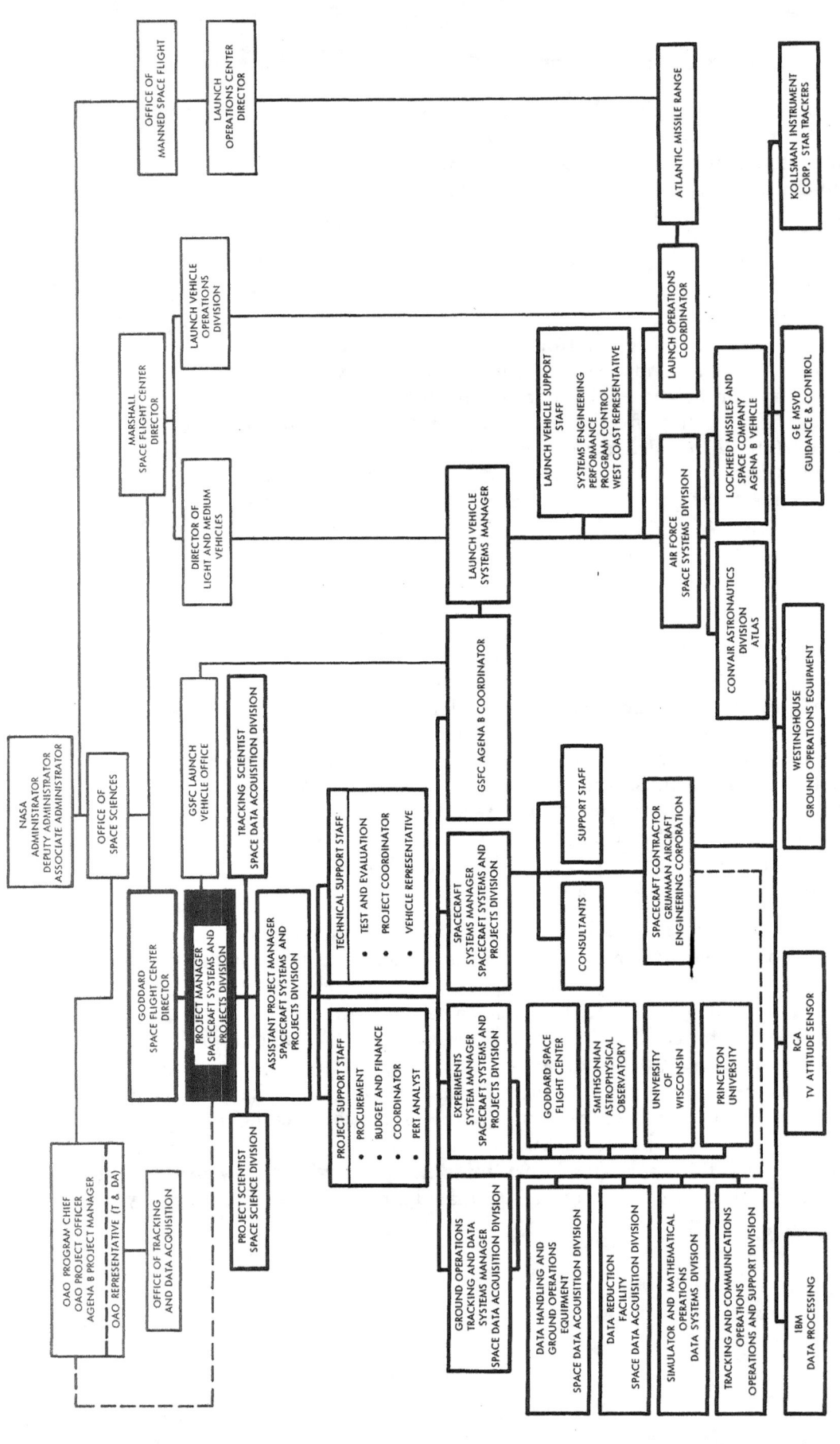

Orbiting Astronomical Observatory project management chart.

The director and key staff elements kept themselves informed through weekly staff meetings and weekly reports, issued every Friday, became "weekend reading material" in preparation for the next staff conference.

But whatever the technique, the management of the Center's complex research and development programs was no easy task; project management developed into a new and important art which affected virtually every level of the organizational strata. Solutions had to be found to such questions as to how NASA Headquarters would deal effectively with the Center; how the Center management would manage the project manager; and how the project manager in turn would manage a multimillion dollar contract with the aerospace industry, again involving a variety of contract managers. The formal management organization at Goddard is shown in accompanying illustrations.

In these early years, a major consideration was the amount of contract assistance which the Center should seek in carrying out its projects. It was obvious that only a relatively small portion could be done in-house. According to Eugene W. Wasielewski, Associate Center Director: "We try to have at least one small satellite under development in-house, at all times. Sometimes we have two in process. We also attempt to do a major share of the work on one of the large satellites. . . . While it is difficult to generalize, we feel we are barely doing enough in-house work to enable us to carry out our programs effectively."[71]

If a major project was to be accomplished through a prime contractor, specifications and requests for proposals were issued. Soon the program involved such prime contractors as Radio Corporation of America (*Relay I* and *II*), Hughes Aircraft Corporation (Syncom), General Electric (Nimbus), Ball Brothers (Orbiting Solar Observatory), Grumman Aircraft Corporation (Orbiting Astronomical Observatory), Thompson Ramo Wooldridge Space Laboratories (*Pioneer V*, Orbiting Geophysical Observatory), and many others. In the area of tracking, data acquisition, and data reduction, there were such industrial giants as International Business Machines, Western Electric, Bell Telephone Laboratories, and Bendix Corporation.

Contracting with industry on a multimillion dollar scale required the Center to seek the highest quality of American scientific and industrial skill, as well as the best capabilities of other Government laboratories. Each experiment in space was characterized by a high degree of attention to individual design and assembly. Even in a series of projects having the same general purpose, the payload packages varied according to the experiments conducted. Seldom, if ever, were any two payloads identical.

Procurement problems were complicated by the fact that in many instances the experiment or spacecraft hardware to be bought had never before been manufactured, indeed had never before been on the drawing board. More often than not, materials of a rare or "exotic" nature and

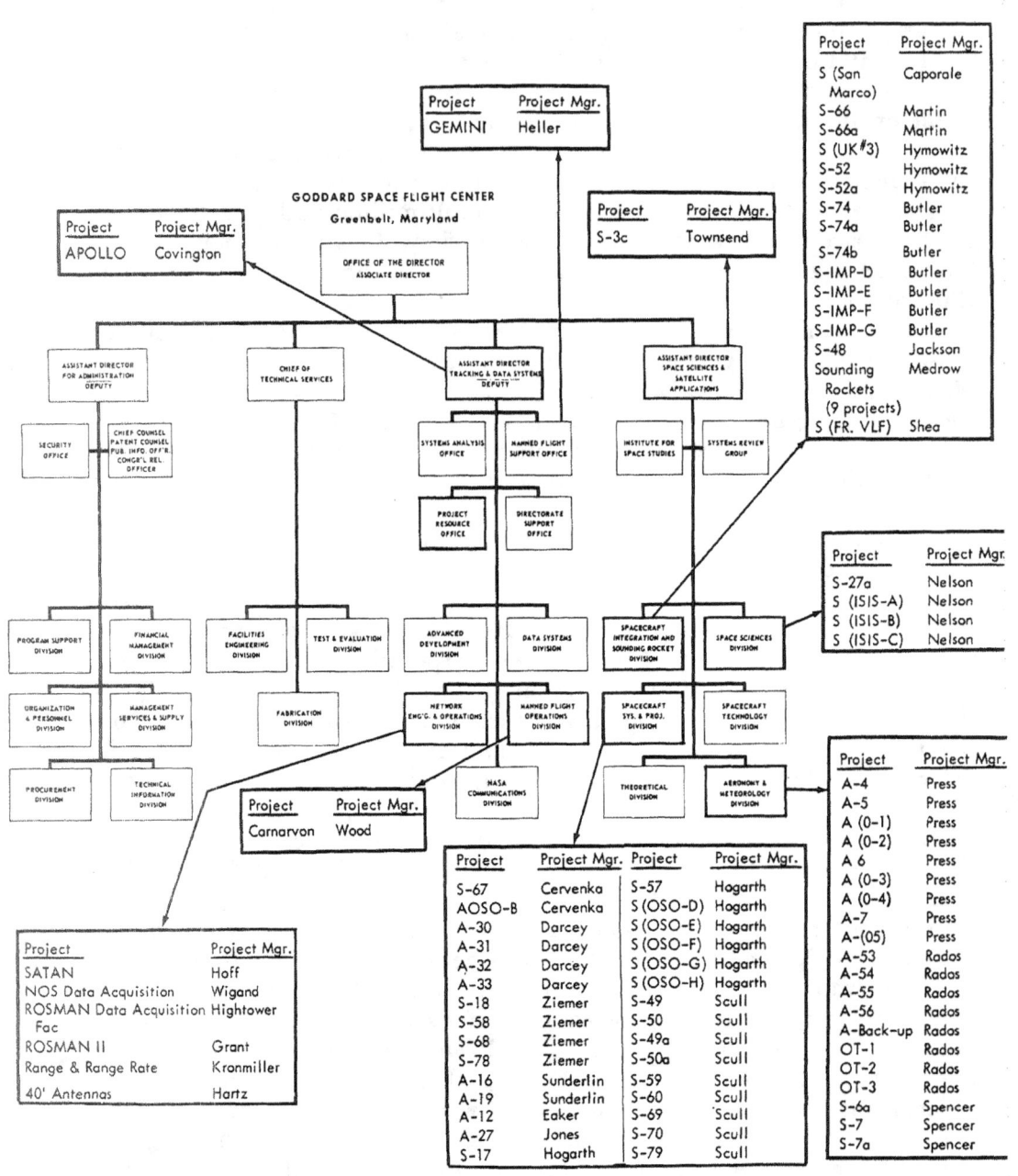

Goddard Space Flight Center project assignments as of September 20, 1963.

Personnel—the Center's most important resource. Here William Cahill discusses Goddard computer operations with a group of new employees.

limited availability were required. This called for extensive knowledge of supply sources, capabilities, and past performance of industrial firms and other vendors.

From January 1, 1960, to June 30, 1960, procurement actions totaled over $61 million; during Fiscal Year 1961, procurement actions totaled $375 million; and during Fiscal Year 1962 the procurement actions totaled over $418 million. (See Appendix E for detailed breakdown.) "Space" became a big and complicated business.

Personnel

The nucleus of Goddard Center personnel was drawn from several NACA laboratories and from the various satellite programs transferred to NASA in 1958 (see ch. 2). But as the Center's missions expanded and the physical plant neared completion, more and more skilled people of various descriptions were needed.

In a labor market which was extremely tight because of the nationwide shortage of scientists and engineers, Goddard had the additional problem of finding interested individuals possessing the specific and unique experience demanded by its programs. Experience with the college recruitment programs was rewarding. Although NASA entrance salaries may have been somewhat below the national average, the challenge of Goddard's mission, facilities, and progressive educational programs attracted high-quality graduates. The critical recruitment was for jobs in which highly specialized experience was necessary: for example, senior people with solid experience in satellite instrumentation, communications systems, systems integration, and spacecraft project management. It was somewhat paradoxical that for this group of personnel the Center was in effect in competition with itself; it competed with the personnel needs of industrial concerns which, with government contracts, were also engaged on space projects. One incentive particularly attractive at Goddard was the opportunity provided by the Center to participate in a major space science project from its inception to completion. The Center's mission frequently called for the "universal type" scientist, capable in areas beyond his immediate scientific discipline, knowledgeable in such fields as aerodynamics, electrical engineering, data transmission, etc.

Rather than establishing positions such as physicist, electrical engineer, etc., NASA categorized and identified positions directly with the nature of the work to be done. Since the areas of academic training did not always correspond with the fields of advanced research and development, the aero-

Aerial view of Goddard Space Flight Center, June 1962.

space technology concept was applied. Under this plan, the title "Aero Space Technologist" was used to cover the broad field of research and development specialties.[72] Specifically, the titles of positions had the symbol AST followed by the specialty.

Five separate categories of employees composed the overall Goddard team: scientists and engineers; "blue collar" craftsmen; technical support personnel; administrators; and clerical personnel. Almost 42 percent of the Center's work force consisted of scientists and engineers, while NASA-wide the ratio was approximately one-third scientists and engineers to two-thirds support personnel.

As previously indicated, the scientists and engineers who had been associated with Project Vanguard and other Government space programs formed the nucleus of the GSFC personnel complement. This group grew rapidly:[73]

1959:   782   (including Space Task Group staff later transferred to Manned Spacecraft Center)
1960:   1,265 (including Space Task Group staff later transferred to Manned Spacecraft Center)
1961:   1,497
1962:   2,850
1963:   3,494 (December 31, 1963)

Sources from which scientists and engineers were recruited during the period of December 1958 to December 1963 were:

Government: 916   (50.8 percent)
Private industry: 495   (27.4 percent)
Schools: 264   (14.5 percent)
Other: 128   (7.07 percent)

Scientists were selected from virtually every source and from many geographic regions. The recruiting program was conducted with the aid of extensive publicity campaigns and through the cooperation of colleges and universities.

For staff personnel, Goddard sponsored graduate study programs and undergraduate cooperative courses with several colleges and universities. Select students who had completed their sophomore year could attend school one semester and work at Goddard the next, alternating in this way until they got their degrees. Under another plan, graduate students could take three-quarter-credit courses at a local university, wherein they worked 3 days a week at Goddard and attended classes on alternate days.

Likewise, scientists and engineers were encouraged to augment their education in one of several local graduate school programs. Because the field of space technology was unique and developed so rapidly, Goddard had a program of seminars, colloquia, and specialized courses.

Goddard also assisted employees who, having completed their studies at the master's degree level, were striving for greater competence and stature. Each year a limited number of carefully selected scientists and engineers were offered an opportunity to spend up to one year in research and study fellowship programs at institutions of their choice. This program enabled an employee to conduct advanced study and to do research under the direction of men with international reputations.

Location plan of Center, June 1962.

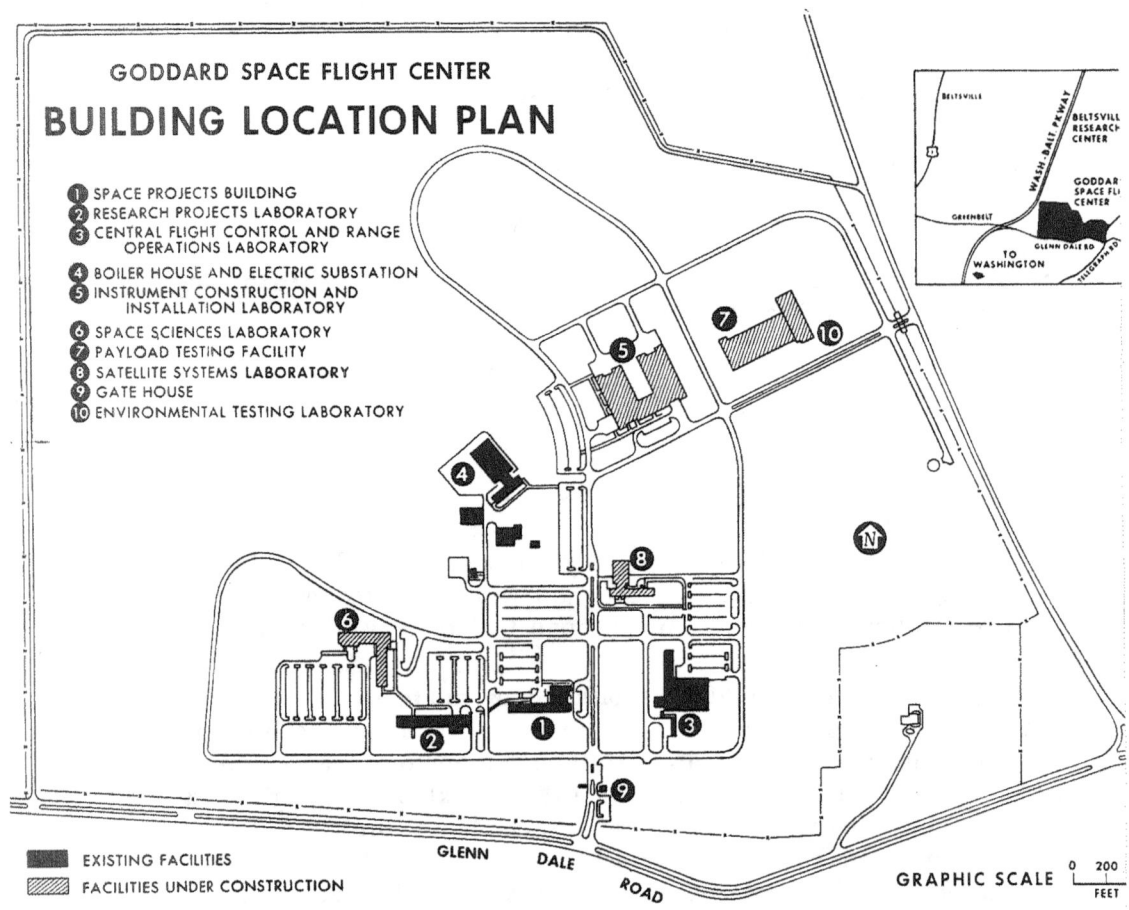

ORGANIZING FOR SPACE SCIENCE

Physical Plant

The physical plant of Goddard Space Flight Center was established with an eye to immediate and future requirements. An engineering master plan was developed by Voorhees, Walker, Smith, Smith & Haines of New York City. It envisioned a "campus type" layout, conducive to effective management and creative activity.

The first construction contract was let on April 10, 1959, to Norair Engineering Corporation, Washington, D.C. This contract called for construction of Buildings 1 and 2, together with access roads and parking areas. The first construction began early on the morning of April 24,

Location plan of Center, 1963 estimates.

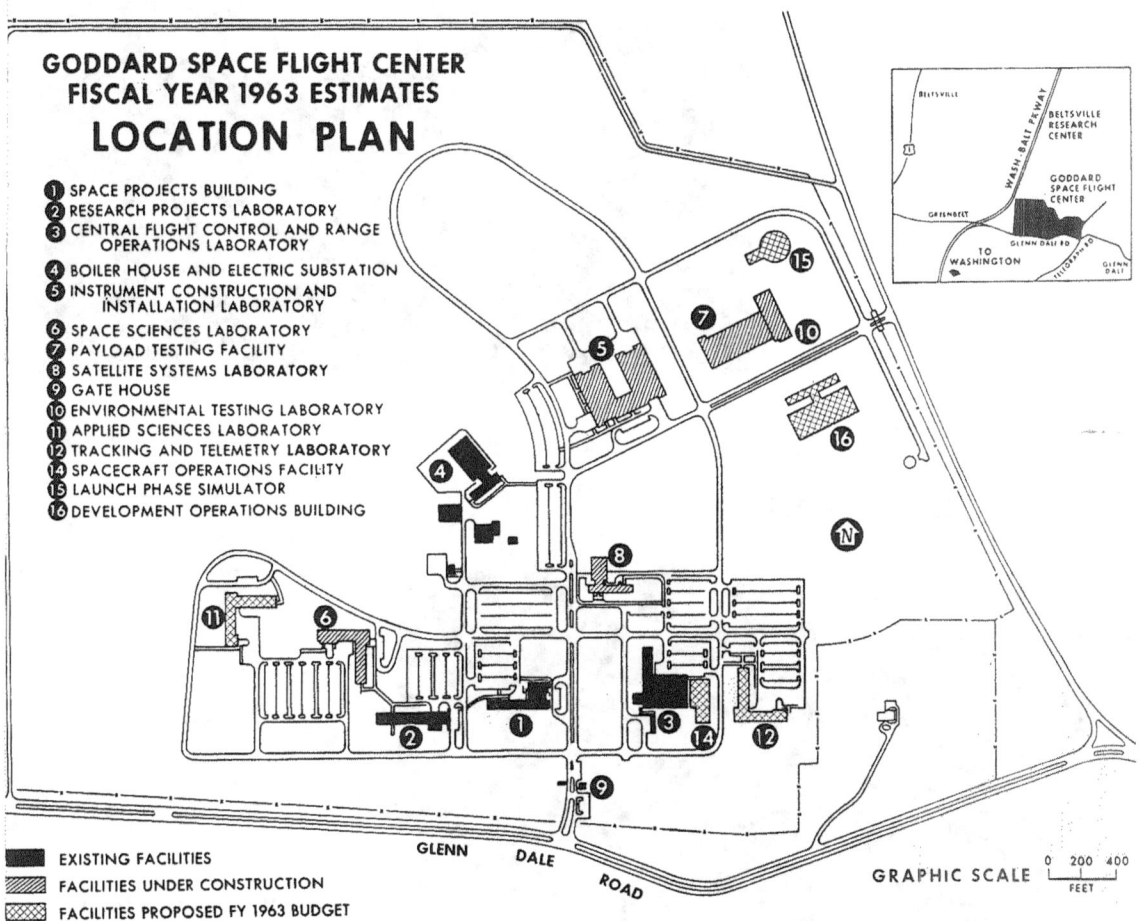

1959, when brush and trees were cleared in the area which was to become the Center's main entrance.

The computer and switchboard rooms were occupied on April 28, 1960. By July of the same year, the remainder of Building 1 was completely occupied, although steam for operation of the heating system and refrigeration compressors for air conditioning were provided by a temporary boiler outside the building. September 16, 1960, saw the full occupation of Building 2.

Site for **Building 1**, June 1959.

Building 1 under construction, October 1959.

Building 1.

Humphreys & Harding, Inc., began construction of the Central Flight Control and Range Operations Building (Building 3) on September 21, 1959. Installation of computer equipment was completed on March 1, 1961, while other portions of the computer and communications area were occupied in November and December of the same year.

Building 4, housing service shops, central power-plant, refrigeration plant, cooling tower, emergency power generators, and office areas, was started on May 23, 1960, under contract with Norair Engineering Corporation. Parking lots and roads were also included under this contract. Steam service lines, temporary boiler, and service shops were completed in November; office space, parking lot, and boilers were placed in operation on December 20, 1960. Construction and installation of refrigeration equipment were completed on May 29, 1961.

The Instrument Construction and Installation Laboratory, Building 5, under contract with Norair Engineering Corporation, was started on November 26, 1960. Initial phases of this structure accommodated many administrative and scientific personnel formerly housed in temporary quarters. In early 1962 it was necessary to modify the machine shop and upper floor areas to house personnel pending completion of Buildings 6, 8, and 11. Building 5 was completed on March 20, 1962.

Arthur Venneri Co. was low bidder on a contract to build the Space Science Laboratory, Building 6. Construction was begun on November 19,

1960. By February 1962, the lower floors were completed and sections A and B were ready for occupancy. The remaining portion of the building required one additional month, and the staff took possession on March 2, 1962.

The contract for Buildings 7 and 10 went to United Engineers and Constructors, Inc., on January 31, 1961. Notice to proceed with the construction of Building 7 was given in May 1961; construction started May 22, 1961. Occupancy by the Test and Evaluation Division, formerly housed in Building 4 and in numerous trailers, began April 28, 1962, and was completed a month later. Construction of Building 10 was started on October 19, 1961, by United Engineers and Constructors, Inc., with a scheduled contract completion date of September 1, 1962. Sufficient portions of the building, together with the overhead crane, were completed by March 2, 1962, so that installation of the Space Environmental Simulator and Dynamic Test Chambers by Minneapolis-Honeywell Corporation could begin.

The Satellite Systems Laboratory, Building 8, under contract with Arthur Venneri Co., was begun on September 16, 1961, and was targeted for occupancy by spring 1963. Featuring a 500-seat auditorium, it also included provisions for a presentation-type stage, multipurpose projection

Aerial view of Buildings 1, 2, 3.

booth, and wide-range sound facilities. Building 8 would house the Director, Associate Director, Assistant Directors, and supporting administrative services offices of the Center.

Building 11, an Applied Sciences Laboratory, was begun by the Norair Engineering Corporation of Washington, D.C., on August 16, 1962, and completed during September 1963. A contract for the construction of Building 12 was awarded to the Piracci Construction Company of Baltimore. This building, a Tracking and Telemetry Laboratory, was to augment such facilities in Building 3. Construction was begun on October 22, 1962, and was completed during November 1963. Each of the buildings have laboratory and office space for approximately 350 employees.

The assignment of space for the most effective administration of the Center continued to be a critical problem, and occupancy in many areas was on a temporary basis. This particularly applied to Buildings 1 and 5. Many of the administrative officers which eventually were scheduled to be located elsewhere at the Center were housed in rented space in the Jackson Building, Bladensburg, Maryland; at Lawrence St., Bladensburg, Maryland; at Litton Industries, College Park, Maryland; at Beltsville; and in the Colemont Building, Silver Spring, Maryland.

Building 8 under construction, July 1962.

Occupancy by Directorate and Location, July 12, 1962

Building [a]	1	2	3	4	5	6	7	8	Jac	Law	CFO	Litton	ARC	Cole	Litton 2A	Cape	NYC
Dir. and Asst. Dir. OA—585:																	
84 Dir., Asst. Dir.—OA	25				41		1	5	12								
82 Fin Mgmt	12				70												
81 O and P	2				79												
57 Mgmt Ser	39				18												
219 P and S	16								100	103							
62 TID	30				32												
Office of Tech. Serv.—456:																	
2 Chief OTS							2										
140 Fac Eng				115													
188 Test and Eval				30			158				25						
126 Fab Div		4			113	5	4										
Track and Data Sys.—725:																	
17 Asst. Dir.			17														
130 Track Sys			125										5				
116 OP and Sup			94				22										
219 Data Sys			219														
164 Spa Data Acq			6			20						138					
79 MSFS			19		60												
Space Sci. and Sat. App.—1,123:																	
11 Asst. Dir.						11											
284 Spa Sci		213				10											
185 Spa Sys and Proj					50	105									49		
175 Aero and Mete	16	137			22												
326 Spac Tech		5		16										12		30	
142 Theo	86					305											56
	226	359	480	161	485	456	187	5	112	103	25	138	5	12	49	30	56

[a] Jac—Jackson St. CFO—Construction Field Office Cole—Colemont NYC—Institute for Space Studies, N. Y.
Law—Lawrence St. ARC—Agricultural Research Center Cape—Cape Canaveral

Organizational Growth

A review of the organization charts (see Appendix F) gives an indication of how the Center grew while, at the same time, adhering to its original concepts. The manned satellite function shown in the first two charts was transferred from Goddard to the Manned Spacecraft Center at Houston, Texas. The other four major elements—Administration (formerly Business Administration), Tracking and Data Systems, Space Science and Satellite Applications, and Technical Services—remained basically the same with further refinements. The Office of Technical Services expanded to include such divisions as Facilities Engineering, Test and Evaluation, and Fabrication.

The Office of Space Science and Satellite Applications, while retaining essentially the same functions, recognized the need for the creation of a separate division for aeronomy and meteorology. The responsibilities of the Office of Tracking and Data Systems grew to the point where the original breakdown of a Theory and Analysis Staff, a Tracking Systems Division, and an Operations Division no longer was adequate. It expanded to include a Space Projects Integration Office, an Operations and Support Division, a Tracking Systems Division (with a plans office), a Space Data Acquisition Division, a Data Systems Division (with a theory and analysis office), and a Manned Space Flight Support Division.

Goddard Institute for Space Studies

To provide a point of contact between the national space program and an area rich in universities and scientific talent, the Goddard Institute for Space Studies was established in New York City in May 1961 as a part of the Goddard Space Flight Center.

The Institute's primary mission was to assist in the analysis and interpretation of data from NASA probes. It was concerned with basic theoretical research in a broad variety of fields, including the structure of the earth, the moon, and other planetary bodies in the solar system; the atmospheres of the earth and other planets; the origin and evolution of the solar system; the properties of interplanetary plasma; sun-earth relations; and the structure and evolution of stars.

The second major mission of the Institute was to arouse the interest of university scientists and students in the space program and to enlist their participation in some of the theoretical problems of space research. With its location in New York, the Institute had a unique opportunity for direct contact with the metropolitan university community. In its first year it developed associations with Princeton University, Yale University, Columbia University, New York University, the City College of New York, and Brooklyn Polytechnic Institute.

Drs. Robert Jastrow, Jackson Herring, Hong Yee Chiu, and Albert Arking of the Goddard Institute for Space Studies in New York.

The Institute was originally designated as a New York office of the Theoretical Division of the Goddard Space Flight Center. Dr. Robert Jastrow was named Director and also continued as Chief of the Theoretical Division. In July 1962, the Institute was separated from the division.

Goddard Launch Operations

Much of the early success of the Goddard satellite program has stemmed from the efforts of its launch teams, organized under the Directorate for Space Science and Satellite Applications, with personnel at both the Eastern and Western Test Ranges (ETR and WTR). The ETR team provided the management and technical direction of the field efforts involving the successful Delta launch vehicle. Under the Field Projects Branch, Goddard scientists and engineers closely supervised, monitored, and directed the launch vehicle preparations, launch operations, and coordination for spacecraft checkout. A similar team at WTR was responsible for the field integration of the spacecraft and launch vehicle and for coordinating and directing mission operations originating from the Pacific coast.

Fabrication and Testing

A scientific satellite was not a mass-produced item, but usually a one-of-a-kind spacecraft containing delicate scientific and electronic equipment. It required the best engineering talents to build and test these spaceborne laboratories. Manufacture of these space vehicles was accomplished either in-house by the Center's staff of skilled fabricators, under contract with American industry, or by a combination of both. The Center's fabrication staff was composed of a small but highly specialized team of engineers and technicians skilled in machining, forming, optics, electronics, satellite assembly, etc.

In its lifetime, each spacecraft would have to survive environmental stresses during ground handling, launching, and then operate effectively in space for its expected lifetime. The task was not a simple one, since these spacecraft contained equipment heretofore used only under the ideal environment of a laboratory. In space, these instruments had to operate reliably at distances and under conditions where they were subjected to solar radiation, space vacuum, extreme temperatures, radiation belts, and solar flares. Unlike a laboratory, there could be no experienced experimenter in attendance, making adjustments and taking readings.

Goddard's Spacecraft Test Facility served as a large-scale laboratory to

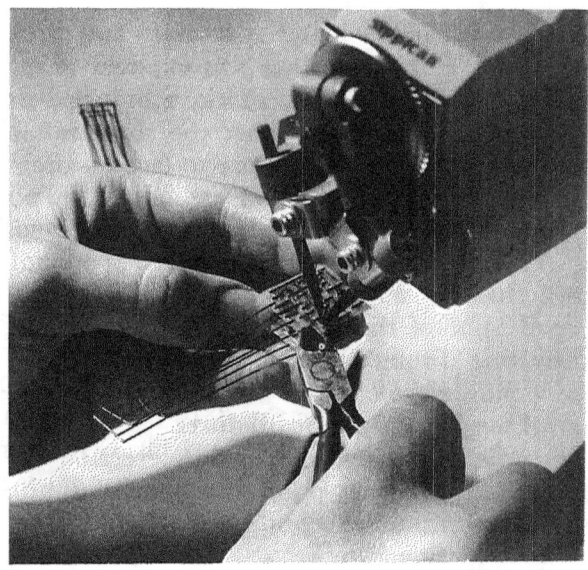

The welding of circuitry for use in a scientific satellite.

Space environment simulator under construction at Goddard.

test Goddard-developed spacecraft and probes. It was the Center's conscience. Here a satellite was exposed to man-made conditions of extreme temperature, humidity, shock, vibration, structural loadings, and various combinations to assure that the spacecraft and equipment could take the punishment which they were to face on their orbital mission. These facilities were capable of handling one 4,000-pound payload, plus two 1,000-pound loads simultaneously, measuring up to 25 feet.

Two major items in the Center's Spacecraft Test Facility were a dynamic test chamber and a space environment simulator. These chambers simulated some of the forces that converge on a spacecraft from initial ground handling through launch and flight. The dynamic test chamber consisted of a stainless-steel structure $33\frac{1}{2}$ feet in diameter and 58 feet high. Powerful mechanical pumps reduced inside pressure to 0.1 mm mercury. Here, dynamic balancing, solar paddle erection, spacecraft orientation, etc., could be tested. In the space environment simulator, a spacecraft could be exposed to simulated conditions of cold outer space, extreme vacuum, and solar radiation.

A review of scientific satellite failures detected by means of environmental test programs was made for the calendar year 1962. Five satellites were selected for this review, all of which were launched and successfully operated in space during 1962. These satellites were chosen to represent several factors that might influence their complexity. For example, weights varying from less than 100 to over 300 pounds and three launch vehicles were represented. The scientific discipline represented by the onboard experiments covered electron density; galactic noise; corpuscular, solar, and cosmic radiation; magnetic fields; ionospheric relations; and communication experiments. The telemetry systems were typically pulse-frequency modulated, although one system included traveling-wave tubes. Only one of the systems used batteries exclusively; the other four included solar cells for power. The satellites reviewed included those developed by NASA, by industry, and through international cooperation. They all, however, were tested under the same philosophy.

The ratio of electrical to mechanical failures was 4:1 (80 versus 20 percent). The mechanical problems were chiefly concerned with antenna designs, subsystem mounting, and local resonances. Stronger and stiffer designs, together with damping (often by potting), were general solutions to these problems. Electrical problems were erratic and spurious, requiring much troubleshooting. Solid-state components often were found to be faulty. Local overheating was often corrected by providing improved heat sinks and heat conduction paths. The failure distribution seemed reasonably consistent between the satellites. Nearly one-half of the failures reviewed occurred during the thermal-vacuum test, which simulated space conditions. However, nearly one-sixth of the failures occurred during

Failure Distribution by Spacecraft

Spacecraft	Weight, lb	Vehicle	Failures during test					
			Electrical		Mechanical		Total	
			No.	Percent	No.	Percent	No.	Percent
A	94	Scout	10	71	4	29	14	12
B	170	Delta	15	83	3	17	18	16
C	86	Delta	18	78	5	22	23	20
D	150	Delta	42	86	7	14	49	43
E	310	Thor-Agena	6	60	4	40	10	9
Total			91	80	23	20	114	100

Failure Distribution by Test Condition

Failure category	Failure during test [a]														
	Electrical							Mechanical						Total	
	A	B	C	D	E	Total		A	B	C	D	E	Total		
						No.	Percent						No.	Percent	No.	Percent
Checkout	---	2	3	5	2	12	13	---	---	1	4	1	6	26	18	16
Vibration	7	5	3	4	1	20	22	4	3	3	1	3	14	61	34	30
Temperature	---	1	1	---	1	3	3	---	---	---	---	---	---	---	3	3
Vacuum	---	1	3	1	---	5	5	---	---	---	---	---	---	---	5	4
Thermal-vacuum	3	6	8	32	2	51	56	---	---	1	2	---	3	13	54	47
Total	10	15	18	42	6	91	100	4	3	5	7	4	23	100	114	100

[a] Test conditions for spacecraft A, B, C, D, and E in table above.

checkout, and about one-third during vibration. One observation made from these data was the importance of completing the entire system and checking it out early in the project life. One-sixth of the errors noted were primarily indicative of the interaction of subsystems and the many interface problems. Cabling and connectors were particular offenders at this stage of checkout. Each of these failures was detected, corrected, tested, and evaluated. The final result—in space flight—was a successful satellite.[74]

Tracking, Data Acquisition, and Data Reduction
6

THE FIRST FUNCTIONAL TRACKING SYSTEM to be constructed for satellites was the Minitrack network. This network grew directly out of arrangements originally made by the United States with agencies abroad as part of the program for the International Geophysical Year. Among the overseas stations tied in with the satellite tracking network were Antigua, West Indies Federation; Quito, Ecuador; Lima, Peru; Antofagasta and Santiago, Chile; Woomera, Australia; and Esselen Park, Union of South Africa. These countries, in a program originally established in 1957 by the U.S. Naval Research Laboratory in cooperation with other agencies here and abroad, were all part of Minitrack.[75]

On January 10, 1959, representatives of NASA and DOD met to coordinate the separate requirements of the two agencies, and arrived at an agreement for a "National Program to Meet Satellite and Space Vehicle Tracking and Surveillance Requirements for FY 1959 and FY 1960." The agreement, signed by Secretary of Defense Neil H. McElroy and NASA Administrator T. Keith Glennan, established respective responsibilities and mutual use of tracking data wherever possible and led to the formation of the continuing NASA–DOD Space Flight Tracking Resources Committee.

The basic responsibilities of the network included: tracking, orbit computation, data acquisition (environmental and scientific telemetry), and data reduction.

The network consisted of three major functional parts. The first, the Minitrack Net, has been used to track all U.S. satellites containing a suitable beacon since the beginning of the space programs in 1957 and 1958.

The Minitract network comprised an organization of fixed ground stations, located throughout the world, to provide precision tracking, command, and telemetry reception from satellites and space probes together with a communications system to transmit this information to a computing facility.

A large percentage of the original stations were located along the 75th

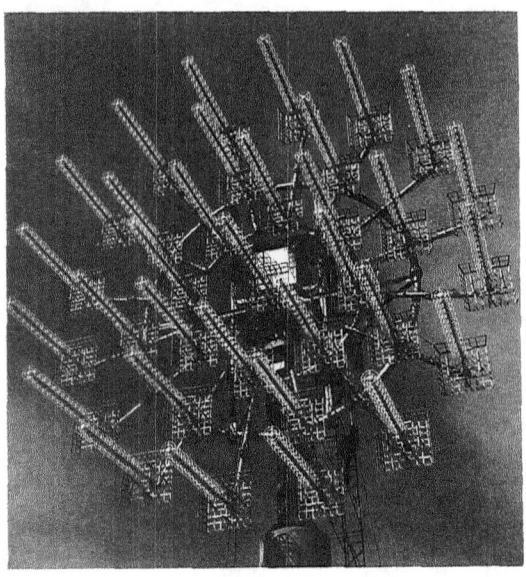

A Wallops Island helical antenna that receives video signals from the Tiros weather satellites.

meridian to intercept satellite orbits with inclinations of less than 45 degrees. New stations were added in higher latitudes to cope with more nearly polar orbits. Furthermore, ten of the stations were supplemented with additional antennas aligned specifically for polar orbit. (As of December 1962, the Minitrack system included the following stations: Blossom Point, Maryland; Fort Myers, Florida; Quito, Ecuador; Lima, Peru; Antofagasta, Chile; Santiago, Chile; Woomera, Australia; Esselen Park, South Africa; Goldstone, California; St. John's, Newfoundland; East Grand Forks, Minnesota; Fairbanks, Alaska; and Winkfield, England.)

The prime Minitrack satellite tracking system consisted of radio interferometers [76] operating in conjunction with a transmitting beacon in the payload itself. Since the establishment of the network, certain enhancements have been added to the original station equipment to provide tracking capability by optical and Doppler means as well. While the original tracking equipment operated on or near 108 megacycles (Mc), the frequency assigned for IGY activities, additional equipment has been provided, tunable over the 136–137-Mc region.

Many of the satellites launched by NASA used very wide bandwidths for transmission of data from the satellite to the ground stations. Since the receiver and sky noise in the telemetry link is proportional to the bandwidth used for reception, either a very high transmitter power or a very high antenna gain, or both, had to be used in a wideband telemetry link to achieve good signal-to-noise ratios. Since transmitter powers were restricted for technical reasons, it became necessary to develop very high gain antennas at the ground stations for receiving the wideband telemetry signals. The

Fairbanks, Alaska, tracking station.

antenna that best satisfied the requirements for high gain and multiple frequency operation was a parabolic antenna 85 feet in diameter.

The first satellite to require a large data acquisition facility for wide bandwidth reception was the Nimbus weather satellite. Nimbus was also one of the first NASA satellites that was to have a polar orbit. So the first station for wideband data acquisition was constructed on Gilmore Creek, 12 miles north of Fairbanks, Alaska; it was completed in May 1962. A contract was awarded the University of Alaska for operation of the station, to provide coverage for 70 percent of the passes of a satellite in a polar orbit. A contract for construction of a second station located near Rosman, North Carolina, was placed in July 1962; this station picked up an additional 20 percent of the passes of a polar satellite. Thus these two stations formed a network which provided coverage of 90 percent of the orbits of a satellite with a very high inclination.

The main antenna for the Alaskan station was an 85-foot-diameter paraboloid of revolution with a focal length of 36 feet. Its surface consisted of double-curved aluminum sheet panels. The surface was separate from the reflector structure so it could be independently adjusted. The antenna reflector was mounted on an X–Y-type mount designed specifically for tracking satellites. It was capable of tracking at rates from 0 to 3 degrees per second, with accelerations up to 5 degrees per second per second.[77]

Tracking Project Mercury [78]

The first Mercury-Redstone flight occurred December 19, 1960. In this flight, the unmanned capsule reached a peak altitude of 135 miles, a range of 235 miles, and encountered 5½ minutes of zero gravity. The flight was a success. The capsule control system, retrorockets, separation rockets, communications equipment, and recovery equipment functioned properly. The capsule was recovered soon after landing by a helicopter dispatched from an aircraft carrier.

Tracking, data acquisition, and communications for this project were the responsibility of the Goddard Space Flight Center. For the Project Mercury flights, Goddard operated a worldwide tracking network, spanning three oceans and three continents. In their location and equipment configuration, these tracking sites were prescribed by the character of the onboard electronics systems and by facilities existing throughout the world, which were used to the maximum extent possible. This maximum utilization of existing facilities was mandatory if the rapid pace set for the project was not to outstrip the development of the ground tracking system.[79]

Since the major requirement was one of safety, a highly reliable command system for backup of the astronaut functions by ground command was installed at strategic points around the earth. This requirement also made it mandatory that the onboard spacecraft systems be carefully monitored during all phases of the flight; this was accomplished by providing real-time telemetry display data at the sites.

Goddard was the focal point for receipt of real-time radar data from the sites. Two IBM 7090 computers provided launch and orbital computing during the flight; real-time display data were then transmitted to the Mercury Control Center at GSFC. An air-to-ground communications system was established together with remote site-to-control-center voice and teletype communications, to maintain network contact with the astronaut during all phases of the flight.

A special facility for testing and evaluation was established at Wallops Island, Virginia. Here a typical site was constructed during the early phases of equipment procurement to evaluate the performance of the systems, determine the interface problems, develop detailed equipment testing procedures, establish calibration techniques and equipment, and perform early training exercises. This proving ground was invaluable in providing a rapid evaluation of contractor-developed equipment and testing procedures. Here also the criteria were established for ultimate acceptance testing of the equipment as it was to be installed at each of the remote sites.

Late in 1961, an industrial team headed by the Western Electric Com-

TRACKING, DATA ACQUISITION, AND DATA REDUCTION

pany turned over this $60 million global network to NASA. Other team members were Bell Telephone Laboratories, Inc.; the Bendix Corporation; Burns & Roe, Inc.; and International Business Machines Corporation. The Lincoln Laboratory of the Massachusetts Institute of Technology had advised and assisted on special technical problems related to the network. The contract had involved extensive negotiations with Federal agencies, private industry, and representatives of several foreign countries in the establishment of tracking and ground instrumentation.

This worldwide network consisted of tracking and instrumentation sites (sixteen land-based sites and two ships), a control center, and a computing and communications center. The network was capable of performing real-time analysis of both the powered phase and orbiting flight. From orbital insertion until landing, the network provided continuous prediction of the capsule location, monitored the status of the capsule and astronaut, and initiated the command functions necessary for the mission.

In view of the fact that the computing system, located at GSFC, required a reliable input of tracking data to assure accurate location of the capsule at all times, two types of radar were incorporated. Because the spacecraft was

The two tracking ships of the Mercury network.

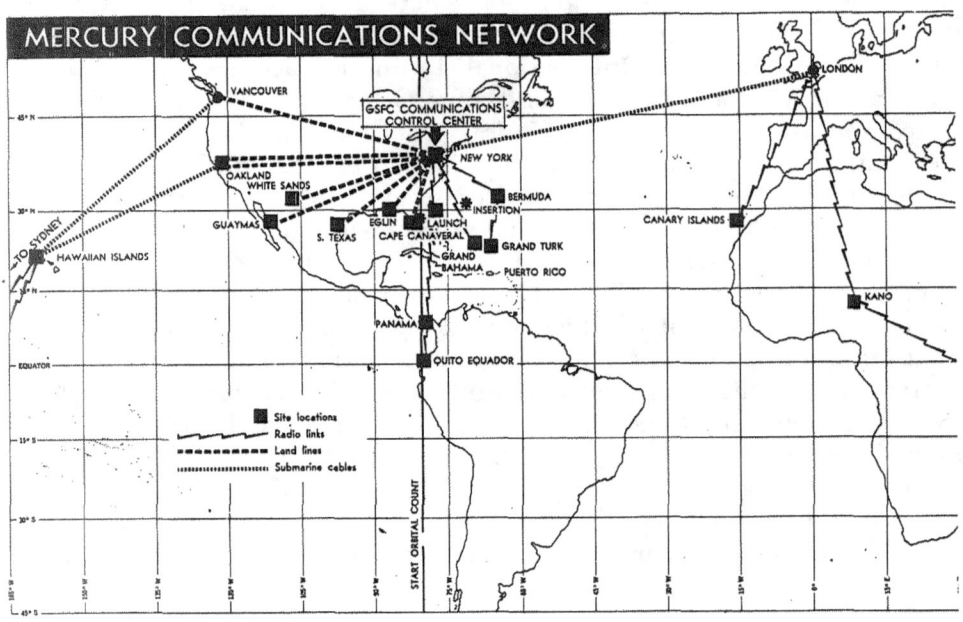

Mercury

of such a size that skin tracking with conventional radar would not be entirely reliable, two radar beacons were placed on board the spacecraft. The frequencies, selected on the basis of the available existing tracking facilities, had C-band as well as S-band tracking capabilities. This provided a degree of redundancy in case one of the onboard beacons failed during an orbital flight.

Of almost equal importance in the early consideration of the orbital flight was the capability of backing up the astronaut by ground command. At strategic sites throughout the network, dual FRW–2 command systems were installed. As it turned out, the command system was required more often during unmanned ballistic and unmanned orbital flights.

For intelligent ground command, a high degree of real-time ground monitoring capability had to be provided for the flight controllers located at the various sites. The spacecraft was designed to incorporate a dual telemetry system operating at separate carrier frequencies. Consequently the ground system had to have the capability of receiving two separate telemetry carriers with the attendant demodulation equipment associated with each.

To assure high reliability of spacecraft-to-ground communications under unknown conditions which might be experienced at approximately 100 miles altitude, again a dual communications system was incorporated into the spacecraft. This consisted of both ultra-high-frequency and high-fre-

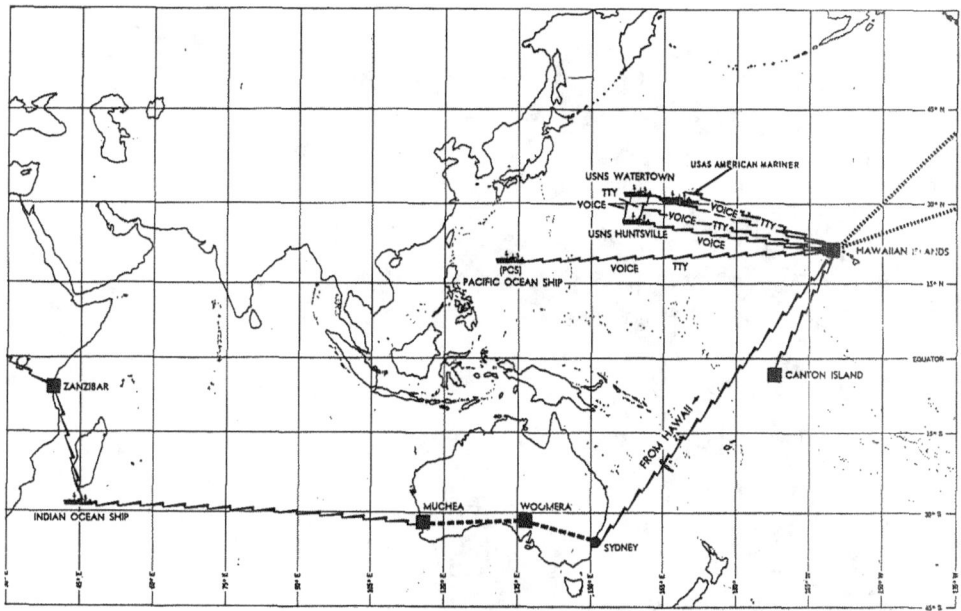

network map.

quency communications systems. Each ground tracking site had associated receiving and transmitting equipment for these frequencies.

To interconnect the network tracking sites with a reliable intercommunications system, two means of communication between sites and with the control center were established. The first and basic type consisted of the teletype communications facilities. Not only did the teletype communications system have to carry the load of communications among the flight controllers located at various sites around the network during the mission, but this system also was designed to deliver the radar tracking data from the radar tracking sites to the central computers. The second type of ground communications network provided the capability to communicate by voice between the control center and all sites.

To further enhance the reliability of the tracking systems, such as radar, command, communications, and telemetry, early and reliable acquisition of the spacecraft as it approached each tracking site was of major importance. Therefore a great deal of emphasis was placed on the system which could reliably provide pointing information to the various tracking antennas as the spacecraft appeared above the horizon. In addition, the provision of accurate radar tracking data, the reliable transmission of commands at a predetermined time, and tagging the received telemetry data with an accurate time reference made it necessary that a universal time system be installed at all sites. This system utilized the time signals of radio stations

Goldstone antenna.

WWV, Beltsville, Maryland, and WWVH, Hawaii, operated by the National Bureau of Standards, for calibration.

The Ground Communications Network was an automatic communications system connecting the Mercury sites around the world with GSFC. All intelligence pertaining to the Mercury capsule, except on life support equipment, passed through GSFC. The system carried telephone, teletypewriter, and high-speed data (1,000 bits per second) information to and from the worldwide network on a real-time basis. It accepted a message from a distant site and delivered it to the final destination, regardless of location, in a little over one second. The voice communications system was essentially a private-line telephone system which terminated in SCAMA (Switching, Conferencing, and Monitoring Arrangement) at GSFC, where it could be interconnected or switched over various lines.

Altogether the Mercury system involved approximately 60,000 route-miles of communications facilities to assure an integrated network with worldwide capability for handling satellite data. It comprised 177,000 actual circuit-miles—102,000 miles of teletype; 60,000 miles of telephones; and over 15,000 miles of high-speed data circuits.

TRACKING, DATA ACQUISITION, AND DATA REDUCTION

The various tracking and telemetry stations throughout the world were integrated into a coordinated network through a communications system terminating in the GSFC Space Operations Control Center.

The Space Operations Control Center had multiple functions:

- Control the operation of all tracking, command, data acquisition, and data transmission facilities utilized in support of scientific space vehicles.
- Coordinate the operation of other ground instrumentation facilities utilized in support of scientific space vehicles, with the exception of certain launch site installations.
- Ensure that operational activities required in support of any spacecraft were properly executed according to the operations plan. In case of inability to fulfill the plan, the control center was to recommend alternative courses of action to the project manager and make certain his decision was properly implemented.
- Provide facilities for monitoring the status of the network and the space vehicle at all times.
- Ensure that the project manager was informed of any departures from normal in the status of the network or of the satellite which might affect the conduct of the operation.
- Coordinate data reduction facilities, both for reduction of tracking data for determination and refinement of the orbit, and also for performance of such computations on the telemetered data as were requested by the project manager. The control center was directly responsible for converting tracking data into formats suitable for data processing.
- Schedule network activities to ensure that the requirements imposed were within the operational capability of the network and avoid conflicts between individual projects insofar as possible.
- Provide facilities in which interested officials could follow the critical phases of specific operations and rapidly obtain information on the status of any satellite during its useful lifetime.

In the performance of these functions the control center utilized a number of special facilities:

Communications.—Ten telephone toll lines, two local voice loop circuits, two special external point-to-point circuits, and ten dial intercom positions provided voice contact between external and internal groups performing functions essential to a given operation. There was also a special voice line to other agencies to pass on satellite orbital data. In addition to the routine in-house routing system for printed messages, the center was tied into the worldwide teletype network via three quasi-real-time data lines and two monitor lines providing minimum delay on circuits of operational importance.

Displays.—Three large edge-lit plexiglass boards were available for display of status, schedules, and graphical data of general interest and importance. An opaque projector presented teletype messages and infor-

In the photograph above, James Donegan (seated second from left), Goddard Mercury Operations Director, and staff monitoring the progress of the MA-6 (John Glenn) flight. Below, the GSFC communications area.

TRACKING, DATA ACQUISITION, AND DATA REDUCTION

mation of interest on a screen. Digital clocks displayed time (GMT): time to lift-off (countdown), and elapsed time of a hold. The clocks could be preset and controlled to indicate the status of proceedings. A world-map display indicated the nominal orbit on a Mercator projection; and active sites were illuminated to indicate status, acquisition, etc. Doppler launch data were superimposed on a nominal curve on a plotting board, indicating launch vehicle performance after lift-off.

Computation.—Three small computers were available: the CDC–160, the LPG–30, and an RPC–4000. These machines processed data of various types and of particular interest into a form for optimum presentation. For example, teletype data were edited and stored on IBM-style magnetic tape for rapid and easy handling. These machines lent themselves to expeditious solution of last-minute data changes. They were also used to perform various studies to aid in operational planning.

The communications network centering on GSFC included 36 full-period leased teletype lines serving 50 continental and foreign stations in the Minitrack and Deep Space networks, other data acquisition and command stations for scientific satellites, and other agencies in the scientific community engaged in the exploration of space. The 36 circuits (later expanded to 42-line capacity) terminated in a Western Union 111B switching center which combined tape relay, message-switching, and circuit-switching capabilities. Circuit combining facilities permitted the interconnection of

During Project Mercury's MA–6 mission.

Goddard computer room.

any station in the network in any combination for direct conference or real-time data exchange as needed. In addition, six full-period off-net lines were brought into the center. Three TWX, RCA, and Western Union commercial refile services were available. These circuits were used to carry all types of administrative and logistics information, satellite tracking data, satellite prediction and orbital data, and certain types of telemetry data.

All equipment and circuits in the Operations Room were arranged on a patch panel which permitted complete flexibility in interchange or substitution of equipments and links. One of the page printers was equipped with a keyboard for use in keyboard-to-keyboard coordination of data runs with remote stations in the network. Each of the circuits could be directly connected to any station in the network by leg-combining repeaters under switch control. An off-line 19/14 teletype set was provided for special tape preparation.

In the Control Room, two page printers were provided for monitoring purposes. Any lines in the network could be monitored upon request, and outbound traffic from the Control Room was carried by courier to the communications room. Voice communication in the Control Room was handled by standard telephone equipment. Each of 12 operating positions had the capability of using an outside exchange line, local interposition extensions, or a general conference loop; selected positions had the capability of point-to-point connection for immediate contact to facilities such as the communications room. The Operations Director, the Project Coordinator, and the Network Controller were able to select any of the lines in the room. Headset transceivers were generally used, leaving the operators' hands free.

Data Reduction Center.—A data processing center was established for determining the orbital parameters of earth satellites and reducing the scientific data obtained by the experiments contained in these satellites. Computer facilities available at the center included four large-scale general-purpose scientific computers (IBM 7090s) with associated peripheral equipment, as well as a variety of small digital computers, such as the CDC–160 and LPG–30. Scientific data obtained by the satellites were processed by special-purpose equipment developed to handle the many different telemetry formats in use. Underlying principles stressed semiautomatic operation with optimum improvement in the signal-to-noise ratio. Conversion to digital format was a one-step operation, allowing rapid entry into a large-scale computer for the complete reduction and analysis phase. Quick-look facilities in the form of tabulation or strip charts gave the experimenters the opportunity to evaluate their data prior to the final reduction and analysis and thereby determine optimum handling procedures to suit their individual needs.

The objective of the GSFC orbital calculation programs was to determine and predict the orbits of satellites on the basis of observational data. Initially the programs were written by the IBM Vanguard group for the Vanguard IGY project. These were subsequently replaced by programs written by GSFC personnel incorporating improvements in methods and efficiency. Orbit prediction could be done in four ways: Hansen theory, Keplerian ellipse, Brouwer theory, and numerical integration of equations of motion. The output of these orbit-prediction programs could be supplied in a variety of local frames at arbitrary times and in any form required. Output forms used included geodetic latitude, longitude, and height above subsatellite points; azimuth, elevation, and range; local hour angle, declination, right ascension, and direction cosines—all with reference to an arbitrarily selected site on the earth's surface.

Data acquired at the many remote stations had to be processed and reduced. Some of these data were not uniform in quality; therefore, the central data reduction facility had to be extremely flexible and capable of taking into account operational errors. The central facility also had to be able to handle many categories of data and the many variations created by a lack of standardization of all missile ranges and data acquisition networks. Almost without exception, telemetry data were recorded at the stations on magnetic tape.

The first step in the operation was an inspection and evaluation of the tapes received. Selection for further processing was based on the general requirements that the tapes had signals with adequate signal-to-noise ratio and usable timing, standard frequency, and tape-speed controls. The data were then digitalized and stored on digital magnetic tape. In cases where experimenters demanded analog records such as film or strip-chart recordings, these were supplied. In some instances, these were the only means by which the data could be presented. But since digitalizing did allow rapid entry of the data in a computer, it was usually suggested as the preferable method.

The final reduction and analysis of the data in a computer allowed insertion of calibration, linearization, smoothing, selection, correlation, and other data and, finally, merging of the data with orbit and aspect data. Mathematical operations on the data, such as insertion into equations relating measured values to other quantities, were easily performed; and the final output to the experimenter could be in the form of tabulation or graphs as desired.

Goddard-Managed Satellites and Space Probes
7

WHEN THE GODDARD SPACE FLIGHT CENTER was established, its basic mission within the NASA structure was the scientific exploration of space—"to meet the gaps in current knowledge . . . and research in space in all the scientific disciplines as dictated by the expanding knowledge of space phenomena."[80] By December 31, 1963, Goddard had launched some thirty satellites and space probes and performed experiments with over three hundred sounding rockets.

Specific emphasis was placed on experimentation in several scientific areas: atmospheric structure, electric and magnetic fields, astronomy, energetic particles, gravitation, ionospheric structure and behavior, and satellite meteorology, and communications. As a result of investigation in these areas, man's knowledge of the upper atmosphere would be increased and the groundwork would be laid for operational systems of meteorological and communications satellites.

Although much can already be written about the results of the explorations carried out at Goddard before 1964, it must be emphasized that the story here is incomplete. At this point in time, the interpretation of space research is an open-end task. It has not yet been possible to fully interpret all of the mass of data that has already been transmitted. Some satellites are still orbiting and transmitting data that will undoubtedly alter present concepts. Finally, discussion continues among scientists about the significance and interpretation of the data that have already been studied. For these reasons, intrinsic in the nature of scientific advances, the following account of satellites, probes, and sounding rockets must necessarily be incomplete.[81]

Early Satellites Related to the Center Mission

The Center considers *Explorer VI,* launched in August 1959, as its first satellite. Prior to that time, the United States had placed eight satellites in orbit and launched one major space probe as a result of pre-NASA programs. Although not directly connected with the work at Goddard Space Flight Center, they have been listed here as precursors of the Center's program.

Explorer I

This satellite was launched on January 31, 1958, under the project direction of the Army Ballistic Missile Agency and was the first U.S. satellite placed in earth orbit. Its IGY experiment was responsible for the discovery of the Van Allen radiation belt—believed by many to be the most significant finding of the International Geophysical Year. It demonstrated the feasibility of temperature control by satellite surface treatment and showed that micrometeoroids are not necessarily a major hazard in space navigation near the earth. The satellite ceased transmission on May 23, 1958.

Vanguard I

Launched March 17, 1958, *Vanguard I* was a test of the Vanguard launch vehicle and satellite ejection mechanism under the management of the Naval Research Laboratory. In an earth orbit, it determined atmospheric density at great altitudes and conducted geodetic measurements. It revealed that the earth is slightly pear-shaped, and the extensive information gained from it was useful in correcting geophysical map errors. It revealed much about the pressure of solar radiation and pioneered the use of photocells as a solar power source. Its lifetime, originally estimated at 200 years, is now believed to be about 2,000 years. One of its transmitters functioned until January 1965.

Explorer III

Explorer III, launched by the Army on March 26, 1958, went into an orbit slightly more elliptical than planned. It yielded valuable data on radiation belts and micrometeoroid impacts as well as external and internal temperatures and transmitted for about two months before reentry on June 28, 1958.

Explorer IV

The Advanced Research Projects Agency of the Department of Defense launched *Explorer IV* on July 26, 1958, for the purpose of studying radiation belts detected by *Explorers I* and *II* and to measure artificial radiation created by previous experiments. It collected data that helped to establish detailed spatial relations and many of the properties of artificial radiation;

SATELLITES AND SPACE PROBES

and it aided in analysis of the earth's magnetic field. Transmitting until October 1958, it decayed about a year later.

Project Score

Launched December 18, 1958, also under the direction of the Advanced Research Projects Agency, Project Score tested the Atlas ICBM as a launch vehicle. It also tested the feasibility of voice and teletype relay via satellite. For the first time a human voice (President Eisenhower's) was beamed from outer space. It was the first known satellite to be guided into orbit by a radio-inertial system. Its beacon signal terminated December 19, 1958, and the voice signal stopped on December 31, 1958. It reentered the atmosphere on January 21, 1959.

Vanguard II

This satellite, launched under NASA direction on February 17, 1959, was placed in orbit for the purpose of studying the earth's cloud cover. While the satellite was successfully placed in orbit, a wobble which developed in the satellite's orientation prevented interpretation of the cloud-cover data from the two optical telescopes. The payload configuration consisted of a sphere with a shell of highly polished silicon-monoxide-coated magnesium. The two transmitters functioned for 19 days.

Discoverer I

Discoverer I was launched by the Air Force under the direction of the Advanced Research Projects Agency on February 28, 1959. Its objective was to demonstrate the orbital capability of the Discoverer satellite with the Thor-Agena booster and the capability of the ground-support equipment. The satellite was the first to be placed in near-polar orbit. Difficulty with stabilization caused tumbling, which hampered continuous tracking. Transmitters included telemetry and a tracking beacon. The satellite reentered the atmosphere and decayed in mid-March 1959.

Pioneer IV

Launched by NASA on March 3, 1959, this instrumented probe was placed in an earth-moon trajectory to measure radiation in space, to test the photoelectric sensor in the vicinity of the moon, to sample the moon's radiation, and to test long-range tracking.

Pioneer IV, the first U.S. solar satellite, achieved an earth-moon trajectory, and yielded important radiation data in space. An injection below the planned velocity caused it to pass 37,300 miles from the moon and prevented near-lunar experiments. The configuration of this spacecraft was conical with a shell of gold-washed fiber glass, which served as a conductor and antenna. It was tracked for 82 hours to a distance of about 407,000 miles. It is in orbit around the sun.

Discoverer II

Launched on April 13, 1959, by ARPA, its objectives were to maintain a life-supporting temperature and oxygen environment and provide data on propulsion, communications, recovery techniques, and the measurement of cosmic radiation. It achieved a near-circular polar orbit and stabilization was controlled. The timer malfunctioned and caused a premature capsule ejection, preventing a recovery attempt.

Discoverer II was the first satellite to carry a recoverable instrument package. Telemetry was received until April 14, 1959, and the tracking beacon functioned until April 21, 1959. The capsule made impact in the vicinity of the Spitsbergen Islands (Arctic Ocean) on April 26, 1959, and was lost.

Goddard Satellites Leave the Launching Pads

With the launching of *Explorer VI* on August 7, 1959, Goddard Space Flight Center began a series of NASA satellite launchings that was to provide an accumulation of new data so significant and detailed that it would give man new perspective on his environment and solidify the foundation for operational weather and communications satellites. The following presentation of Goddard launchings is primarily chronological, but summary discussions have been inserted in appropriate places to give a clearer picture of some of the achievements of satellite-derived scientific research.

Explorer VI

Explorer VI, launched by an Air Force Thor-Able booster on August 7, 1959, was the first scientific satellite under the project direction of Goddard Space Flight Center. In addition to Goddard personnel, several universities and one private laboratory participated in equipping the satellite with experiments and interpreting the data received.[82]

The Universities of Chicago and Minnesota and the Space Technology Laboratories (STL) developed the instruments for measurement of the Van Allen radiation belt. STL also provided instrumentation for measuring the earth's magnetic field and for a type of one-line television scanning of the earth's cloud cover. NASA and the Air Force Cambridge Research Laboratories provided micrometeoroid experiments. Stanford University provided all equipment for the observation of very-low-frequency radio signals.

Explorer VI, launched from AMR August 7, 1959.

This "paddle wheel" satellite stopped transmitting on October 6, 1959, and reentered the earth's atmosphere sometime before July 1961. During its active lifetime, *Explorer VI* transmitted much valuable data. The most significant results can be briefly summarized.

The University of Chicago, using a triple coincidence telescope—an electronic device for measuring particles—reported that there appeared to be high-energy radiation of 10–20 million electron volts (MeV) on the inner side of the Van Allen radiation belt. It is apparently a narrow proton band, 330 miles thick, at about 1,240 miles from the earth. The total counting rate maximum was about 1,400 counts per square centimeter per second. They detected no protons with energies greater than 75 MeV, or electrons with energies greater than 13 MeV in the vast outer low-energy radiation region. There was some indication that radiation intensity varied with time.

The University of Minnesota also made measurements on the radiation belt but used an ionization chamber and Geiger counter. These measurements were compared with measurements of the University of Chicago and with those made by other spacecraft. Although instruments differed, the University of Chicago and the University of Minnesota measurements showed the same region of hard radiation. At times the radiation intensities at great distances dropped to a level about 5,000 times less than those measured by *Pioneer IV* and 10 times lower than *Pioneer III*.

Variations of radiation with respect to time were noted. In August, radiation at great distances had increased to a point consonant with the

Pioneer III measurements. In general, the measurements showed that the radiation belt had a complicated and variable structure. There appeared to be some correlation between the variations in pockets of radiation at large distances and solar outbursts.

The Space Technology Laboratories installed five instruments on *Explorer VI* that measured various kinds of radiation and magnetic fields and televised the earth's cloud cover. Although the intensity of the radiation fields did not increase with solar activity in readings taken August 16, August 20 readings seemed to indicate correlation between solar activity and increased radiation. One hundred and forty traverses of the radiation belt region indicated that the low-energy radiation zone has a gross structure similar to the radiation belts reported by Van Allen, but fluctuations in intensity indicated that both the inner and outer zones are much more complicated than had been previously indicated.

The data from the magnetic field experiment were quite involved, but there were no unexpected findings. The television scanner data gave only a crude image of the cloud cover, but they did correlate roughly with data from meteorological maps and are of historical significance in that they formed the first satellite-originated, complete, televised cloud-cover picture.

The micrometeoroid counting rate, as conducted by NASA and the Air Force Cambridge Research Laboratories, was somewhat lower than that found in earlier experiments. Stanford University's very-low-frequency detecting experiments showed that the Navy's 15.5-kilocycle-per-second transmitting station was received clearly below the D-region of the ionosphere and dropped out after passage through the D-region at 43.4 miles altitude.

Vanguard III

Vanguard III was launched on September 18, 1959, marking the end of the Vanguard launching activities. The 50-pound satellite achieved the desired orbit of 2,329 miles apogee and 319 miles perigee. The satellite was equipped by the Naval Research Laboratory to measure solar x-radiation and by Goddard Space Flight Center to measure magnetic fields, micrometeoroid impacts, and satellite temperatures. The ionization chambers from the Naval Research Laboratory were saturated most of the time because of the high apogee of the satellite. But the information enabled scientists to refine their determinations of the lower edge of the Van Allen radiation belt.

GSFC's magnetometer worked well; it showed there were systematic variations from the predicted fields. The micrometeoroid experiment showed from 4 to 15 impacts of particles 10^{-9} g or larger per square meter per hour. The thermistors showed that the temperature of the satellite varied between 6° and 27° C. The satellite ceased transmitting on December 11, 1959; it is expected to remain in orbit about 40 years.[83]

Explorer VII

The seventh Explorer satellite was launched by an ABMA Juno II booster on October 13, 1959. The 91.5-pound satellite achieved a successful orbit with an apogee of 680 miles and a perigee of 342 miles. It was the last ARPA originated satellite under NASA cognizance. The Naval Research Laboratory experiment measured Lyman-alpha and solar x-rays. The Research Institute for Advanced Study of the Glenn L. Martin Company installed Geiger tubes for heavy primary cosmic rays; total cosmic-ray counts were made by the University of Iowa. The radiation balance in the earth's atmosphere was measured by the University of Wisconsin. Goddard Space Flight Center conducted micrometeoroid and temperature experiments. Information was received about solar x-ray and Lyman-alpha radiation, heavy cosmic radiation, and cosmic radiation counts. Of special interest was the intense radiation recorded on one orbit as the satellite passed through an aurora.

Some of the experiments were coordinated to show correlations between atmospheric and solar activity. The Geiger-tube radiation counts indicated a correlation during periods of high activity between solar activity and optical emissions from the lower atmosphere. University of Wisconsin experiments indicated correlation between the earth's atmospheric temperature and space temperature.

The heavily instrumented satellite was spin-stabilized and carried two sets of radiation detectors, one an integrating ionization chamber and a Geiger-Mueller tube combined in a single package, and the other a proportional counter telescope. The vehicle also carried a magnetometer to measure the component of the ambient magnetic field perpendicular to the spin axis of the vehicle. A micrometeoroid detector and a device for determining the attitude of the vehicle relative to the sun were also included, but failed to function properly.

The Goddard instrumentation recorded the first penetration of a sensor in flight by a micrometeoroid. The tracking beacon became inoperative on December 12, 1959. Several of the instruments, although designed for short-term transmission, were still transmitting data in 1961.[84]

Pioneer V

One of the most dramatic space launchings in 1960 was that of *Pioneer V*, on March 11. This 94.8-pound interplanetary probe was placed into a successful orbit around the sun by an Air Force Thor-Able booster. Communication was lost soon after launch as the result of a malfunctioning diode, but was restored on April 24.

Pioneer V set a communications distance record that stood until the fall of 1962 when *Mariner II* set a new record on its flight toward

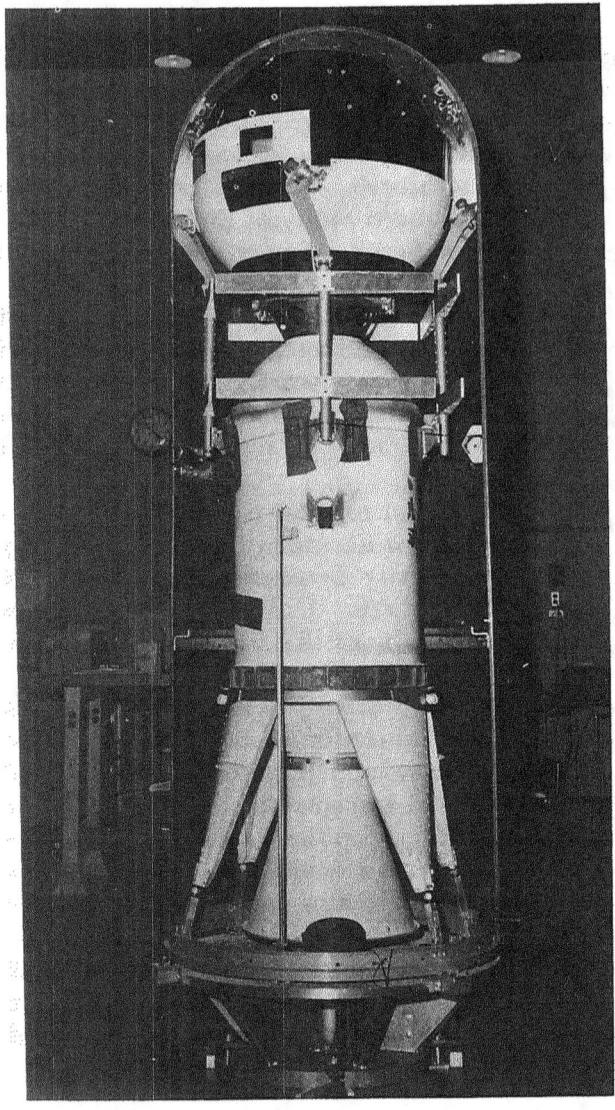

Pioneer V in final checkout.

Venus. *Pioneer V* continued to transmit interplanetary data up to a distance of 17,700,000 miles and its tracking signal was received from a distance of 22,500,000 miles from the earth.

Some of the scientific achievements of *Pioneer V* can be briefly summarized.

Galactic Cosmic Rays.—The frequently observed sharp decreases in the intensity of galactic cosmic rays, called Forbush decreases, are phenomena of solar origin and are not caused by the earth or its magnetic field.[85] This fact was determined by a comparison of data from *Pioneer V* with simultaneous observations obtained from ground-based neutron monitors. The comparison showed that the magnitude of the large Forbush decreases of April 1, 1960, in the vicinity of the earth was almost identical with that observed at the position of *Pioneer V*, 3,100,000 miles (5 million km) closer to the sun.

The mechanism which produces the 11-year variation in cosmic-ray intensity is centered in the sun, and the size of the volume of space in which the intensity is reduced is greater in radius than one astronomical unit during that part of the solar cycle which was covered by the *Pioneer V* measurements.[86] This result is inferred from the fact that the reduction in intensity produced by this mechanism was nearly the same at the earth and at *Pioneer V*.

Solar Cosmic Rays.—Cosmic rays were observed in space, completely free of any effects due to the earth. Instruments aboard *Pioneer V* directly detected particles accelerated by solar flares, which constitute a potential hazard to man in space. Numerous bursts of such particles were observed with the ionization chamber-Geiger tube package. The correlation between these observations and data from ionosondes at Thule and Resolute Bay on the minimum ionospheric reflection frequency indicates that the particles detected in the bursts were solar protons, with energies probably between 10 and 50 MeV, and that very few, if any, solar electrons above 50 MeV result from flares.[87] Many of these proton events were detected by the counter telescope also. However, the threshold of this instrument for proton detection was about 75 MeV, whereas that for the ionization chamber-Geiger tube package was about 20 MeV.

These solar flare particles, which produce ionization in the polar atmosphere for many successive hours, are not stored in the geomagnetic field. This fact was established by a comparison of data received from *Pioneer V* with polar cap absorption data acquired simultaneously.[88]

The energetic electrons in the earth's outer radiation zone arise from an acceleration mechanism within the geomagnetic field rather than from direct injection of energetic electrons into the field. This conclusion was based on the fact that the flux of electrons of energies in the 50-billion-electron volt (BeV) range reached very high levels in the outer radiation zone, according to *Explorer VII* data, while few, if any, electrons of these energies were detected at *Pioneer V*.[89]

Bremsstrahlung-producing radiation (electromagnetic radiation produced by the sudden retardation of a charged particle), evidently accelerated by solar activity, is present in interplanetary space during the periods of such activity.[90] This type of radiation was detected by the proportional counter telescope on numerous occasions.

The Geomagnetic Field.—The termination of the geomagnetic field was observed at about 14 earth radii on the daylight side of the earth, at least during periods of little geomagnetic activity.[91] The distance from the earth of this termination was considerably greater than that predicted by most theories.

An anomaly in the geomagnetic field was observed at about 6 earth radii on the daylight side of the earth.[92] This deviation from the assumed dipolar character of the field was similar to that observed with the earth satellite on the night side of the earth.[93] The anomalous component of the field is believed to be produced by a ring current circling the earth with its axis parallel to the geomagnetic axis.[94]

The existence of rapid fluctuations in the geomagnetic field between 10 and 14 earth radii was confirmed.[95] This phenomenon was first observed with instruments aboard the space probe *Pioneer IV* in October 1959.[96] These fluctuations may be produced by the interaction of the geomagnetic field and the ionized particles therein with the interplanetary medium.[97]

Magnetic Fields in Interplanetary Space.—During periods of low solar activity, the interplanetary magnetic field was found to be about 2.7×10^{-5} gauss and nearly perpendicular to the plane of the ecliptic.[98] During periods of solar activity, fields greater than 50×10^{-5} gauss were observed. The direction of the fields at such times could not be determined. The correlation of these increases in the interplanetary field with solar and terrestrial effects leads the experimenters to conclude that these high fields accompany the plasma ejected from the sun during active periods. Further, these comparatively intense fields are believed to be responsible for the exclusion of galactic cosmic rays from regions of the solar system during Forbush decreases.[99]

Measurement of the Astronomical Unit.—The astronomical unit, expressed in the terms of solar parallax, was found to be 8.79738 ± 0.00082 seconds of arc.[100] This result, determined from the long-range tracking of *Pioneer V*, is in good agreement with the value of 8.79835 seconds of arc obtained from optical observations of the asteroid Eros.

Tiros I

Tiros I, the first meteorological satellite, was launched into a near-circular orbit of 428.7 miles (690 km) perigee and 465.9 miles (750 km) apogee on April 1, 1960.[101] Shaped like a hat box, it was about 21 inches high and 42 inches in diameter and weighed about 270 pounds. The top and sides were covered with solar cells, the primary source of power. Its main sensors were two TV camera systems. When viewing the earth vertically, one camera took pictures about 700 to 800 miles on the side, while the other took more detailed pictures about 80 miles on the side. Since it was spin-stabilized, the cameras could view the earth during only part of each orbit. Tape recorders made it possible to store pictures taken over areas distant from the United States and to read them out as the satellite passed over the command and data acquisition stations at Fort Monmouth, New Jersey, and Kaena Point, Hawaii.

Tiros I had a useful lifetime of 78 days (1302 orbits). On June 16, 1960, a stuck relay in the satellite drained the batteries, and continued operation caused general failures. During its operational lifetime, *Tiros I* provided 22,952 exciting pictures of the earth's cloud cover, of which an estimated 60 percent were of meteorological interest. Weather patterns over the earth lying roughly between 50° N and 50° S latitude were photographed. These weather patterns, compared with data provided by conventional observations, showed that the satellite data provided more complete and more accurate information than had been possible in the past.

Studies of *Tiros I* pictures indicate that distinct cloud vortex characteristics are probably associated with individual storm types. Striking patterns of large spiral cloud formations, some as much as 1,550 miles in diameter, were observed. Jet streams, thunderstorms, fronts, and regions of moist and dry air were discernible in some photographs. In the absence of obscuring clouds, large ice packs were sometimes seen.[101]

Echo I

Echo I was a 100-foot-diameter, aluminized-plastic inflatable sphere that was placed into orbit on August 12, 1960. The initial orbital parameters were: apogee, 1,049 miles (1,689 km); perigee, 945 miles (1,522 km); period, 118.3 minutes; inclination, 47.2 degrees. The Mylar polyester sphere, including subliming powders, weighed approximately 124 pounds and had been designed by the Langley Research Center. Initial inflation was accomplished by the expansion of residual entrained air when the packaged sphere was ejected from the payload. Inflation was maintained by the use of subliming chemicals. The sphere carried two 10-milliwatt tracking beacons, powered by chemical batteries and solar cells. It was the first passive communications satellite.

The specific objectives of the launch were: to orbit a 100-foot-diameter, aluminized-plastic sphere to be used as a passive reflector of electromagnetic waves; to study the effects of the space environment on large area-to-mass-ratio structures; to measure the reflective characteristics of the sphere and the electromagnetic propagation characteristics of space; and to conduct experiments to determine the feasibility of using such satellites as passive relays in worldwide communication systems.

The principal communications experiments were conducted by the Jet Propulsion Laboratory (JPL) station at Goldstone, California; the Bell Telephone Laboratory (BTL) station, Holmdel, New Jersey; and the Naval Research Laboratory (NRL) station, Stump Neck, Maryland. Goldstone and Holmdel carried out the first communication experiment; JPL transmitted at 2,390 Mc and received at 960.05 Mc; at the other end of the

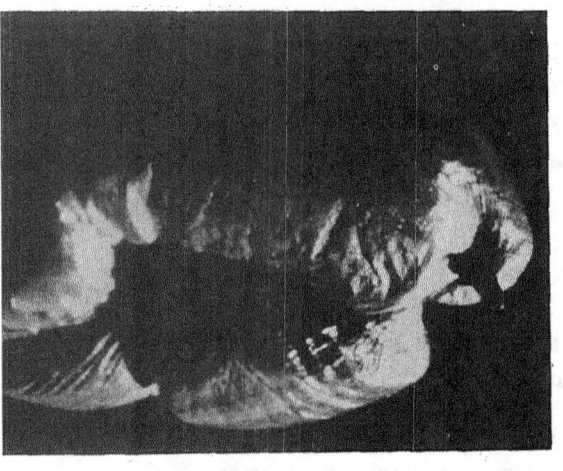
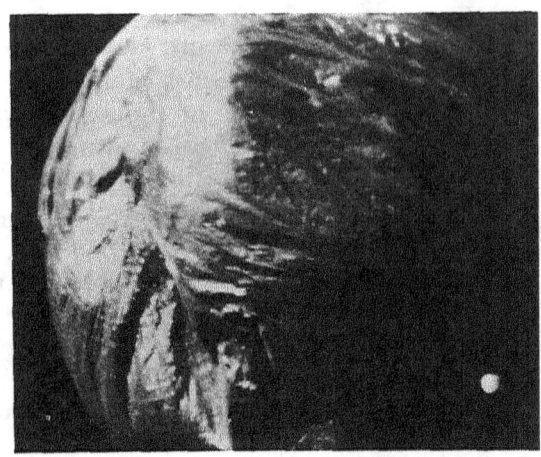
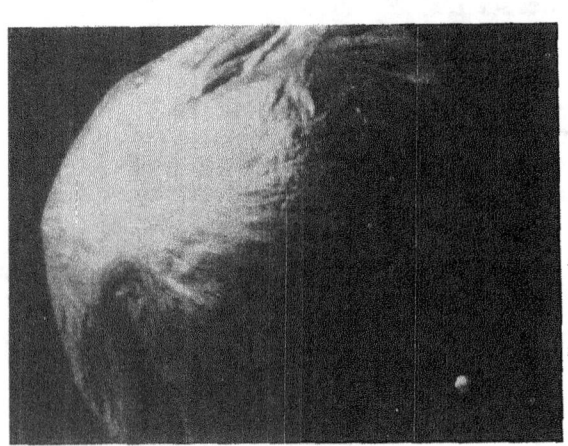
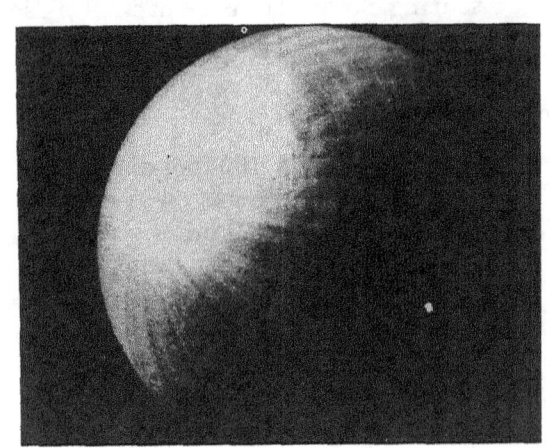

Echo inflation test sequence.

link, BTL received at 2,390 Mc and transmitted at 960.5 Mc. Voice messages were transmitted over the two-way link utilizing wideband frequency modulation with special demodulation techniques developed by BTL. NRL received JPL transmissions at 2,390 Mc, and transmitted at either 2,390 or 2,390.4 Mc, using 2,390 Mc when JPL was not transmitting and 2,390.4 Mc for differentiation from JPL transmissions when JPL was transmitting simultaneously. In addition to experimenting with voice modulation, tests were made with continuous wave "sine wave" modulation, etc., to provide further data for evaluation of the characteristics of the passive satellite transmission media. At a secondary priority on later passes of the satellite, experiments were conducted utilizing narrow-band phase modulation, narrow-band frequency modulation, and single sideband.

SATELLITES AND SPACE PROBES

Immediately after launch, and for a few days thereafter, beacon function was completely satisfactory. Suitable orbital elements were determined; and the highly directional antennas at NRL, JPL, and BTL were directed at the satellite to well within the required accuracies of 0.2°, 0.2°, and 0.4°, respectively. After a period of time, orbital data points became increasingly uncertain because of several causes. The first of these was the failure of the battery system; this limited beacon power to that supplied directly by the solar cells, so the beacons could function only when the satellite was in sunlight. A second cause was the progressive darkening of the epoxy resin with which the solar cells were coated. This darkening was a design feature which would cause the cells to become inactive within a period of approximately 6 months to 1 year, so that beacon transmissions would eventually cease. Finally, *Echo I*'s gradual change in its aspect with relation to the sun meant that it passed through the earth's penumbra for more extended periods. This development precluded beacon function; the Minitrack orbital data were based on fewer data points, with a consequent loss of accuracy. Antenna-pointing accuracy requirements no longer could be met. Minitrack stopped tracking *Echo I* on December 28, 1960.

Significant information has been gained from *Echo I*.[102]

- The use of a sphere as a passive reflector has been effectively demonstrated. System and space transmission losses did not differ substantially from those anticipated.
- With good reflector sphericity there was no appreciable cross-polarization of reflected signals. However, as wrinkling occurred, there was increasing cross-polarization of signal components.
- The uniquely large area-to-mass ratio of *Echo I* has made it possible to determine the effect of solar pressure on its motion, and a quantitative measure of such effects has been applied to orbit predictions.
- *Echo I* provided data for air drag in the upper atmosphere measurements. Thirty-fold diurnal changes in density in the upper atmosphere at the altitude of about 620 miles (1,000 km) and changes produced by solar disturbances on high-altitude atmospheric density have been reported. The density of the atmosphere at the altitude of the *Echo I* orbit has been calculated from the drag and is reported as about 10^{-17} at about 620 miles (1,000 km) altitude and about 10^{-18} at about 992 miles (1,600 km) altitude.
- An interesting use has been made of observations of *Echo I* to acquire information on the ozone distribution in the atmosphere. An experimenter reported he found the maximum of ozone content between 12.4 miles (20 km) and 18.6 miles (30 km) altitude, with the number of molecules per cubic centimeter as $2 \times 10^{12}/cm^3$ and decreasing 10^{11} molecules/cm^3 at the altitude of 40.3 miles (65 km).
- Another experimenter reported that he had used a special radar tracking technique to search for ionization surrounding *Echo I* but that he

found no evidence of such ionization in the case of either *Echo I* or *Sputnik III*.
- The whole process of tracking, determining orbits, and pointing narrow-beam antennas using computed drive tapes has been demonstrated successfully; this means a satellite can be used in synchronized fashion for communication relay between two remotely located sites.
- Radar cross-sectional measurements have been made on the satellite. They have indicated a gradual decrease in average area which, presently, appears to be about one-half of the original value.

At the time of publication, *Echo I* was still in orbit, visible to the naked eye in the night sky, and still capable of reflecting radio signals. Although the satellite was slightly crumpled and had lost much of its original shape as the result of micrometeoroid punctures, its orbit and condition were better than had been anticipated. The satellite was expected to reenter the earth's atmosphere in 1968.

Explorer VIII

This satellite was launched on November 3, 1960, with the objective of studying the temporal and spatial distribution of ionospheric parameters by direct measurement. The 90-pound satellite was launched into an orbit of 50° inclination with a 258-mile (415 km) perigee and a 1,423-mile (2,290 km) apogee. Spin stabilization at 30 rpm was used. Data transmissions stopped on December 27, 1960, the approximate date estimated for exhaustion of the chemical batteries. No solar cells were used.

The instruments included a radio frequency (RF) impedance probe for determination of electron concentration, four ion traps for measurements of positive-ion concentration and mass distribution, two Langmuir probes for measurement of electron temperature, an electric-field meter for determination of satellite charge distribution, and two instruments for determining micrometeoroid impacts. All the instruments operated continuously with the exception of the electric-field meter, which operated only by command.

One of the micrometeoroid detectors consisted of two microphones with a maximum detectable sensitivity of 10^{-4} dyne-second and a dynamic range of 3 decades. This detector reported frequency and momenta of impacts. The second micrometeoroid detector used a photomultiplier tube with a 1000A evaporated layer of aluminum on the window. Light flashes generated by micrometeoroid impacts on the aluminum were translated into pulses of varying length and amplitude by the photomultiplier. The pulses were interpretable in terms of the kinetic energy of the impinging particle. Sensitivity was estimated to be sufficient to detect particles of $<10^{-15}$ g having a velocity of 20 km/sec.

The radio frequency impedance probe experiment could be connected to the antenna upon command, to study ion sheath effects.

Two of the four ion traps were single-grid models and thus sensitive to photoemission. The remaining two ion traps were multiple-grid models not sensitive to photoemission. Comparison of the data from the two experiments gives the magnitude of the photoemission current.

The Langmuir probe measured both electron and positive-ion currents. The electric-field meter was of the rotating-shutter type and capable of measuring fields up to 10,000 volts/meter. Its noise equivalent was less than 5 volts/meter and residual drift less than 5 volts/meter.

Explorer VIII data have disclosed a number of interesting discoveries. It measured the diurnal electron temperatures between altitudes of 244 miles (400 km) and 1,364 miles (2,200 km) and found the daytime temperature to be about 1,800° K and the nighttime temperature to be about 1,000° K. In its orbit it measured an electron concentration of 1.3×10^4 electrons/cm^3. Its ion probe reported the mean mass of the ions as 16 atomic mass units, indicating the predominance of atomic oxygen in the ions. The ion density profile determined by *Explorer VIII* up to 465 miles (750 km) was similar to that determined before, but, in addition, showed the upper atmosphere to be isothermal. Another interesting and important result found by *Explorer VIII* was the ratio of helium to hydrogen ions. Above 496 miles (800 km) the helium ion was found to be an important constituent of the ionosphere, and at the altitude of 1,364 miles (2,200 km) there was a heavy predominance of helium ions over hydrogen ions. This discovery explains the high air densities encountered by *Echo I* in its high-altitude orbit. The instruments carried by *Explorer VIII* also revealed that the spacecraft was charged and that, as the altitude of the craft increased, the negative charge changed to positive.[103]

Tiros II

Launched November 23, 1960, into an orbit of 431 miles (730 km) apogee and 406 miles (625 km) perigee, *Tiros II* was similar to *Tiros I* but carried, in addition, infrared sensors to observe the radiation from the earth and its atmosphere, and a magnetic coil for partial control of orientation. The magnetic coil was included on the basis of experience with *Tiros I*, whose spin axis had moved in an unexpected manner (but fortunately remained more favorable for observations). These motions were caused by the interaction of an induced magnetic field in the satellite with the magnetic field of the earth. The new coil allowed some control of satellite orientation and camera pointing.

Presumably because of a malfunction associated with the lens of the wide-angle camera, the pictures from this *Tiros II* camera failed to show the detail that was so striking in the *Tiros I* photographs. They did disclose large cloud masses or clear areas, and proved useful in day-to-day weather analyses and forecasting. The narrow-angle camera pictures were of excellent quality.

Prior to the launching of *Tiros II*, 21 countries were offered the necessary orbital data if they wished to conduct special meteorological observations to be correlated with the satellite observations. Ten of the 17 countries which indicated a desire to participate chose to proceed with their programs even though the wide-angle camera picture quality proved poorer than expected.

The measurements made by the infrared detectors on *Tiros II* were: temperature of the top of the water vapor layer (6.3 microns (μ)); surface temperatures or cloud-top temperatures (8 to 12 μ), which help to distinguish cloudy areas at night; the amount of reflected radiation (0.2 to 5 μ); the amount of emitted radiations (7 to 30 μ); and low-resolution cloud pictures (0.5 to 0.7 μ).

The magnetic orientation coil, on several separate occasions, functioned as planned. The sensor instrumentation worked properly until mid-January 1961 when a malfunction in the clock control system forced discontinuance of remote wide-angle pictures to reduce the danger of a power drain that would disable the entire satellite.

The operation of *Tiros II* continued during most of 1961, but with a more or less progressive deterioration in the quality of data obtained. The blackbody sensor of the wide-angle radiometer failed in March. Of the scanning radiometer channels, the 6.3-μ channel had degraded early in the year, the 7- to 30-μ channel in early April; and no useful infrared data were obtained after April 23, 1961. The infrared electronics and tape recorder continued to function until the satellite was shut off in early December after more than a year in orbit. Except for temporary suspensions, the cameras had been programed to operate regularly until the *Tiros III* launch on July 12, 1961, and from then until August 8, 1961, on an average of two orbits per day. Cloud pictures of operationally significant meteorological value were obtained in early August. After August 8, camera programing was sporadic because of power limitations. The last pictures, in November 1961, were still not completely useless although obviously degraded. A total of 36,156 pictures was obtained.

The third pair of spin-up rockets was fired with partial success in mid-September 1961. The fourth pair was successfully fired on September 28 after more than 10 months in orbit. The magnetic orientation coil worked well until late November 1961. In early December 1961, the satellite no longer appeared to be responding to orientation coil commands and was so oriented that the sun shining on the base plate produced excessive heating and very little power. The beacons were shut off on December 3, 1961, but the satellite is still in orbit. Case studies have clearly demonstrated the expected correlation between the 8- to 12-μ atmospheric window data and the concurrent patterns of cloud cover and cloud-top altitude.

During the fall of 1961, a *Tiros II Radiation Data User's Manual* was published, along with Volume I of the *Tiros II Radiation Catalog*. Volume I of this catalog contains analyzed grid point data for fifty orbits of the *Tiros II* scanning radiometer data.[104]

Perhaps the most striking of the *Tiros II* picture data were several examples of narrow-angle camera photographs of sea ice in the area of the Gulf of St. Lawrence. One series of such examples was obtained in January 1961; the second in late March. The March pictures included coverage from the Gaspé Peninsula to east of Newfoundland and show significant changes in the ice patterns, particularly in the vicinity of Anticosti Island.

Explorer IX

Explorer IX, a 12-foot inflatable sphere made of Mylar and aluminum foil, weighed 15 pounds. It was launched on February 16, 1961, for the purpose of obtaining atmospheric densities from drag measurements. A project of Langley Research Center with GSFC participation, *Explorer IX* went into orbit with a perigee of 395 miles, an apogee of 1,605 miles, and an inclination of 39°. The tracking beacon carried by the sphere failed to operate after the sphere was placed in orbit, but optical tracking from the ground was successful. By means of this satellite, density of the atmosphere at an altitude of 434 miles was calculated from drag for comparison with values calculated from *Echo I* at about twice this altitude. The sphere was sensitive to changes in the density along relatively small segments of its orbit and was thus able to reveal the effect of solar disturbances on upper-atmosphere density.

The results obtained by *Explorer IX*, combined with those of *Echo I* and other satellites such as *Vanguard I*, showed that the upper atmosphere is a dynamic region of changing density caused by the diurnal variation of sunlight and by the smaller changes of energy associated with solar disturbances.[105] The satellite reentered April 9, 1964.

Explorer X

This spin-stabilized earth satellite, weighing about 79 pounds, was launched on March 25, 1961. It entered a highly eccentric orbit with an apogee of about 186,000 miles located at an angle of 140° to 150° from the sun-earth line, and a perigee of about 100 miles.

The highly eccentric orbit was selected for the purpose of studying the properties of the magnetic field and the solar interplanetary plasma over a region extending from close to the earth out to a point where the effects of the earth's magnetic field should be negligible. The scientific payload consisted of an extremely sensitive rubidium-vapor magnetometer, two fluxgate saturable-core magnetometers to measure the spatial and temporal var-

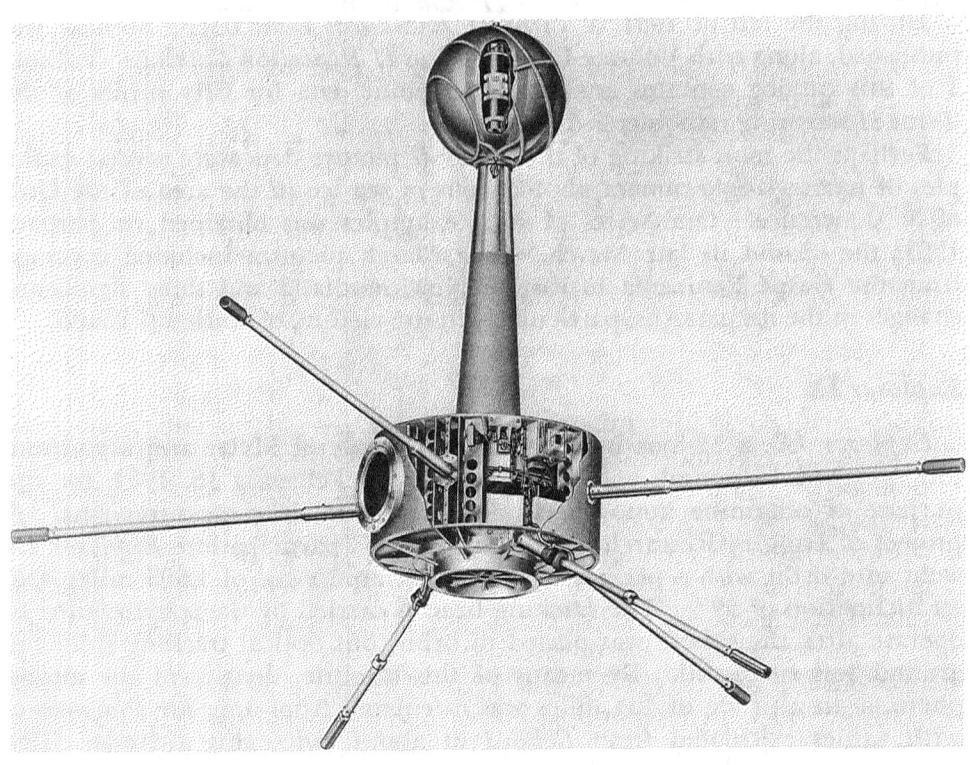

Explorer X, an interplanetary probe, launched March 25, 1961.

iations of the geomagnetic and interplanetary fields, and a multigrid plasma probe to determine the flux, energy spectrum, and directionality of very low energy protons in the plasma. Since the satellite was spin-stabilized, a sun sensor was used to provide information about the orientation of the instruments relative to the sun. Power for the satellite was provided by batteries having an active life of about 60 hours, sufficient to permit continuous measurements to be made on the first outward pass to apogee.

Quiet magnetic conditions prevailed prior to and during the first day of the satellite's outward pass to apogee; solar activity was confined to Class 1 and 1— flares. However, at 10:15 Universal Time on the second day after launch, a Class 3 flare occurred near the east limb of the sun, producing disturbed magnetic conditions in the vicinity of the satellite and at the earth. Significant findings are:

Magnetic Fields.—The measured geomagnetic field between 1.8 and 5 earth radii over the South Atlantic Ocean was found to be less than the computed field.[106] The discrepancy was attributed to the existence of the field source having its maximum strength at an altitude between 1.8 and 3 earth radii at the geomagnetic equator. From 5 to 6.6 earth radii, the measured geomagnetic field was in agreement with the computed field. Superposition of the earth's field and the interplanetary field was detected between 11 and 19 earth radii. Beyond 19 earth radii the earth's field was negligible.

The interplanetary field between 20 and 21.5 earth radii was found to be stable. However, its magnitude was more than anticipated. The field was approximately radial from the sun. An abrupt change in the character of the interplanetary field was detected simultaneously with the first detection of the solar interplanetary plasma at 21.5 earth radii. Large fluctuations both in magnitude and direction were encountered for five hours after the onset of the abrupt change and were attributed to the passage of shock waves.[107]

When the satellite was at 37.1 earth radii, an increase in the interplanetary field intensity occurred at about the same time that a sudden commencement was observed at the earth's surface, indicating that little if any delay was associated with the arrival of the sudden commencement disturbance at the earth's surface from outside the earth's field.

Solar Interplanetary Plasma.—The *Explorer* X satellite provided the first experimental observation of a plasma in interplanetary space.[108] A strongly spin-modulated signal was present at all energies near the earth, from 1.3 to 2.9 earth radii. The interplanetary plasma was first detected at 21 earth radii, and its presence was confirmed out to 38.5 earth radii.

Correlation of the plasma data with the magnetic field data indicated that the presence of the plasma was coincident with a relatively weak field which fluctuated in magnitude and direction. Absence of plasma was associated with strong steady fields directed away from the sun. Large fluctuations in plasma intensity were detected between a minimum detectable value of less than 5×10^6 and about $10^{10}/cm^2$-sec.

Explorer XI

Explorer XI, the gamma-ray astronomy satellite, was launched on April 27, 1961, into an orbit with an apogee of 1,113.2 miles, perigee of 304 miles, an inclination of 28.8°, and period of 108.1 minutes. The objective of this satellite was the detection of extraterrestrial high-energy gamma rays, such as result from the decay of neutral π mesons. The experiment was designed to detect and map the direction and intensity of the galactic gamma rays above the earth's atmosphere. Earlier balloon experiments had been limited by the background radiation produced in the residual atmosphere.

Explorer XI resembled an old-fashioned street lamp with the payload constituting the lamp and the attached burned-out fourth stage the post. Before being put into orbit, the payload and the fourth stage were spun about the longitudinal axis at 6 cps, but the whipping of the external loop antennas and the nutation damper included in the satellite converted the rotation into an end-to-end tumbling that was desired for scanning the entire sky. Storage batteries carried the major portion of the power load.

The instrumentation consisted of a gamma-ray telescope and sensing devices in the forward end of the payload. The latter reported the position of the satellite relative to the earth's horizon and to the sun. The telescope consisted of a sandwich of scintillation crystals and of a Cerenkov detector contained in an anticoincidence shield. The use of crystals with different fluorescent decay rates—namely, CsI(Tl) and NaI(Tl)—permitted differentiation between gamma rays and neutrons. When the instrumentation was not in the anticoincident mode, primary cosmic-ray protons could be observed. These particles have known energy and can be used as a calibration standard for the energy of the gamma rays.

The energy spectrum of the positive particles was found to be peaked at 500 electron volts even though the shape of the energy spectrum showed large variations. The number density of the plasma protons ranged from 6 to 20 cm^3. The plasma arrived from the general direction of the sun.

The aluminum housing of the satellite, serving as a micrometeoroid shield, emitted secondary neutrons and gamma rays, the background counts from which could obscure the results if the intensity of celestial gamma rays was low.

Explorer XI achieved too high an apogee and consequently reached into the inner Van Allen belt, which masked the gamma-ray counts in that portion of the orbit. As a result, useful data were supplied only about five percent of the time in orbit. Preliminary results of the analysis of these data have been given by the Massachusetts Institute of Technology, based on some 23 hours of useful observing time in a period of 23 days.

During this period 127 events which could have been gamma rays occurred. Of these, 105 were shown by analysis to have come from the direction of the earth and were presumably produced in the earth's atmosphere; the remaining 22 came from a variety of directions. The analysis of arrival directions was complicated by the fact that all portions of the sky were not scanned for the same length of time. Therefore use was made of an idealized model of the galaxy. It was assumed to be a disk 100,000 light years in diameter and 1,000 light years in thickness, filled uniformly with a gas of one hydrogen atom per cubic centimeter, and having a cosmic-ray density equal to its value in the vicinity of the earth. By means of this model, "predicted" intensities were used to evaluate an expected number of counts in each of the cells into which the sky was divided. Comparison of the "predictions" with the observations showed a good degree of consistency with regard to spatial distribution and to the number of events. The results are consistent with a source strength of gamma rays in the galaxy of the order of 10^{-24} cm^{-3}-sec^{-1}.[109]

Tiros III

This satellite was launched on July 12, 1961, into an orbit with an apogee of 506.44 miles, a perigee of 461.02 miles, and a 48.2° inclination. Both TV cameras were of the wide-angle type. *Tiros III* was basically the same as *Tiros II* except that a third set of infrared sensors developed by V. Suomi, of the University of Wisconsin, was added; these sensors, very much like his experiment on *Explorer VII*, consisted of two pairs of hemispheres (each pair consisting of one black and one white hemisphere), mounted on mirrors, on opposite sides of the spacecraft.

The satellite was commanded from, and data were acquired at, stations located on St. Nicolas Island, California, and Wallops Island, Virginia. An auxiliary command (only) station at Santiago, Chile, permitted obtaining more hurricane and other cloud-picture data over the tropical Atlantic Ocean than would otherwise have been possible.

Tiros III proved a worthy successor to the earlier satellites in the series, especially with regard to the discovery and tracking of Atlantic hurricanes and Pacific typhoons. The cloud-picture data from these and other weather situations were made available for operational weather analysis and forecasting through internationally disseminated operational nephanalyses.

On more than 50 separate occasions, *Tiros III* photographed tropical cyclones in all stages of development. Five hurricanes (Anna through Esther, inclusive) and one tropical storm were seen in the Atlantic; two hurricanes and a tropical storm were seen in the data-sparse Pacific near Baja California. Typhoons Kathy through Tilda, nine storms in all, were followed in the central and western Pacific.

On a single day (September 11, 1961), *Tiros III* photographed Hurricane Betsy (in a dissipating stage), Carla (as it hit the Texas coast), and Debbie; discovered the tropical storm later designated as Hurricane Esther; and photographed Typhoons Nancy and Pamela.

One of the two *Tiros III* cameras ceased operation on July 27, 1961, apparently because of a stuck shutter, confirming the desirability of redundant wide-angle cameras. Some deterioration of the quality of the TV pictures of the second camera was noted as early as the second week in August, and became progressively worse. Routine preparation of operational nephanalyses was stopped in late November 1961, when the picture quality became too poor. Few nephanalyses were transmitted after that date, and archiving of the pictures was terminated. The tape recorder on the second camera ceased to function on December 5, 1961, preventing further data acquisition in remote mode. It was possible to resume direct picture taking in early January 1962, but quality remained poor and the pictures were of little, if any, practical value. Over 35,000 pictures were obtained through the end of December 1961.

Degradation of the 6.3-μ and 7- to 30-μ infrared channels was noted as early as August 5, 1961. No useful infrared data were obtained after October 30, 1961; for several weeks before then, about 50 percent of the data were lost because of problems with the tape recorder playback mechanism.[110]

Explorer XII

Explorer XII, an Energetic Particles Satellite, was launched on August 15, 1961, in an orbit with an apogee of 47,800 miles, a perigee of 180 miles, and an inclination of 33°. The highly eccentric orbit permitted measurements both in interplanetary space and inside the earth's magnetosphere. About 90 percent of the time of the satellite was spent in the Van Allen belt region. The objectives of this satellite were to describe the protons and electrons trapped in the Van Allen radiation belt; to study the particles coming from the sun, including the occasional very intense bursts of high-energy protons which present a hazard to manned flight; to study the cosmic radiation from outside the solar system; and to correlate particle phenomena with the observed magnetic field in space about the earth.

Explorer XII was octagonal in shape, 19 inches from side to side, and 27 inches long, including a magnetometer boom. It carried four laterally extended paddles, carrying solar cells. Prior to injection into orbit, the satel-

Explorer XII, an energetic particles satellite, was launched on August 15, 1961.

lite and third stage were spun to 150 rpm for stabilization. After burnout of the third stage, a yoyo despin device slowed the rate down to 31 rpm. Further despinning to 18 rpm occurred as the solar paddles were extended just prior to separation of the spacecraft from the third stage.

Seven experiments were carried in the satellite:

(1) A proton analyzer was used to measure the proton flux and distribution of energies in the space beyond 6 earth radii. Although the mass of the particles was not measured, the particles were assumed to be protons, since the latter probably constitute at least 85 percent of the positive-ion population in space.

(2) A three-core flux-gate magnetometer, sensitive to a few gammas, was used to measure the earth's vector magnetic field at distances of 3 to 10 radii.

(3) The trapped-radiation experiment (four Geiger counters and three CdS cells) measured the fluxes and energies of particles emitted by the sun, the galactic cosmic rays, as well as trapped Van Allen belt particles.

(4) The cosmic-ray experiment monitored cosmic rays beyond the effect of the earth's magnetic field during the apogee portion of the orbit. The instrumentation consisted of a double telescope for cosmic rays, a single crystal detector of energetic particles, and a Geiger-Mueller telescope for cosmic rays. The group of instruments was capable of obtaining data on the flux of moderate to very energetic protons 1 to 700 MeV, on the flux of low-energy alpha particles, and on the differential spectrum of proton energy.

(5) An ion-electron detector was carried to measure particle fluxes, types, and energies in and above the Van Allen belt. This device consisted of a photomultiplier tube coated with a powder phosphor, $ZnS(Ag)$, in combination with absorbing screens for the detection of energetic particles and with a scattering block for the detection of electrons. The energy flux could be measured for protons with energies below 1 MeV and for electrons below 100 KeV.

(6) An experiment was carried to determine the deterioration of solar cells resulting from direct exposure to the radiation in the Van Allen belt, and to compare the effectiveness of glass filters in preventing degradation of solar cells.

(7) An optical aspect experiment was carried to determine the orientation in space of the spacecraft as a function of time. Six photodiodes gave the position of the spin axis of the satellite relative to the sun's elevation with an accuracy of about 5° in azimuth and elevation.

The launch occurred as planned, and all experiments functioned normally until December 6, 1961, when the satellite ceased transmitting; telemetry coverage of nearly 100 percent was maintained until then. By September 12 the spin rate of the satellite had increased from the initial value of 27.8 to 28.63 rpm as a result of the solar-radiation pressure on the solar cell paddles. The unprotected cells in the solar cell experiment suffered a major degradation when the spacecraft passed through a point of high proton concentration in the Van Allen belt. Degradation for the test cell patches covered by glass was not noticeable.

The instrumentation carried in this satellite indicated that the level of electron flux in the outer portion of the Van Allen belt was about three orders lower than what had previously been considered to be present. The ion-electron detector, which was capable of measuring protons of low energy, showed that protons are present in the outer Van Allen belt. This region was at one time thought to contain only electrons. The instrumentation on board this satellite detected electrons out to a boundary of 8 earth radii, with a maximum flux at 6 to 7 earth radii. The data from the satellite confirmed existence of a low-energy proton current ringing the earth in an east-to-west direction, perpendicular to perpetual north-south spiraling motion along geomagnetic field lines.[111]

Explorer XIII

The effects of micrometeoroids or cosmic dust in collisions with spacecraft, and the degree of likelihood of collisions, were unknown, but the possibility that the particles constitute a hazard to travel in the space environment was very real and required investigation. The micrometeoroid satellite, *Explorer XIII*, a project of the Langley Research Center with GSFC participation, was designed to provide an improved estimate of the danger of penetration of spacecraft by cosmic dust by securing direct measurements of the puncture hazards in spacecraft structural skin specimens at satellite altitudes. In addition, the satellite carried instruments to measure micrometeoroid flux rates and to obtain data regarding the erosion of spacecraft materials.

The satellite was launched on August 25, 1961, failed to obtain a high enough perigee, and all the data collected during its 2½ days of life were telemetered during 13 minutes of time. It reentered August 27, 1961.

Explorer XIII carried a group of experiments installed around the fourth stage of a Scout launch vehicle. The satellite was cylindrical in shape, 76 inches in length, and 24 inches in diameter; the overall weight was about 187 pounds. Five types of detectors were carried. One consisted of a battery of pressurized cells in which the pressure was released upon puncture. In the second type, foil gauges showed impact of micrometeoroids by a change in resistance. A third type of experiment showed a change of resistance in wire grids when a wire was broken by impact. Another detector used photoelectric cells which detected the light transmitted through the aluminized Mylar sheets. The fifth experiment recorded impacts on piezoelectric crystals.

The satellite was spin-stabilized during the last stage of burning. It was

SATELLITES AND SPACE PROBES

expected that interaction of the satellite with the magnetic field of the earth would cause the original spin to turn into a tumbling motion after 10 days. Two separate telemeters were used for storing and telemetering data. Both used Goddard's Minitrack telemetry and coding system.[112]

P–21 Electron Density Profile Probe

The P–21 space probe was launched on October 19, 1961, from Wallops Island by means of a Scout rocket in a nearly vertical trajectory. The altitude achieved was 4,261 miles. The payload itself was in the form of an eight-sided frustrum and weighed 94 pounds. The heat shield had been selected for low electrical conductivity, to permit an attempt to use the stored antennas during the propulsion period.

The flight of the probe was about 2 hours in duration. It was launched near mid-day at a time when the ionosphere appeared quiet. The payload was spin-stabilized about its axis before fourth-stage separation. Telemetry of the swept-frequency r-f probe data made use of the 73.6-Mc frequency. The instrumentation carried consisted of: (1) a continuous-wave (CW) propagation experiment to measure the ionosphere profile; (2) an r-f probe experiment to measure the ionospheric electron density, especially at altitudes above 620 miles where data are particularly scarce; and (3) a swept-frequency probe to provide information on the power absorbed by electron pressure waves. Two CW signals were transmitted from the rocket to the ground, one at 12.27 Mc and one at 73.6 Mc. They were controlled to an exact 6 to 1 ratio. Comparison of the two frequencies for Faraday rotation gave the columnar electron density in the path traversed by the beam; this experiment was of value only during the ascent of the rocket and only below 2,480 miles. A new instrument was used with the dual purpose of providing information on electron densities in the ionosphere and information concerning the behavior of the probes themselves.

It was planned to calibrate the r-f probe during ascent to 2,480 miles by means of the CW experiment and to use the probe measurements for the remainder of the flight. The r-f probe made use of a capacitor in which the dielectric constant of the gas between the plates was affected by the ionization of the gas. Comparison of the capacity of the probe when in the ionosphere and when outside supplied the information necessary to determine electron density in the ionosphere. The heat shield had been selected for low electrical conductivity to permit an attempt to use the stored antennas during the propulsion period.

Tiros IV

This satellite was launched on February 8, 1962, into an orbit with a 48.29° inclination, an apogee of 525 miles, and a perigee of 471 miles. One of the TV cameras was a standard Tiros wide-angle type; the second carried a Tegea Kinoptik lens not previously used on Tiros. In the scanning radiometer, the 7- to 30-μ sensor had been dropped to provide a channel for picture-timing data. Beacon frequencies had been transferred from the 108-Mc to the 136-Mc band. Otherwise the satellite was basically the same as *Tiros III*. Vidicon picture transmission ceased on June 11, 1962, after some 30,000 pictures were transmitted.[113]

OSO I

The Orbiting Solar Observatory satellite, designed for the study of radiations from the sun with instrumentation located above the earth's atmosphere, was launched on March 7, 1962. *OSO I* weighed about 458 pounds. It reached an orbit with an apogee of 369 miles and a perigee of 343.5 miles, an inclination of 32.8°, and a period of 96.15 minutes. The measurements included ultraviolet radiation studies using narrow-band detectors, a gamma-ray experiment to study solar emission in the 0.1- to 500-MeV region, neutron flux measurements from the earth and the sun, and studies of the time variation of solar ultraviolet, x-ray, and gamma-ray emissions. An instrument obtained spectral and spatial variations in the 1A and 10A radiation region as the instrument's look-angle scanned across the inner corona and the solar disk. Gamma rays in the 0.1- to 5-MeV region were detected by a combination of scintillation counters. An electronic pulse height analysis of the counter pulses yielded the energy spectrum of the rays. To measure gamma rays in the 100- to 500-MeV region, a Cerenkov detector viewed from four directions by multiplier tubes in conjunction with an anticoincident scintillator was used. The neutron experiment, using BF_3 counters, monitored the neutron flux from the earth

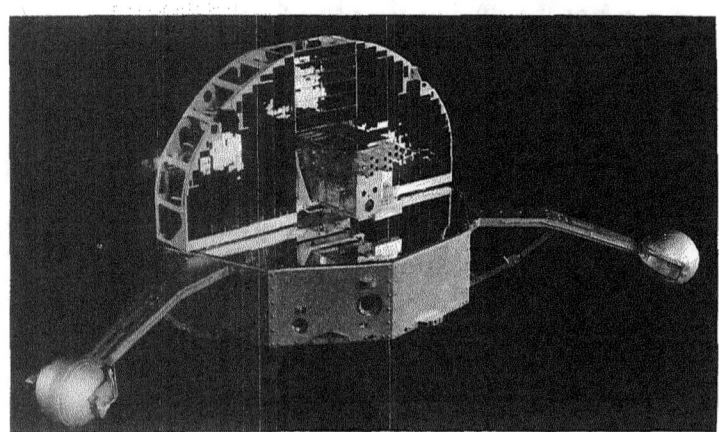

Orbiting Solar Observatory I, launched March 7, 1962.

and the sun. One objective of this experiment was to determine whether the lower radiation belt arises from the decay of neutrons emitted from the earth's atmosphere.

The satellite consisted of a spinning portion for gyroscopically stabilizing the payload in space and a servo-driven instrumentation section providing azimuth and partial elevation control. This instrumentation section was oriented within one minute of arc. It housed about 173 pounds of instrumentation requiring a stabilized view of the sun for operation. Additional instrumentation was carried in the spinning wheel portion of the satellite. Power for the servosystems and the instruments was furnished by solar cells and rechargeable chemical storage cells.

After May 22, 1962, contact was lost with *OSO I.* Then contact was reestablished on June 24, 1962, and satisfactory data were received until August 6, 1963, thereby exceeding the expected life of the satellite by almost a year. The satellite is still in orbit.

All experiments transmitted as programed, resulting in a significant increase in knowledge about the composition and behavior of the sun. Data revealed tentative evidence that solar flares may be preceded by a series of microflares whose sequence and pattern may be predictable. *OSO I* reported at least four of these series during a year in orbit.[114]

P–21a Electron Density Profile Probe

A second electron density profile probe, P–21a, was launched on March 29, 1962. (See the account of the launch of P–21 on October 19, 1961, for more details on the mission and instrumentation of the electron density probes.) In addition to equipment carried on P–21, P–21a had a swept-frequency radio-frequency probe and a positive-ion detector. Like P–21, the second probe was to measure electron density profile and intensity of ions in the atmosphere. However, it was launched so the measurements could be made at night. The probe achieved an altitude of 3,910 miles and transmitted good data. The 12.3-Mc transmission failed, but Faraday rotation data were obtained by 73.6 Mc to give electron density profile at nighttime. As a result of P–21a data, it was concluded that the characteristics of the ionosphere differ drastically from the daytime state when the temperature of the ionosphere is much cooler.

The data from the P–21 and P–21a probes and from satellites were combined with sounding rocket data to enable scientists to map in greater detail the structure of the upper atmosphere. These data show that there are two transition regions (from O^+ to He^+ and from He^+ to H^+) in the upper ionosphere rather than a single transition from O^+ to H^+ as was previously believed. The O^+/He^+ transition, i.e., $O^+/He^+=1$, was between 800 and 1,400 kilometers, depending on the atmospheric temperature. The measured temperature in the upper ionosphere was found to be constant with altitude within a few percent and consistent with a previously developed empirical relation which predicts the temperature as a function of diurnal time and of solar activity. The determined altitudes of the ion transition levels were in good agreement with a theoretical model which describes these altitudes as a function of atmospheric temperature.

Ariel I

On April 26, 1962, the first international satellite was launched as a joint United Kingdom-United States project. The satellite was designated *Ariel I*, the International Ionosphere Satellite. The spacecraft transmitted regularly until November 1963, when its transmission became intermittent.

The satellite had an overall diameter of 23 inches, was 22 inches high, and weighed 136 pounds. It was spin stabilized at a rate of 12 to 36 rpm. Solar cells and nickel-cadmium batteries supplied the power. Data were stored on a 100-minute tape recorder. Orbital parameters were: apogee, 754.2 miles; perigee, 242.1 miles; inclination, 53.86°; period, 100.9 minutes; eccentricity, 0.057.

This project developed from proposals made in 1959 to NASA by the British National Committee on Space Research. These proposals were in response to a United States offer to the Committee on Space Research (COSPAR) of the International Council of Scientific Unions to launch

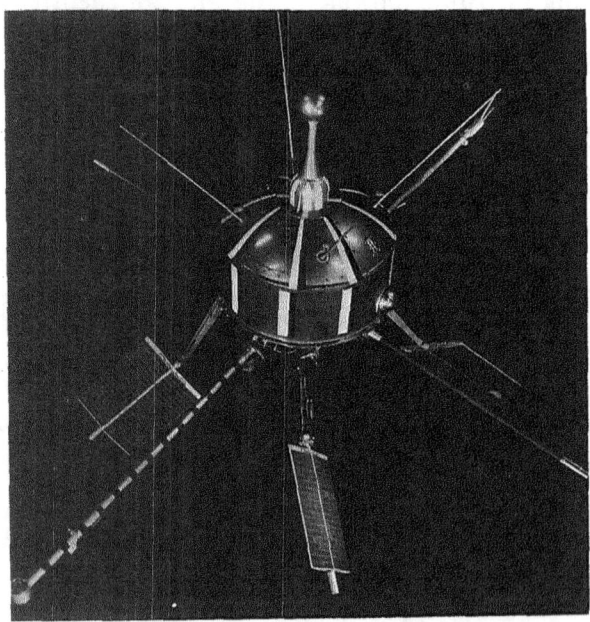

Ariel I, launched April 26, 1962.

scientific experiments or complete satellites prepared by scientists of other nations. The content of the program and the division of responsibility between NASA and the British Committee were agreed to during discussions that took place in late 1959 and early 1960. Subsequently the NASA Administrator assigned project responsibility for the United States to the Goddard Space Flight Center.

This assignment included the design, fabrication, integration, and testing of the spacecraft structure, power supply telemetry, command receiver, thermal control, and data storage. GSFC supplied the vehicle, was responsible for launch, performed data acquisition via the worldwide Minitrack network, and provided data processing. The United Kingdom had the responsibility for the design, fabrication, and testing of all flight sensors and their associated electronics up to the telemetry encoder input. The United Kingdom also was responsible for data analysis and interpretation.

The University College, London, carried out a number of experiments that included an electron temperature and density determination based on Druyvesteyn's modification of the Langmuir probe to determine the electron density and temperature near the satellite. Ion mass, composition, and temperature instrumentation was essentially the same as the electron temperature experiment with another method of temperature measurement, a solar Lyman-alpha emission measurement. This last experiment measured two parts of the solar spectrum to enable simultaneous and nearly continuous observations of the state of the ionosphere and of the solar atmosphere. In addition, University College installed instruments to measure solar x-ray emission, latitude and longitude of the sun with respect to the satellite, and satellite spin rate.

Imperial College, London, installed instrumentation to make accurate measurements of the primary cosmic ray energy spectrum and the effects of interplanetary magnetic field modulation of this spectrum. The University of Birmingham conducted ionosphere electron density measurements with instruments that differed from those of the University College, London, to provide checks on the measurements.

Goddard Space Flight Center provided subsystems that accommodated telemetry, data encoder, tape recorder, power system, and spacecraft parameters (housekeeping) system.[115]

Tiros V

Tiros V was launched from the Atlantic Missile Range on June 19, 1962. The infrared sensor had not functioned in a prelaunch test, but the launch was not delayed to repair the sensor; the satellite was needed for observations, during the August-September tropical storm season. The orbit, with a 367-mile perigee and a 604-mile apogee, was more elliptical than planned, but good picture transmission was still possible.

Tiros V is mated to second stage of launch vehicle.

Tiros V liftoff from Cape Canaveral, June 19, 1962.

SATELLITES AND SPACE PROBES

The first camera ceased transmitting data on July 6, 1962, and the second on May 4, 1963. During its 10½ operational months, *Tiros V* transmitted 57,857 pictures, about 80 percent of which were usable for cloud cover analysis.

Of the ten major tropical storms in the 1962 season, *Tiros V* observed five of them. In addition, it supplied data that were used in weather analysis for the orbital flights of Project Mercury Astronauts Walter Schirra and Gordon Cooper. *Tiros V* data were used by Australia in the first instance of international use of weather satellite data. On September 6, 1962, data from *Tiros V* satellite were sent to France via the *Telstar I* communication satellite.[116]

Telstar I

A dramatic new era in world communications was inaugurated on July 10, 1962, when a Goddard Space Flight Center team launched the world's first active communications satellite. *Telstar I* was a product of private industry, American Telephone & Telegraph Company, launched for AT&T by NASA on a reimbursable basis. Here was a satellite which enabled a whole continent to "see" across oceans. Television programs from and to Europe, for instance, brought new, real-time sights and sounds into the homes of millions. Even though Telstar's "mutual visibility"—the time during which signals could be sent and received—was relatively short (approximately 15 to 20 minutes), the portents of this new communications medium was immediate. With an elliptical orbit that crossed the Van Allen belts, *Telstar I* taught engineers a great deal about radiation damage to communications equipment.

Telstar I, launched July 10, 1962.

A legislative debate soon ensued on Capitol Hill as to how this new communications system was to be used operationally—by private industry, by a public utility, or by a Governmental agency. On August 31, 1962, President John F. Kennedy signed Public Law 87-624, the "Communications Satellite Act of 1962" (Exhibit 16). This law created a "communications satellite corporation for profit which will not be an agency or establishment of the United States Government, but which would have government representation on its Board of Directors and have many of its activities regulated by Government." A space-age development now became a new business enterprise and marked a new form of Government-business collaboration.[117]

Tiros VI

Tiros VI was launched on September 18, 1962. Good coverage of a large portion of the earth's cloud cover was possible because the satellite went into an almost circular orbit with an inclination of 58.3°. The launch was originally scheduled for November but was pushed forward to September so that *Tiros VI* could serve as a backup for *Tiros V* during the last half of the tropical storm season. It also was used for weather observations for the Project Mercury flights of Astronauts Walter Schirra and Gordon Cooper.

Transmission was such that *Tiros VI* was able to aid in the detection of hurricanes and typhoons in both the 1962 and 1963 tropical storm seasons. Data from this satellite were used by the U.S. Weather Bureau in daily forecasts.

On December 1, 1962, the medium-angle camera ceased to function, and it was announced on October 17, 1963, that the satellite was no longer transmitting data. During its 13 months of active lifetime, *Tiros VI* transmitted over 67,000 pictures.[118]

Alouette I

On September 29, 1962, a second international satellite, the *Alouette I*, was launched. A United States-Canadian project, it was NASA's first satellite launched from the Pacific Missile Range and into polar orbit.

Alouette I was a project of the Canadian Defence Research Board, and a part of NASA's Topside Sounder Program. The primary objective was to examine the structure of the ionosphere from above in a manner similar to that being used by ground-based sounding stations. In particular, information was desired about the ionosphere in the region above the maximum electron density of the F layer, usually about 188 to 250 statute miles above the earth's surface.

The topside sounder was carried in a satellite traveling at 638-mile apogee and 620-mile perigee. It was launched with a Thor-Agena vehicle. The satellite provided radio transmissions (downward and via the Echo

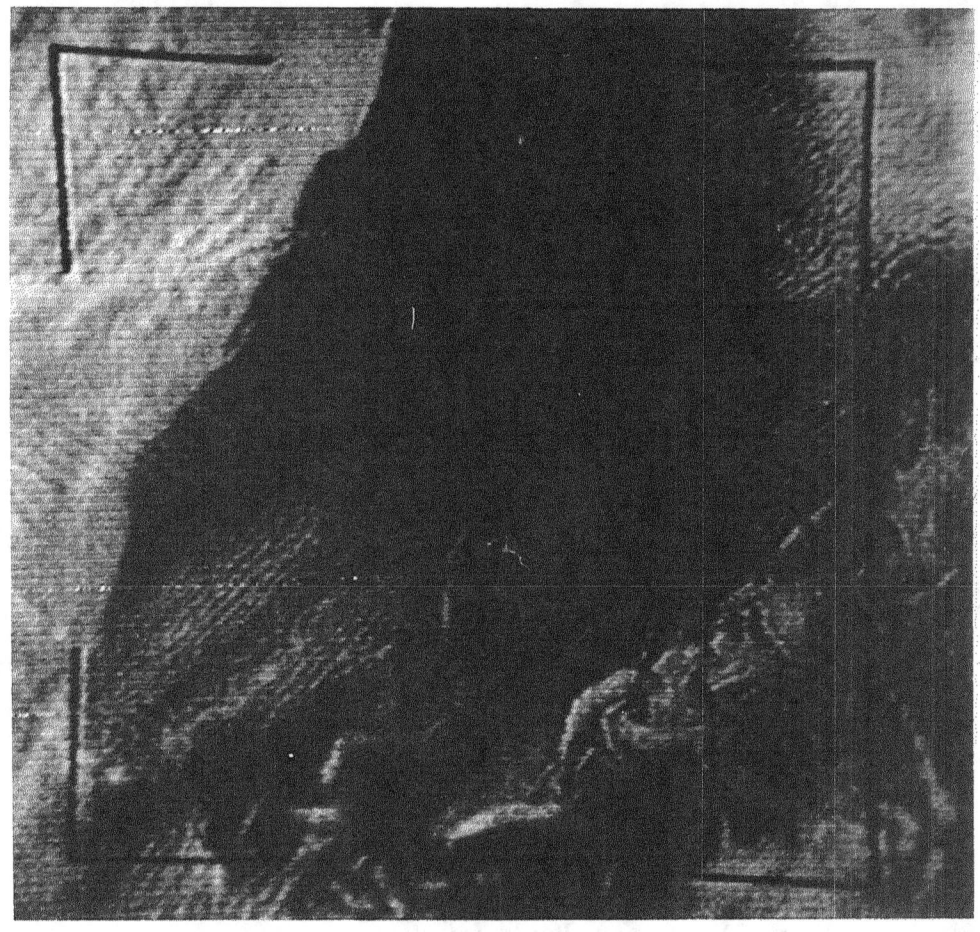

Tiros VI photographed Cape Blanc, Africa, the clearly discernible ocean, and eddy pattern from Canary Islands, April 23, 1963.

satellite) over a frequency range of about 2 to 15 Mc. The data were telemetered to Canadian, United States, and United Kingdom sites.

After having been in orbit for one year, all four experiments—ionospheric sounder, energetic-particle counters, VLF receiver, and cosmic-noise-intensity equipment—continued to provide data without degradation of quality since launch. Energetic photons and corpuscles, together with micrometeoroids, were gradually decreasing the efficiency of the solar cells. The solar-cell charging power was down to 58 percent of its initial value, but the efficiency was declining at a much slower rate, confirming predicted rate of decrease.

Canadian scientists check *Alouette I*, which was launched September 29, 1962.

The control center for the *Alouette I* satellite was at the Defence Research Telecommunications Establishment (DRTE) in Ottawa, Canada. The magnetic tape was processed at DRTE and topside ionograms were filed at the World Data Center, Boulder, Colorado.

From preliminary analysis, it appeared evident that the *Alouette I* topside sounder not only clarified many of the earlier concepts about the structure of the topside ionosphere, but at the same time raised a number of new questions concerning the relative importance of solar, magnetic, and corpuscular control of the topside ionosphere.

The analysis of *Alouette I* data has led to the publication of numerous scientific papers. About two-thirds of these were of Canadian origin, the remaining one-third were by scientists of the United Kingdom and United States.[119]

Explorer XIV

This Energetic Particles Satellite, launched on October 2, 1962, contained a cosmic-ray experiment, an ion detector experiment, a solar cell experiment, a plasma probe and analyzer, a trapped radiation experiment, and a magnetometer experiment. Its objectives were to describe the trapped

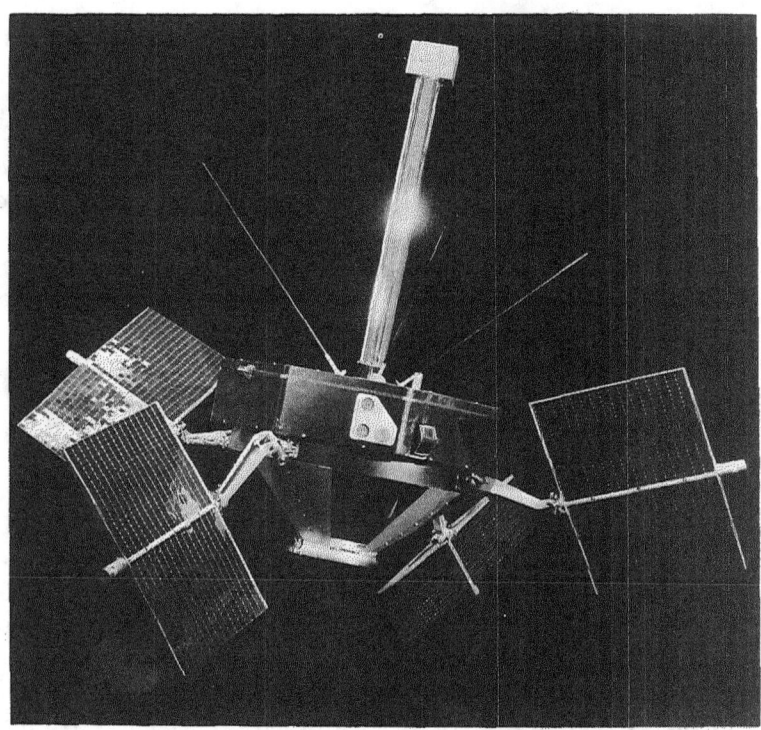
Explorer XIV.

corpuscular radiation, solar particles, cosmic radiation, and the solar wind and to correlate the particle phenomena with the magnetic field observations. This satellite was to continue and extend the energetic-particles study undertaken by *Explorer XII*.

The 89-pound *Explorer XIV* satellite was launched into a highly elliptical orbit with an apogee of 61,090 miles and a perigee of 174 miles. The satellite transmitted data from the six experiments and tracking beacon regularly until January 1963, when stabilization difficulties caused the loss of about 15 days of data. On January 29, transmission was resumed and continued until mid-August when all usable signals except the tracking beacon ceased. In 10 months the satellite had transmitted 6,500 hours of usable data.

Instruments enabled scientists to chart the boundaries of the magnetosphere with more precision than had been previously possible, and it was found that it flared away from the earth in a definite shape.

The effects of magnetic fields on particles and radiation, and variations in the magnetic field during the day-night cycle were observed. There was possible confirmation of *Explorer VI*'s claim of a ring current on the night side of the earth. Dr. James Van Allen said that *Explorer XIV* and *Explorer III* data indicated the radiation from high-altitude nuclear tests that was trapped in the ionosphere would remain much longer than he had previously estimated.[120]

Explorer XV

This satellite was launched October 27, 1962, for the purpose of studying the new artificial radiation belt created by nuclear explosions. Instrumentation was similar to *Explorer XII* and included p-n junction electron detectors, scintillator detectors, a scintillator telescope, and a triaxial fluxgate magnetometer.

The satellite went silent in February 1963, after transmitting 2,067 hours of data. Digitized data were sent to the five experimenters, who used it to determine more exactly the intensity and location of the artificial radiation.[121]

Relay I

Relay I, NASA's first active communications satellite, was launched by Goddard Space Flight Center on December 13, 1962. The objective of the 172-pound satellite was to investigate wideband communications between ground stations by means of a low-altitude satellite. *Relay I* was placed in an orbit with an apogee of 4,612 miles and perigee of 819 miles. Although *Relay I* was primarily a communications satellite, the major portion of its instrumentation was designed to evaluate the satellite's circuitry and equipment and transmit this information to earth. Seven instruments were designed to measure energetic particles and the effects of these particles on *Relay I*'s instrumentation.

Communications signals to be evaluated were an assortment of television signals, multichannel telephone, and other communications. Wideband stations used in the experiment were located at Rumford, Maine; Pleumeur-Bodou, France; Goonhilly Downs, England; and Weilheim, West Germany. Narrowband stations were located at Nutley, New Jersey, and Rio de Janeiro, Brazil.

Some of the most noteworthy of the transmissions made on *Relay I* were the first three-way link between North America, South America, and Europe; first simultaneous TV transmission to London, Paris, and Rome (March 11, 1963); first known color transmission (March 19, 1963); and transmission of the coverage of President Kennedy's tragic death (November 22 to 26, 1963).

Transmission difficulty was experienced during the week after launch and for about a week during March 1963, but long-term performance of the satellite was considered excellent. Although all planned experiments had been completed by March 1963, *Relay I* was still transmitting at the end of 1963. At that time it had completed 2,880 orbits, performed 1,330 wide-

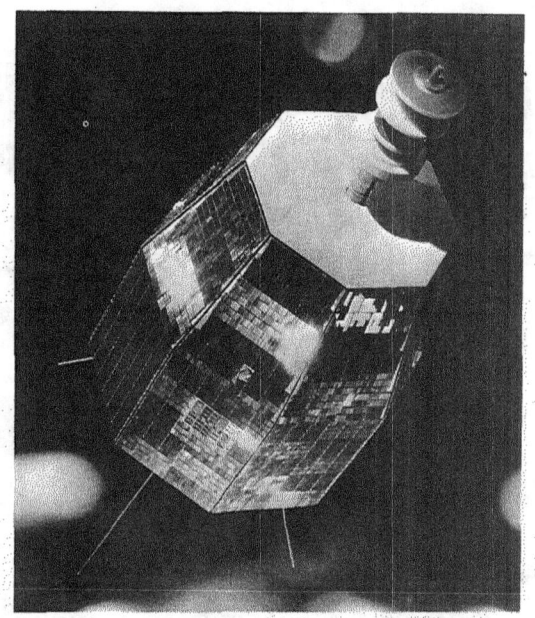

Relay I, which was launched December 13, 1962.

band experiments, 720 narrowband experiments, and 157 demonstrations (TV and narrowband). The transponder had been operated for 288 hours over a period of 720 operations.[122]

Syncom I

Syncom I was launched on February 14, 1963. The objective was to place in orbit a 24-hour (synchronous) active communications satellite with the Delta rocket. The booster functioned well, but 20 seconds after the apogee motor fired to place the satellite in a near-synchronous orbit, all communications were lost. Optical sightings of the satellite were made after some days; it was found to be in a near-synchronous orbit traveling eastward at about 2.8° per day. The firing of the apogee motor may have damaged the satellite.[123]

Explorer XVII

This Explorer satellite, launched on April 3, 1963, was to make studies of the earth's upper atmosphere. Specifically, it was designed to make direct samplings of atmospheric constituents such as helium, nitrogen, and oxygen. Measurements were made with two mass spectrometers, four vacuum-pressure gauges, and two electrostatic probes. Telemetry was performed with a new pulse-code-modulation system—a solid-state system

Syncom I.

providing output power of 500 milliwatts and capable of supplying 40 separate channels of information in digital form.

After a few days of operation *Explorer XVII* had more than tripled all previous direct measurements of the neutral gases in the earth's upper atmosphere. Among the data was confirmation that the earth is surrounded by a belt of neutral helium at an altitude of 150 to 60 miles. The satellite became inactive on July 10, 1963.[124]

Telstar II

On May 7, 1963, *Telstar II* was launched into orbit from Cape Canaveral. This medium-altitude, active communications satellite was designed to provide additional information on TV, radio, telephone, and data transmission, as well as experiments on the effects of radiation on the on-

Explorer XVII readied for launch.

board communications equipment. It was a project of AT&T launched by a Goddard Center launch team. The Center also participated in some of the experiments. It was successfully used for transmitting TV, color TV, and voice messages between the United States, France, and England. The transmitter failed after some 60 days in orbit.[125]

Tiros VII

Tiros VII was launched June 19, 1963, into an orbit designed to provide maximum Northern Hemisphere hurricane coverage for the 1963 season. The satellite was equipped with two vidicon cameras with wide-angle lenses, a five-channel medium-resolution radiometer to measure infrared radiation, an electron temperature probe, and a magnetic attitude coil.

Coverage extended to 65° N and 65° S latitudes, and included Hurricanes Arlene through Ginny during the 1963 hurricane season. The electron temperature probe malfunctioned 26 days after launch, but the two cameras and the infrared subsystems remained active for over two years. Spacecraft reliability had truly made great strides and the Goddard team and its contractors had laid the foundation for an operational system of weather satellites.[126]

Syncom II

Syncom II communications satellite was launched July 26, 1963, and by September had been maneuvered into a near-perfect synchronous orbit. Firings of hydrogen peroxide jets on August 11 slowed *Syncom II* from 7° drift per day to 2.7° drift per day, and on August 12 the drift was reduced to 1.2° per day. By September 7 the satellite was in orbit over Brazil and the South Atlantic Ocean at an altitude of more than 22,000 miles in an orbit that varied from an absolute circle by no more than 4.5 miles. Orbital period was 23 hours, 55 minutes and 54 seconds—only 0.09 second shorter than the mean sidereal day. The satellite had a drift of about one degree per month that was corrected by a periodic figure-eight carrying the satellite along the 55° meridian to points 33° north and south of the equator. NASA Administrator James E. Webb called completion of the positioning maneuvers the culmination of "one of the outstanding feats in the history of space flight."

The satellite was equipped with a spin-stabilized active repeater consisting of a 7,200-Mc receiver and an 1,800-Mc transmitter with an output of 2 watts. A vernier velocity-control system was installed for orientation of spin axis and adjustment of the orbit. In addition, onboard instrumentation could measure the effect of radiation on the solar cells that powered the spacecraft. Measurement of power loss resulting from radiation damage confirmed the desirability of changing the next Syncom satellite to n/p cells with 0.012-inch quartz cover slides.

A telephone conversation between President John F. Kennedy and Nigerian Prime Minister Sir Abubaker Tafawa Balewa on August 23, 1963, was the first transmission of the satellite. The success of this project confirmed the feasibility of earth-synchronous satellite systems, a technical achievement of major significance.[127]

The success of Goddard's work on the synchronous satellite development was to pave the way for the world's first commercial communications satellite, *Early Bird*. After its NASA research and development role was completed, *Syncom II* continued its useful service. NASA transferred operation of the satellite to DOD on January 1, 1965; DOD used it for communications with the armed forces in Vietnam.

Explorer XVIII

Explorer XVIII, an Interplanetary Monitoring Platform (IMP), was launched on November 27, 1963. The satellite carried ten experiments designed to study the radiation environment of cislunar space and to monitor this region over a significant portion of a solar cycle. Special emphasis was placed on the acquisition of simultaneous data to aid in the determination of interdependent effects of magnetic and ion fields. In addition, it was hoped that knowledge could be gained for the further development of a simple, relatively inexpensive, spin-stabilized spacecraft for interplanetary investigation.

An elliptical orbit of 121,605-mile apogee and 122-mile perigee, with

Useful Lifetimes of Goddard Satellites

Project	Other name	Useful life, days	Remarks
Explorer VI	S-2	68	
Explorer VII	S-1a		
Explorer VIII	S-30	55	
Explorer X	P-14	2	Satellite achieved estimated useful lifetime.
Explorer XI	S-15	180	
Explorer XII	S-3	112	Transmission ceased abruptly.
Explorer XIII	S-55a	2	
Explorer XIV	S-3a	310	
Explorer XV	S-3b	95	
Explorer XVII	S-6	100	Power supply designed for 3 months.
Explorer XVIII	IMP-I	300	Survived 9-hour shadow at 160 days.
Echo I	A-11	250	Number of days useful as a communications satellite; number of days useful for scientific information indefinite.
Tiros I	A-1	77	Number of days TV data meteorologically useful.
Tiros II	A-2	231	
Tiros III	A-3	145	Number of days TV system useful.
Tiros IV	A-9	146	Number of days IR system provided useful meteorological data.
Tiros V	A-50	330	Number of days TV system useful.
Tiros VI	A-51	388	
Tiros VII	A-52	920[a]	
Tiros VIII	A-53	740[a]	
Ariel I	S-51	320	Some damage from Starfish event (74 days); useful data continued at a decreasing rate.
Alouette I	S-27	1,185[a]	
Relay I	A-15	924	
Syncom I	A-25	0.25	
Syncom II	A-26	900[a]	
Vanguard III		85	Achieved estimated useful lifetime.

[a] Still functioning December 31, 1965.

an orbital period of 93 hours, was achieved. All experiments and equipment operated satisfactorily except for the thermal ion experiment, which gave only 10 percent usable data. The satellite was the first to accurately measure the interplanetary magnetic field and shock front, and to survive a severe earth shadow of 7 hours and 55 minutes. One year later, on December 15, 1964, Dr. Norman F. Ness, speaking at a Goddard Scientific Symposium, discussed some of the findings produced by the IMP satellite.

Dr. Ness compared the earth to a comet, explaining the presence of a long magnetic tail extending to an unknown distance out beyond the dark (night) side of the earth.

The new theory was drawn from the results of the first detailed mapping of the earth's magnetic field on the nighttime side of the magnetosphere by this satellite. Earlier, scientists had believed the earth's magnetic field on its dark side was draped out far beyond the earth in a massive closed teardrop. Under the new theory countless magnetic lines of force stretch out like the tail of a comet to a still-to-be-determined distance in space, possibly beyond the moon. Within this vast comet-like tail the lines of force in the Northern Hemisphere are directed toward the sun; in the Southern Hemisphere, away from the sun. In between there is a neutral zone.

Dr. Ness characterized this neutral zone—which had been hypothesized but never before detected—as a thin sheet, which is a permanent part of the earth's environment and virtually void of magnetic activity. Though the neutral zone's exact role was unknown, it may be responsible for speeding particles into the earth's polar region, either directly or via the Van Allen belt to cause aurora. The neutral zone may even play a major role in the development of the Van Allen belt. Because of its location, it may give rise to gegenschein, a slight but noticeable increase in the luminosity of the night sky.[128]

Tiros VIII

The eighth Tiros weather satellite was launched on December 21, 1963. The satellite went into a successful orbit, and proved for the first time the feasibility of automatic picture transmission (APT) for direct facsimile readout from the satellite. This system allowed weathermen around the world, using inexpensive ground equipment, to receive an almost real-time photo of the weather in their area.

Recognizing the continuing success of the Tiros weather satellite program, the U.S. Weather Bureau in late 1962 began discussions with NASA to continue this program on an operational basis. This follow-on effort was named Tiros Operational Satellite System (TOSS). On May 23, 1963, TOSS was formally implemented when the Weather Bureau issued a $9,132,000 purchase order to NASA providing for three Tiros spacecraft, two Thor-Delta launch vehicles, plus associated launch, data acquisition programing, and data analysis services. The first two TOSS spacecraft would be identical to *Tiros VI* with its two wide-angle cameras. The third would embody a "cart wheel" configuration producing vertically oriented pictures taken by cameras looking out from the spacecraft rim.[129]

Boosters and Sounding Rockets
8

SATELLITES WERE THE BACKBONE of the Goddard Space Flight Center research program, but no account of the Center would be complete without discussing the boosters that placed these satellites into orbit and the sounding rockets with their complementary research.

Delta

By far the most important of the launch vehicles has been the Delta. It was important to the entire U.S. space program because of the frequency of its use and its proven reliability. It is also important in a study of the Goddard program because its development was closely associated with the efforts of the Goddard Space Flight Center. Like the original core of Goddard personnel, the Delta evolved, in part, from the Vanguard project at the Naval Research Laboratory.

In September 1955, NRL was given a green light for the development of a rocket capable of launching a satellite into orbit. This project, known as Project Vanguard, was to be part of the International Geophysical Year and was not to interfere with the Department of Defense rocket program. The development of the Vanguard vehicle and the orbiting of Vanguard satellites has already been discussed (see ch. 2).

Although the Delta is generally considered the progeny of the Vanguard, there is another thread of the Delta development that must be picked up. Near the end of 1955, the U.S. Air Force awarded Douglas Aircraft Company a contract to develop a 1,500-mile Intermediate Range Ballistic Missile (IRBM). While this vehicle, known as the Thor, was still in development, it was augmented by mating the Thor with two upper stages that were essentially the upper stages of the Vanguard. The new vehicle was fired as the Thor-Able in April 1958 and was widely used in the early period of the Air Force space program.

Just one year later a NASA contract was signed for the Delta, a rocket with satellite launching capability. This rocket, which was patterned after the Air Force Thor-Able, first successfully launched a satellite on August 12, 1960.

The Delta was originally conceived as an "interim launch vehicle" for satellite launching through 1961 or 1962—a rocket to serve until larger-thrust rockets were developed. But the Delta proved so reliable that it became an "off the shelf" item. Although some programs, such as manned space flight, had to wait until larger rockets were available, most of NASA's satellite needs could be met satisfactorily and economically with a Delta-size vehicle. The Delta was continuously modified so that it kept pace with the demands made on a medium-weight satellite launch vehicle. It thus achieved exceptional reliability and versatility under the guidance of its Goddard project managers.

These improvements not only included innumerable modifications but a general upgrading of the capabilities of the rocket. The Delta rockets used through 1960 and 1962 lifted payloads of less than 300 pounds into relatively low orbits. Later Deltas were capable of placing satellites of over 800

DELTA CONFIGURATION

Stages: 3

Propellants: 1st stage, liquid oxygen and kerosene; 2d stage, unsymmetrical dimethylhydrazine and inhibited red fuming nitric acid; 3d stage, solid

Thrust: 1st stage, 170,000 lb at sea level; 2d stage, 7,700 lb; 3d stage, 2,800 lb

Maximum Diameter: 8 ft, excluding fins

Height: 88 ft, less spacecraft

Payload: 800 lb in 350 n.mi. orbit; 130 lb escape

---

pounds into orbits of around 1,000 nautical miles. Changes included stretching the second stage propellant tank by three feet in 1962 and replacing the X-248 solid propellant third stage in 1963 with the higher performance X-258.

It is ironic that Delta's remarkable flight record started with a failure. In its debut at Cape Canaveral on May 13, 1960, Delta No. 1 failed to put a 100-foot-diameter Echo balloon into orbit. A circuitry problem in the second stage was diagnosed as the problem. The circuitry subsystem was redesigned, more severely tested, and installed in the second stage for another flight. On August 12, 1960, 3 months after the first failure, Delta No. 2 successfully launched a backup Echo passive communications satellite into orbit. Delta No. 2 was the start of a successful launch string that would last for 2½ years and include 21 successful space missions.

Among the satellite "firsts" boosted into orbit by Delta were: the first passive communications satellite, *Echo I*; the first international satellite,

Delta Growth (1962 to 1963)

Delta configuration	Earth orbit (300 miles)	Escape
DM19 (1960)	525 lb	70 lb
DSV-3A and 3B (1962) 3-foot longer second stage tanks	800 lb[a]	95 lb
DSV-3C (1963) X-258 motor replaced X-248 motor	800 lb[a]	115 lb

[a] Structural limits of the second stage limit spacecraft weight to 800 pounds.

Delta Launch Vehicle Record, 1960 to 1963

Vehicle No.	Mission	Results	Launch	Weight, pounds
1	Echo	Failed	May 13, 1960	132
2	Echo I	Successful	August 12, 1960	200
3	Tiros II	"	November 23, 1960	280
4	Explorer X	"	March 25, 1961	80
5	Tiros III	"	July 12, 1961	280
6	Explorer XII	"	August 16, 1961	90
7	Tiros IV	"	February 8, 1962	280
8	OSO I	"	March 7, 1962	500
9	Ariel I	"	April 26, 1962	160
10	Tiros V	"	June 19, 1962	300
11	Telstar I	"	July 10, 1962	171
12	Tiros VI	"	September 18, 1962	280
13	Explorer XIV	"	October 2, 1962	89
14	Explorer XV	"	October 27, 1962	98
15	Relay I	"	December 13, 1962	172
16	Syncom I	"	February 14, 1963	150
17	Explorer XVII	"	April 3, 1963	150
18	Telstar II	"	May 7, 1963	410
19	Tiros VII	"	June 19, 1963	300
20	Syncom II	"	July 26, 1963	150
21	Explorer XVIII (IMP I)	"	November 27, 1963	138
22	Tiros VIII	"	December 21, 1963	265

Delta on launch pad; Scout launches *Explorer XVI*.

Ariel I; the first privately owned satellite, *Telstar I*; the first synchronous orbiting satellite, *Syncom II*; the Orbiting Solar Observatory; the Tiros weather satellites; and *Explorer XVIII*, the first Interplanetary Monitoring Platform.

The Delta was used to launch 21 of the 33 satellites discussed in chapter 7. The first launch (the attempted Echo launch on May 13, 1960) was the only failure that the Delta experienced in the 22 launches. With its outstanding record of reliability and extensive use, the Delta was truly the workhorse of the NASA scientific satellite program.

In addition to the Delta, other boosters have been used to launch Goddard satellites, including the Delta precursor, the Vanguard, for the launch of *Vanguard III*, and the four-stage Scout. This Scout solid-fuel rocket, managed by NASA's Langley Research Center, was used to launch smaller satellites and two space probes. Several Department of Defense rockets were also available for Goddard's use. The Juno II, an Army Redstone-derived vehicle, was used for three of the Explorer satellites. The Air Force's Thor-Able launched three early Goddard satellites and its Thor-Agena was used to launch *Alouette I*.[130]

Sounding Rockets

In addition to its satellite program, Goddard-managed sounding rocket experimentation made many contributions to the space science program. By December 31, 1963, the Center had fired 292 sounding rockets. Of these, 203 were considered successful, 31 partially successful, and 58 failures —either because the rocket failed or the payload was not recovered.[131] (It should be pointed out that classification of a sounding rocket flight as successful, partially successful, or failure, involves some arbitrary decision, and that the above totals are subject to further interpretation.) This program supported astronomy, solar physics, energetic particles and fields, ionospheric physics, and planetary atmosphere and meteorology. Many sounding rocket launchings also served to flight-test equipment intended for use on satellites.[132]

The sounding rocket program also enabled many foreign countries to participate in atmospheric and space research. Frequently the Goddard Center assisted other nations in these cooperative efforts.

While the sounding rocket program has never caught the public imagination as have the more dramatic satellite programs, these activities have played an important role in Goddard Center's scientific investigation of space, and no account of the early years at Goddard would be complete without some discussion of them.

As a practical matter, satellites cannot orbit below 100 miles because of atmospheric drag. Balloons and aircraft were not effective above about 20 miles. A device was needed to take measurements in the upper atmosphere, particularly in the zone between 20 and 100 miles. This was the early impetus to the development of sounding rockets. A statement made by Dr. Homer E. Newell, Jr., before a Senate committee serves as a brief background to the Goddard sounding rocket program:

> The United States has been using sounding rockets for upper air research and rocket astronomy since the close of World War II. WAC Corporal, V-2, Viking, Aerobee, Aerobee-Hi, Nike-Deacon, Nike-Cajun, Nike-Asp, and Rockoons were used. Altitudes attained were below 200 miles for the most part. Many hundreds of rockets were fired prior to the start of the International Geophysical Year; an additional 200 were

Sounding rocket readied for launch at Fort Churchill, Canada.

On April 26, 1962, the Japanese electron temperature experiment was launched from Wallops Island by a Nike-Cajun sounding rocket.

BOOSTERS AND SOUNDING ROCKETS

fired as part of the International Geophysical Year program. Current rate of rocket soundings is somewhat below 100 per year. Higher altitude rockets are being introduced into the work to extend the atmospheric observations to one to several thousands of miles altitude. Launchings have been carried out at White Sands, N. Mex.; Wallops Island, Va.; San Nicolas Island, Calif.; Cape Canaveral, Fla.; Fort Churchill, Canada; Guam; and from shipboard in the North Atlantic, the Mid-Pacific and South Pacific, and the vicinity of Antarctica.[133]

Most Goddard sounding rockets were launched from Wallops Island, Virginia, and Fort Churchill, Canada.

Immediately after the end of World War II, the United States began an upper atmosphere research program using V-2 rockets. As the supply of V-2s ran out and the need for rockets specifically designed for research purposes became evident, the development of the Viking, the Aerobee, and the Nike-Deacon rockets was undertaken. The latter two rockets played prominent roles in NASA's sounding rocket program.

The Nike-Deacon consisted of a Nike-Ajax first stage and a Deacon second stage. From 1947 on, one or another version of the Aerobee served as the principal sounding rocket workhorse. The Aerobee came in two configurations: the Aerobee 150 and the Aerobee 300. The Aerobee 150 was a two-stage system consisting of a solid-propellant booster and a liquid-propellant sustainer stage. Both burned for the duration of the launch and the booster merely provided assistance at takeoff. The Aerobee 300 was the Aerobee 150 propulsion system with the addition of a third stage. While the Aerobee 150 had a three-fin configuration, the Aerobee 150A was four-finned. There was also an Aerobee 300A, which used a 150A second stage.

Nike-Apaches and Nike-Cajuns were perhaps the most heavily used rockets in the United States sounding rocket program. Identical in appearance, the Apache propellant provided more power than the Cajun and thus took a given payload to a higher altitude.

The Javelin was the largest sounding rocket used with any frequency. This four-stage rocket was designed primarily for the researcher who wished to place a payload experiment of between 90 and 150 pounds at altitudes between 500 and 650 miles. The largest scientific sounding rocket was the Journeyman. It lifted a payload of between 50 and 150 pounds to altitudes between 900 and 1,300 miles, although its usefulness could be extended to heavier payloads and higher altitudes.

Other sounding rockets included: the Astrobee 200, a two-stage solid-propellant rocket, similar to the Aerobee 150 but with a higher acceleration to the payload; the Astrobee 1500, a two-stage solid-propellant rocket capable of reaching 1,500 miles altitude; and the Black Brant II built by Canadian Bristol Aerojet Limited.

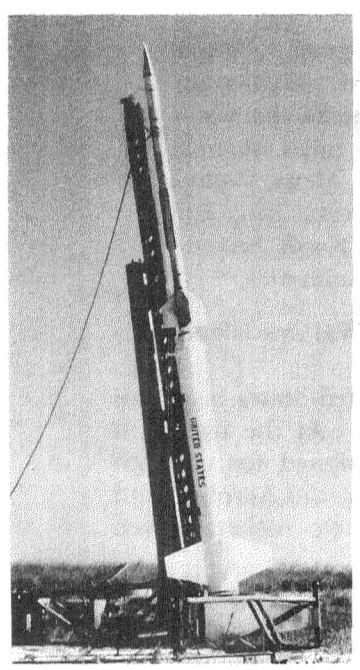  

Nike-Cajun; Javelin; Nike-Apache.

A closeup of the sounding rocket used in the first joint flight effort by the United States and Japan being checked by a Japanese scientist.

BOOSTERS AND SOUNDING ROCKETS

Major Sounding Rockets

NIKE-APACHE/NIKE-CAJUN

Overall:
  Total length: 28 ft
  Gross weight: 1,600 lb
  Propellant: Solid fuel
  Payload weight:
    Minimum: 40 lb
    Nominal: 60 lb
    Maximum: 120 lb
First stage: Nike (M5E1) booster
  Length: 12.4 ft
  Principal diameter: 16.5 in.
  Thrust: 42,500 lb
  Burning time: 3.5 sec

Second stage: Apache (TE 307–2)
  Length: 8.9 ft
  Principal diameter: 6.5 in.
  Thrust: 5,000 lb
  Burning time: 6.4 sec
Second stage: Cajun (TE 82–1)
  Length: 8.9 ft
  Principal diameter: 6.5 in.
  Thrust: 8,500 lb
  Burning time: 4 sec

JOURNEYMAN

Overall:
  Stages: 4
  Total length: 62 ft
  Gross weight: 14,079 lb
  Payload weight:
    Minimum: 75 lb
    Nominal: 125 lb
    Maximum: 175 lb
First stage: Sergeant
  Maximum diameter: 31 in.
  Thrust: 50,000 lb
First stage booster:
  2 Recruits
  Thrust: 36,000 lb (each)

Second and third stages: Lance
  Length: 16 ft
  Maximum diameter: 15 in.
  Thrust: 44,000 lb
  Burning time: 6.4 sec
Fourth stage: X–248
  Length: 6 ft (plus 3-ft payload)
  Maximum diameter: 19 in.
  Thrust: 3,000 lb
  Burning time: 42 sec

JAVELIN

Overall:
  Total length with nominal payload:
    Approximately 49 ft
  Gross weight less payload:
    Approximately 7,500 lb
  Fuel: Solid
  Payload weight:
    Minimum: 20 lb
    Normal: 125 lb
    Maximum: 175 lb
First stage: Honest John (M–6) booster
  Diameter: 22.9 in.
  Length: 16 ft
  Thrust: 82,000 lb
  Burning time: 5 sec

Second and third stages: Nike (M5E–1) motor
  Diameter: 16.5 in.
  Length: 11.2 ft
  Thrust: 42,500 lb
  Burning time: 3.3 sec
Fourth stage: X–248 rocket motor
  Diameter: 19 in.
  Length: 6 ft (plus 2.4-ft payload)
  Thrust: 3,000 lb
  Burning time: 42 sec

Goddard Space Flight Center fired about 100 sounding rockets per year from 1959 through 1963, so it is impractical to discuss each firing. This program was a significant element of the Center's scientific and technological endeavor. Not only do sounding rockets continue to fulfill the early purpose of taking measurements in the upper atmosphere but they also serve a variety of specialized scientific and meteorological purposes at other altitudes. Furthermore they are economical testing devices for equipment and systems that will later fly in expensive satellites.

Goddard Looks to the Future
9

THE CENTRAL IDEA of the scientific exploration of space has been expressed by the first Director of Goddard Space Flight Center (1959 to 1965), Dr. Harry J. Goett:

> The characteristic of space science is such that it spreads across many disciplines, and a very broad segment of the scientific fraternity is helping unravel the meaning of the new scientific data being brought back from outer space. You must recognize the efforts of the orbital mechanicians, physicists with various areas of specialization, astronomers, geologists, and geodesists in the analysis of the results. . . . Each of these disciplines is finding that it has a new frontier. New areas of research are being created by the data which rockets and satellites provide. We surely have just started to realize their potentials.
>
> The job of putting together the cosmic jigsaw of space from the bits and pieces of data obtained from our satellites is one that engages the efforts of many people throughout the scientific community; and this jigsaw puzzle goes together so gradually that there are no singular events which merit a headline.[134]

At the end of 1963, many pieces of the jigsaw puzzle had been fitted together and scientists were able to plan for the future. The Vanguard and Explorer satellites were the forerunners of complex "second generation" observatory spacecraft containing as many as 30 individual experiments. Soon the data flow from these space-borne laboratories would be measured literally in miles of magnetic tape daily. By 1964 some 15 to 20 satellites were being interrogated every day, producing some 50 miles of data tapes daily and tens of thousands of data bits per second. Somehow the future had arrived sooner, much sooner, than engineers, scientists, and managers had dared to expect. The chain reaction of space-related scientific inquiry was to continue.

Vanguard II (SLV-4), launched in February 1959, gave impetus to the Tiros weather satellites, which within the brief span of 4 years would develop into the Tiros Operational Satellite System (TOSS). The Nimbus weather satellite program would promise further breakthroughs in space-borne meteorology.

Rapid strides were also made by NASA in the area of improved communication techniques. Technological advances produced by AT&T's Telstar, NASA's Echo, Relay, and Syncom systems soon found further applications. In 1965 the control of *Syncom II* and *Syncom III* would be transferred to the Department of Defense for operational communications and for study in design of military communications systems. *Early Bird*, the world's first commercial communications satellite, would be built by the Syncom contractor, Hughes Aircraft Co., for the Communications Satellite Corporation and would be closely patterned on the earlier Syncom.

Facilities and experience gained in space tracking, especially in support of Project Mercury, would provide a readily adaptable base for the tracking of Projects Gemini and Apollo.

Years ahead of man's first travels to the moon and to other planets, Goddard communication links from far-flung tracking stations would carry the first closeup photographs of the moon and Mars.

Splash crater on the moon.

Goddard's top management staff, 1962. From left to right: Dr. M. J. Vaccaro, Assistant Director for Administration; J. T. Mengel, Assistant Director for Tracking and Data Systems; E. W. Wasielewski, Associate Director; Dr. H. J. Goett, Director; Dr. J. W. Townsend, Assistant Director for Space Science and Satellite Applications; and L. Winkler, Chief of Technical Services.

The vast investment of talent and money had begun to pay off. New scientific knowledge and technology poured in as returns to a Nation which some five years earlier had embarked on a new national purpose—the exploration of space. Now a new debate developed: How fast? How much?

NASA Administrator James E. Webb, sensing the Nation's feeling, remarked: "In the years ahead, we can expect continuing and necessary debate on the rate and 'mix' of the space investment."[135] He considered it extremely important that the country strive to maintain a well-balanced effort, duly recognizing the potential returns from manned exploration, scientific investigations, practical applications, and possible military uses of space, with a substantial share of attention to basic research in each area.

What was required, in the national interest, was a judicious evaluation of the Nation's new opportunities produced by the Space Age. If the United States was to ensure its security and position as leader of the Free World and gain the scientific and economic benefits which space would surely produce, it had an opportunity it could not afford to neglect. At Greenbelt, Maryland, a team of 3,500 engineers and scientists was ready to contribute to this endeavor.

The Center's management staff in 1965. From left to right: Robert E. Bourdeau, Assistant Director for Projects; Herman E. LaGow, Assistant Director for Systems Reliability; John T. Mengel, Assistant Director for Tracking & Data Systems; Eugene W. Wasielewski, Associate Director; Dr. John F. Clark, Director; Dr. John W. Townsend, Jr., Deputy Director; Dr. Michael J. Vaccaro, Assistant Director for Administration and Management; William G. Stroud, Chief, Advanced Plans Staff; Daniel G. Mazur, Assistant Director for Technology; Dr. George F. Pieper, Assistant Director for Space Sciences.

Goddard's past and its hopes for the future were summed up by the Center's second director, Dr. John F. Clark:

> We who are engaged in the hectic task of space exploration have little time to reminisce about past accomplishments and little inclination to speculate about future achievements beyond, at the most, a few years.... Looking back at our early years of space exploration, one fact becomes paramount: We have telescoped time and emerged with the means to explore space, to use it for the benefit of not only this nation, but of the world. The capabilities that we have built up have not just been placed in space, but rather have been anchored on the solid earth in laboratories, launch facilities, and in the dedication of the men and women who make up this cooperative team.
>
> One can speculate—but it is only speculation—about what the future may bring. There are many avenues of exploration open: the moon, the space environment near earth, the planets, and even the galaxies. But without knowing what constraints we may encounter in the availability of people, dollars or objectives, such speculation can be a rather academic exercise. It is certain, however, that the years immediately

ahead will be filled with intense activity in space, just as they have been in the past decade. We shall be expanding our knowledge and operational capability constantly. Only the rate of progress is uncertain.[136]

Somehow the dream of a lonely New England professor had become the commitment of a new generation.

Footnotes

Chapter 1

1 For a discussion of the relative role of Dr. Goddard among the space pioneers see: Willy Ley, *Rockets, Missiles and Space Travel*, New York: 1961; Eugene M. Emme, ed., *History of Rocket Technology*, Detroit: 1964.

2 Milton Lehman, *This High Man: The Life of Robert H. Goddard*, New York: 1963, pp. 14, 22–23.

3 E. R. Hagemann, "Goddard and His Early Rockets," *Journal of the Astronautical Sciences*, Summer 1961, pp. 51–52; Shirley Thomas, "Robert H. Goddard," *Men of Space*, Philadelphia: 1960, I, 23.

4 Robert H. Goddard, "An Autobiography," *Astronautics*, April 1959, p. 27.

5 *Ibid.*

6 Lehman, pp. 40–50.

7 Lehman, pp. 56–70.

8 G. Edward Pendray, "Pioneer Rocket Development in the United States," Hagemann, pp. 53–54. In *History of Rocket Technology*, p. 22. For the best account of Dr. Goddard's relationship with the Smithsonian Institution, see Bessie Zaban Jones, *Lighthouse of the Skies: The Smithsonian Astrophysical Observatory: Background and History, 1846–1955*, Washington: 1965, pp. 241–276. Dr. Charles G. Abbott of the Smithsonian in 1958 still considered Goddard's 1916 proposal "'the best presentation of a research in progress that I have ever seen.'"

9 Lehman, pp. 96–97.

10 *Ibid.*, pp. 102–112; Hagemann, pp. 54–56; Jones, pp. 254–258.

11 Lehman, pp. 139–144.

12 Eugene M. Emme, "Yesterday's Dream . . . Today's Reality," *The Airpower Historian*, October 1960, pp. 219–220; Jones, pp. 266–272.

13 Robert H. Goddard, *Rocket Development: Liquid Fuel Rocket Research, 1929–1941*, New York: 1948.

14 Lehman, pp. 341–353.

15 *Ibid.*, pp. 378–390. See Arthur C. Clarke, *Man and Space*, New York: 1964.

16 *Ibid.*, pp. 395–399.

17 The Center was formally dedicated on March 16, 1961.

Chapter 2

18 The pre-World War II activities of the American Rocket Society are briefly discussed in G. Edward Pendray, "Pioneer Rocket Development in the United States," in Eugene M. Emme (ed.), *The History of Rocket Technology*, Detroit: 1964, pp. 10–28.

19 James A. Van Allen, John W. Townsend, Jr., and Eleanor C. Pressley, "The Aerobee Rocket," in Homer E. Newell, Jr. (ed.), *Sounding Rockets*, New York: 1959, pp. 54–56.

20 John P. Hagen, "The Viking and the Vanguard," in Emme, *op. cit.*, pp. 122–141.

21 Van Allen, Townsend, and Pressley, *op. cit.*, pp. 54–70.

22 Milton Rosen, *The Viking Rocket Story*, New York: 1955; Homer E. Newell, Jr., "Viking," in *Sounding Rockets*, pp. 235–242.

23 Newell, "Viking," pp. 239–242. The upper atmospheric research group of the Naval Research Laboratory consisted largely of people who were to transfer to NASA a decade or so later—Dr. Homer E. Newell, Milton W. Rosen, Dr. John W. Townsend, Leopold Winkler, Daniel G. Mazur, and others. (Statement of T. E. Jenkins, formerly Administrative Officer of the Beltsville Space Center, July 15, 1963.)

24 Wernher von Braun, "The Redstone, Ju-

piter, and Juno," in Emme, *op. cit.*, pp. 108–109; Robert L. Perry, "The Atlas, Thor, Titan, and Minuteman," in Emme, *op. cit.*, pp. 144–145.

[25] Frank J. Malina, "Origins and First Decade of the Jet Propulsion Laboratory," in Emme, *op. cit.*, pp. 63–65.

[26] R. Cargill Hall, "Early U.S. Satellite Proposals," in Emme, *op. cit.*, pp. 67–93.

[27] *Ibid.*, pp. 28–33.

[28] *Ibid.*, pp. 35–37.

[29] *Ibid.*, pp. 40–41.

[30] Kurt R. Stehling, *Project Vanguard*, Garden City, N. Y.: 1961, pp. 37–42.

[31] The origins of the IGY are discussed in Walter Sullivan, *Assault on The Unknown; The International Geophysical Year*, New York: 1961, pp. 20–35.

[32] R. Cargill Hall, "Origins and Development of the Vanguard and Explorer Satellite Programs," Goddard Historical Essay Winner for 1963, published in *The Airpower Historian* (October 1964), pp. 101–112.

[33] *Documents on International Aspects of the Exploration and Use of Outer Space, 1954–1962*, Staff Report Prepared for the Committee on Aeronautical and Space Sciences, U.S. Senate, 88th Congress, 1st Session, Washington, D.C.: 1963, p. 27.

[34] The activities of the Stewart Committee are reported in Hall, "Origins and Development of the Vanguard and Explorer Satellite Programs," *The Airpower Historian* (October 1964), pp. 101–112.

[35] Stehling, *op. cit.*, p. 50. The Office of Naval Research actually was a cosponsor with the Army of the "Project Orbiter" proposal, and also of the NRL proposal.

[36] Hagen, pp. 439, 449–450.

[37] On Dr. Hagen, see: Stehling, pp. 69–71.

[38] Hagen, p. 440.

[39] *Ibid.*, pp. 440–441.

[40] Hall, "Origins and Development of the Vanguard and Explorer Satellite Programs," *op. cit.*, 101–112.

[41] Hagen, pp. 441–442.

[42] Hagen, pp. 442–444; Milton W. Rosen, "Placing the Satellite in its Orbit," *Proceedings of the XLIV IRE*, XLIV (1956), 749.

[43] *Ibid.*, pp. 444–446; John T. Mengel, "Tracking the Earth Satellite, and Data Transmission, By Radio," *Proceedings of the IRE*, XLIV (1956), 755.

[44] Stehling, p. 79.

[45] *Ibid.*, pp. 80–81.

[46] Hagen, p. 447.

[47] *Ibid.*

[48] *Ibid.*, pp. 447–448; Stehling, pp. 123, 142, 143.

[49] Stehling, *op. cit.*, pp. 25, 181.

[50] David S. Akens, *Historical Origins of the George C. Marshall Space Flight Center*, Huntsville, Alabama: 1960, pp. 44–47; Stehling, pp. 142–143; Wernher von Braun, "Redstone, Jupiter, and Juno," in Emme, *op. cit.*, pp. 107–121, and Eric Bergaust, *Reaching for The Stars*, Garden City, N.Y.: 1960. Studies of Air Force rocket technology during this same period include: Robert L. Perry, "The Atlas, Thor, and Titan," *Technology and Culture* (Fall 1963), pp. 466–477; and John L. Chapman, *Atlas: The Story of a Missile*, New York: 1960.

[51] Hagen, *op. cit.*, pp. 448–449.

[52] Eugene M. Emme, *Historical Sketch of NASA*, NASA EP-29. Washington, D.C.: 1965, pp. 5–13; *Documents on International Aspects of the Exploration and Use of Outer Space, 1954–1962*, pp. 66–80.

[53] See: *First Semiannual Report to the Congress of the National Aeronautics and Space Administration*, Washington, D.C.: 1959, 18–19; *Second Semiannual Report of the National Aeronautics and Space Administration*, Washington, D.C.: 1960, pp. 13–15.

[54] Hagen, pp. 449–450.

[55] *Ibid.*, p. 451.

Chapter 3

[56] "NASA General Directive No. 1," September 25, 1958. (Appendix H, Exhibit 2.)

[57] "Executive Order 10783," October 1, 1958. (Appendix H, Exhibit 4): "White House Press Release," October 1, 1958. (Appendix H, Exhibit 3.)

[58] "NASA Release," October 1, 1958 (Appendix H, Exhibit 5.)

59 Senator J. Glenn Beall, "Press Release," August 1, 1958. (Appendix H, Exhibit 1.)

60 "NASA General Notice No. 1," January 15, 1959. (Appendix H, Exhibit 6.)

61 "NASA General Notice," January 22, 1959. (Appendix H, Exhibit 7.)

62 Thomas E. Jenkins, "Memorandum for the Record," February 15, 1959. (Appendix H, Exhibit 8.)

Thomas E. Jenkins, "Memorandum for All Concerned," March 6, 1959. (Appendix H, Exhibit 9.)

63 T. Keith Glennan, "Memorandum from the Administrator," May 1, 1959; Abe Silverstein, "Memorandum to Assistant Directors and Division Chiefs," May 1, 1959; NASA Release No. 59–125, May 1, 1959; GSFC Release No. 3–10–61–5, March 12, 1961; GSFC Release No. 3–14–61–1, March 14, 1961. (Appendix H, Exhibits 10 through 14.)

64 The Advanced Research Projects Agency (ARPA) of the Department of Defense, created in February 1958, was the first organizational response to the challenge of Sputnik. This agency was directly in charge of the entire U.S. space effort between the time of its creation and the activation of NASA in October 1958. When NASA was activated, ARPA turned over to the new agency Project Vanguard and several other projects in a germinal stage. Many of NASA's most successful programs, such as Tiros, originated in ARPA. The role ARPA played in the origins and development of the U.S. space program has not yet been fully analyzed. Such an analysis must be undertaken before the history of the U.S. effort in space can be considered complete.

65 This and the following quotations from speeches at the dedication are taken from the respective speakers' texts.

Chapter 4

66 Address of Dr. T. Keith Glennan, Administrator, National Aeronautics and Space Administration, to Science, Engineering, and New Technology Committee, Oregon State Department of Planning and Development at Portland, Oregon, on October 12, 1960.

67 James M. Grimwood, *Project Mercury: A Chronology*, NASA SP–4001, Washington, D.C.: 1963, p. 120.

Chapter 5

68 Sir Eric Ashby, *Daedalus*, XCI (Spring 1962), 269.

69 H. Doc. No. 174, 87th Congress, 1st Sess., p. 11.

70 Speech at Rice University, Houston, Texas, Sept. 12, 1962.

71 Eugene Wasielewski, text of speech delivered at the National Rocket Club, Washington, D.C., December 19, 1962.

72 For the concept and origin of this personnel series, see Robert L. Rosholt, *An Administrative History of NASA, 1958–1963*, NASA SP–4101, pp. 141–144.

73 When President Eisenhower assigned responsibility for the development and execution of a manned space flight program, the National Aeronautics and Space Administration was in the process of being organized. Studies and plans for the manned satellite program were presented to Dr. T. Keith Glennan, Administrator, NASA; and on October 7, 1958, he gave orders to proceed with them. In November 1958, the Space Task Group was officially established to conduct the manned space flight program to be known as Project Mercury. The Space Task Group was organized under the Goddard Space Flight Center but was administratively supported by the Langley Research Center and physically located there. It later became evident that the scope and size of the manned space flight program required an entirely separate center, which subsequently led to the creation of the new Manned Spacecraft Center at Houston, Texas. Responsibilities for the Project Mercury worldwide tracking and data complex remained with the Goddard Space Flight Center.

74 J. C. New, "Scientific Satellites and the Space Environment," NASA Technical Note D–1340, June 1962; J. H. Boeckel, "The Purposes of Environmental Testing for Scientific

FOOTNOTES

Satellites," NASA Technical Note D-1900, 1963; A. R. Trimmins and K. L. Rosette, "Experience in Thermal-Vacuum Testing Earth Satellites at Goddard Space Flight Center," NASA Technical Note D-1748, 1963.

Chapter 6

75 William R. Corliss, "The Evolution of STADAN," GSFC Historical Note No. 3, 1967, pp. 13-33.

76 The radio interferometer measures two of the three direction cosines of a line from the center of the station to a satellite as a function of time while the satellite passes through the beam pattern of the receiving antennas. The third direction cosine is thus defined, and the angular position of the satellite is determined. From a series of independent angle measurements from various ground stations, satellite orbital elements may be computed.

Phase comparison techniques are used to measure the differences in arrival time of the wavefront of a distant point source at pairs of receiving antennas separated by known distances in wavelengths of the transmitted frequency. Measurement of this radio path difference is accomplished by a comparison of the phase angle of the signal received at one antenna to that received at another.

The antennas are aligned along baselines in the east-west and north-south directions. Since the phase measurement system is capable of indicating phase difference to a small fraction of a wavelength, two pairs of antennas are aligned along orthogonal baselines many multiples of a wavelength long to obtain good angular resolution. These are termed *fine* antennas. As a radio source passes through the antenna pattern, the relative phase will cycle from 0 to 360 degrees for each wavelength added to the radio path difference. The phase meters repeat their readings every 360 electrical degrees, so a number of different space angles will produce identical phase readings during a satellite transit. This ambiguity is resolved by employing several progressively shorter baselines, which produce fewer integral numbers of wavelength change as the radio source passed through the antenna beam. These are termed *medium* and *coarse* antennas. Ambiguity antenna information determines the number of full wavelengths to be added to the relative phase angle measured at the fine antennas to define a data point. See Corliss, *op. cit.*, pp. 10 ff.

77 *Ibid.*, pp. 50-60.

78 For the overall story of Project Mercury, see James M. Grimwood, *Project Mercury: A Chronology*, NASA SP-4001, Washington, D.C.: 1963; Loyd S. Swenson, Jr., James M. Grimwood, and Charles C. Alexander, *This New Ocean: A History of Project Mercury*, NASA SP-4201, Washington, D.C.: 1966.

79 For the story of Mercury network development and equipment, see William R. Corliss, "The Beginnings of Manned Space Flight Tracking," unpublished GSFC Historical Note No. 4, 1967.

Chapter 7

80 Memorandum for the Record, Feb. 16, 1959. (Appendix H, Exhibit 8.) See Appendix A on the objectives of the United States space program.

81 This compilation is based primarily on the following annual reports to the Committee on Space Research (COSPAR): *Goddard Space Flight Center Contributions to the COSPAR Meeting, May 1962,* Washington, D.C.: National Aeronautics and Space Administration, 1962 (TN D-1669); *United States Space Science Program Report to COSPAR, Sixth Meeting, Warsaw, Poland, June, 1963,* Washington, D.C.: National Academy of Sciences-National Research Council, 1963; *Goddard Space Flight Center Contributions to the COSPAR Meeting, June, 1963,* Washington, D.C.: National Aeronautics and Space Administration, 1963 (G-545); *United States Space Science Program Report to COSPAR, Seventh*

Meeting, Florence, Italy, May, 1964, Washington, D.C.: National Academy of Sciences-National Research Council, 1964. In addition to the COSPAR reports, much information in Goddard Space Flight Center's *Goddard Projects Summary: Satellites and Sounding Rockets* has been used. Also see Appendix B, "Chronology of Major NASA Launchings, October 1, 1958 through December 31, 1962," in House Committee on Science and Astronautics, *Astronautical and Aeronautical Events of 1962*, Washington: 1963, pp. 299-305.

[82] For results of *Explorer VI*, see, among others, C. Y. Fan, P. Meyer, and J. A. Simpson, *Journal of Geophysical Research*, LXVI, No. 9 (September 1961); A. Rosen and T. Farley, *Journal of Geophysical Research*, LXVI, No. 7 (July 1961); A. Rosen, T. Farley, and C. P. Sonett, in *Space Research, Proceedings of the First International Space Science Symposium, Nice, 11-16 January 1960*, ed. by H. K. Kallmann Bijl, Amsterdam: North-Holland Publishing Co., 1960, pp. 938-980; C. P. Sonett, E. J. Smith, D. L. Judge, and P. J. Coleman, Jr., *Physical Review Letters*, IV, No. 4 (February 1960), 161-163.

[83] "Geomagnetic-Field Studies Using Earth Satellites," *IGY Bulletin*, XCVI (April 1961), 6-12; J. P. Heppner, J. C. Cain, I. R. Shapiro, and J. D. Stolarik, "Satellite Magnetic Field Mapping," NASA TN D-696, May 1961; "IGY Satellite 1959 Eta," *IGY Bulletin*, XXVIII (October 1959), 10-14; H. E. LaGow and W. M. Alexander, "Recent Direct Measurements of Cosmic Dust in the Vicinity of the Earth Using Satellites," *Space Research, Proceedings of the First International Space Science Symposium, Nice, 11-16 January 1960*, ed. by H. K. Kallmann Bijl, Amsterdam: North-Holland Publishing Co., 1960.

[84] "IGY Satellite 1959 Iota," *IGY Bulletin*, XXIX (November 1959); W. C. Lin, "Observation of Galactic and Solar Cosmic Rays, October 13, 1959 to February 17, 1961 with Explorer VII (Satellite 1959 Iota)," SUI-61-16, Department of Physics and Astronomy, State Univ. of Iowa (August 1961); G. H. Ludwig and W. A. Whelpley, "Corpuscular Radiation Experiment of Satellite 1959 Iota (Explorer VII)," *Journal of Geophysical Research*, LXV, No. 4 (April 1960), 1119; J. A. Van Allen and W. C. Lin, "Outer Radiation Belt and Solar Proton Observations with Explorer VII During March-April 1960," *Journal of Geophysical Research*, LXV, No. 9 (September 1960), 2998.

[85] C. Y. Fan, P. Meyer, and J. A. Simpson, *Physical Review Letters*, V, No. 6 (September 1960), 269.

[86] *Ibid.*, p. 272.

[87] R. L. Arnoldy, R. A. Hoffman, and J. R. Winckler, unpublished communication.

[88] C. Y. Fan, P. Meyer, and J. A. Simpson, *Journal of Geophysical Research*, LXV, No. 6 (June 1960), 1862.

[89] R. L. Arnoldy, R. A. Hoffman, and J. R. Winckler, *Journal of Geophysical Research*, LXV, No. 9 (September 1960), 3004.

[90] C. Y. Fan, P. Meyer, and J. A. Simpson, *Physical Review Letters*, p. 269.

[91] P. J. Coleman, Jr., C. P. Sonett, D. L. Judge, and E. J. Smith, *Journal of Geophysical Research*, LXV, No. 6 (June 1960), 1856.

[92] *Ibid.*

[93] C. P. Sonett, E. J. Smith, D. L. Judge, and P. J. Coleman, Jr., *Physical Review Letters*, IV, No. 4 (February 1960), 161.

[94] E. J. Smith, P. J. Coleman, Jr., D. L. Judge, and C. P. Sonett, *Journal of Geophysical Research*, LXV, No. 6 (June 1960), 1858.

[95] J. P. Coleman, Jr., C. P. Sonett, D. L. Judge, and E. J. Smith, *Journal of Geophysical Research*, LXV, No. 1 (January 1960), 1856.

[96] C. P. Sonett, D. L. Judge, A. R. Sims, and J. M. Kelso, *Journal of Geophysical Research*, LXV, No. 1 (January 1960), 55.

[97] C. P. Sonett, *Physical Review Letters*, V, No. 2 (July 1960), 46.

[98] P. J. Coleman, Jr., L. Davis, Jr., and C. P. Sonett, *Physical Review Letters*, V, No. 2 (July 1960), 43.

[99] P. J. Coleman, Jr., C. P. Sonett, and L. Davis, Jr., *Journal of Geophysical Research*, LXVI, No. 7 (July 1961), 2043.

[100] J. B. McGuire, E. R. Spangler, and L. Wong, *Scientific American*, CCIV, No. 4 (April 1961), 64.

[101] For this and other Tiros satellites, see John Ashby, "A Preliminary History of the Evolution of the Tiros Weather Satellite Program," unpublished GSFC Historical Note No. 1, 1964; Richard Chapman, "Tiros-Nimbus: Administrative, Political, and Technological Problems of Developing U.S. Weather Satellites," unpublished study in the Inter-Univer-

sity Case Study Program, Inc., Syracuse, N.Y., 1966. For *Tiros I* results, see *IGY Bulletin,* No. 51 (September 1961), pp. 14-17, 23-24; "Roundup on *Tiros I," Astronautics,* V (June 1960, 32-44.

[102] For results of *Echo I,* see R. Bryant, *Journal of Geophysical Research,* LXVI (1961), 3066-3069; *IGY Bulletin,* No. 39 (September 1960) pp. 13-17; L. Jaffe, "Project Echo Results," *Astronautics,* VI, No. 5 (May 1961), 32-33, 80; W. C. Jakes, Jr., "Project Echo," *Bell Laboratories Record,* XXXIX, No. 9 (September 1961), 306-311.

[103] For *Explorer VIII* results, see "Ionosphere Direct Measurement Satellite," *IGY Bulletin,* No. 42 (December 1960), pp. 10-13; M. Melin, "Observing the Satellites," *Sky and Telescope,* XXI (January 1961), 11-12; R. E. Bourdeau and J. E. Donley, "*Explorer VIII* Measurements in the Upper Ionosphere," NASA TN D-2150, June 1964.

[104] The *Manual,* the *Catalog,* and copies of the magnetic tape data tabulations are available from the National Weather Records Center. For general information on the Tiros program, see footnote for *Tiros I.* For results of *Tiros II,* see W. R. Bandeen, R. A. Hanel, John Licht, R. A. Stampfl, and W. G. Stroud, "Infrared and Reflected Solar Radiation Measurements from the *Tiros II* Meteorological Satellite," *Journal of Geophysical Research,* LXV, No. 10 (October 1961), 3169-3185; *Proceedings of the International Meteorological Satellite Workshop, November 13-22,* 1961, Washington: NASA and U.S. Department of Commerce, Weather Bureau, 1962; "The *Tiros II* Cloud-Cover and Infrared Satellite." *IGY Bulletin,* No. 43 (January 1961), pp. 9-13.

[105] For *Explorer IX* results, see *IGY Bulletin,* No. 46 (April 1961), pp. 12-16; *STL Space Log,* I, No. 5, June 1961; W. J. O'Sullivan, C. W. Coffee, and G. M. Keating, "Air Density Measurements from the *Explorer IX* Satellite," *Space Research III,* ed. by W. Priester. New York: John Wiley and Sons, Inc., 1963, pp. 89-95.

[106] J. P. Heppner, N. F. Ness, T. L. Skillman, and C. S. Scearce, *Goddard Space Flight Center Contributions to 1961 Kyoto Conference on Cosmic Rays and the Earth Storm,* Washington: NASA, 1961. The previous field computations are to be found in H. F. Finch and B. R. Leaton, *Monthly Notices Royal Astronomical Society, Geophysical Supplement,* VII, No. 6 (November 1957), 314.

[107] For other *Explorer X* results, see *IGY Bulletin,* No. 48 (June 1961), pp. 1-4; *STL Space Log,* I, No. 5 (June 1961).

[108] H. S. Bridge, C. Dilworth, A. J. Lazarus, E. F. Lyon, B. Rossi, and F. Scherb (Massachusetts Institute of Technology), "Plasma Measurements from *Explorer X," 1961 Kyoto Conference on Cosmic Rays and the Earth Storm.*

[109] For *Explorer XI* results, see *IGY Bulletin,* No. 50 (August 1961), pp. 10-13; *STL Space Log,* I, No. 5 (June 1961).

[110] For general information on the Tiros program, see footnote to *Tiros I.* For results of *Tiros III,* see *IGY Bulletin,* No. 51 (September 1961), pp. 14-17; *Sky and Telescope,* XXII, No. 3 (September 1961), 143-145.

[111] For *Explorer XII* results, see Air Force Special Weapons Center Report No. TN-61-34; State University of Iowa Report No. 61-23.

[112] For *Explorer XIII* results, see "Explorer 13 Micrometeoroid Satellite," *IGY Bulletin,* No. 50 (October 1961), pp. 14-16; *STL Space Log,* I, No. 6 (September 1961); "The Micrometeoroid Satellite, Explorer 13 (1961 Chi)," NASA TN D-2468; "Micrometeoroid Satellite (Explorer XIII) Stainless Steel Penetration Experiment," NASA TN D-1986 (October 1962).

[113] For Tiros program information, see footnote to *Tiros I.* For *Tiros IV* results, see *IGY Bulletin,* No. 58 (April 1962) and No. 62 (August 1962); *U.S. Space Science Program* (Report to COSPAR), National Research Council, 1 May 1962; *Sky and Telescope,* XXIII (May 1962), 256-259; F. Bartko, V. Kunde, C. Catoe, and M. Halev, "The TIROS Low Resolution Photometer," NASA TN D-614 (September 1964).

[114] For *OSO I* results, see F. P. Dolder, O. E. Bartoe, R. C. Mercure, Jr., R. H. Gablehouse, and J. C. Lindsay, "The Orbiting Solar Observatory Spacecraft," *Space Research III,* ed. by W. Priester, New York: John Wiley and Sons, Inc., 1963; W. A. White, "Solar X-Rays: Slow Variations and Transient Events," GSFC Report No. X-614-63-195, presented at the

4th International Space Science Symposium, Warsaw, Poland, June 1963; W. M. Neupert, W. E. Behring, and J. C. Lindsay, "The Solar Spectrum from 50 Angstroms to 400 Angstroms," GSFC Report No. X–614–63–196, presented at the 4th International Space Science Symposium, Warsaw, Poland, June 1963; W. M. Neupert and W. E. Behring, *Solar Observations with a Soft X-Ray Spectrometer*, NASA TN D–1466, September 1966; J. C. Lindsay, "Scientific Results of the First Orbiting Solar Observatory," *Transactions of the American Geophysical Union*, XLIV (September 1963), 722–725.

[115] For *Ariel I* results, see "Ariel-Joint United Kingdom-United States Ionosphere Satellites," *IGY Bulletin*, No. 59 (May 1962), pp. 1–5; Elliott, Quenby, Mayne, and Durney, "Cosmic Ray Measurements in the U.K. Scout 1 Satellite," *Journal of British Institute of Radio Engineers*, XXII (September 1961), 251–256; M. O. Robins, "The *Ariel 1* Satellite Project and Some Scientific Results," paper presented at 9th Anglo-American Conference, held at Cambridge, Mass., October 16–18, 1963, and Montreal, Canada, October 21–24, 1963; *Ariel 1: The First International Satellite*, NASA SP–43, 1963 (revised 1964).

[116] For general information on the Tiros program, see the footnote for *Tiros I*. For results of *Tiros V*, see *IG Bulletin*, No. 62 (August 1962); "Weather Satellite Systems," *Astronautics and Aerospace Engineering*, I (April 1963).

[117] For *Telstar I* results, see *IG Bulletin*, No. 62 (August 1962); *Space/Aeronautics*, XXXVIII (January 1963), 41; "Project Telstar: Communications Experiment," *Journal of the Society of Motion Picture and Television Engineers*, LXXII (February 1963), 91–96; D. S. Peck, R. R. Blair, W. L. Brown, F. L. Smits, "Surface Effects of Radiation on Transistors," *The Bell System Technical Journal*, XLII (January 1963), 95–129.

[118] For general information on the Tiros program, see the footnote for *Tiros I*. For results of *Tiros VI*, see *IG Bulletin*, No. 66 (December 1962), 584–585.

[119] For *Alouette I* program information, see Jonathan Casper, "History of Alouette: NASA Case-Study of an International Program," unpublished NASA Historical Note No. 42, 1965; for *Alouette I* results, see J. O. Thomas, "Canadian Satellite: The Topside Sounder Alouette," *Science*, CXXXIX (January 18, 1963), 229–232; E. S. Warren, "Sweep-Frequency Radio Soundings of the Topside of the Ionosphere," *Canadian Journal of Physics*, XL (1962), 1692.

[120] For *Explorer XIV* results, see Franck, Van Allen, Whelpley, and Craven, "Absolute Intensities of Geomagnetically Trapped Particles with *Explorer XIV*," State University of Iowa Report No. 62–31 (December 1962), and *Journal of Geophysical Research*, LXVIII (March 1963), 1573–1579; "Collected Papers on the Artificial Radiation Belt from the July 9, 1963 Nuclear Detonation," *Journal of Geophysical Research*, LXVIII (February 1, 1963), 605ff.; "*Explorer XIV* Energetic Particles Satellite," *IG Bulletin*, No. 66 (December 1962), p. 585; H. Meyerson, "Energetic Particles Satellite, S–3a, Spacecraft Description and Preliminary Project Results," NASA Report N–90–013 (February 1963).

[121] For *Explorer XV* results, see *IG Bulletin*, No. 68 (February 1963); *Study of the Enhanced Radiation Belt*, Goddard Space Flight Center, Greenbelt, Md., 1962.

[122] For *Syncom I*, see "Syncom 1," STL "Communications Satellites," paper presented at the First World Conference on World Peace Through Law, Athens, Greece, June 30–July 7, 1963; R. C. Waddel, "Radiation Damage to Solar Cells on *Relay I* and *Relay II*," *Radiation Effects on Solar Cells and Photovoltaic Devices*, I, Proceedings of the Fourth Photovoltaic Specialists Conference, NASA-Lewis Research Center, Cleveland, Ohio, June 2, 1964; R. E. Warren and J. R. Burke, "Project Relay," *British Communications and Electronics*, VIII (August 1962), 582–583.

[123] For *Syncom I*, see "Syncom I," STL *Spacelog* (June 1963), pp. 31–33, (September 1963), pp. 41–42; "Syncom Lost and Found," *Sky and Telescope* (April 1963), pp. 210–212; Donald D. Williams, "Control of the 24-hour Syncom Satellite," *Missiles and Space* (February 1963), pp. 14, 15, 58.

[124] For *Explorer XVII* results, see "Preliminary Results of Explorer 17 Announced," NASA News Release No. 63–79, April 18, 1963; Spencer, Newton, Reber, Brace, and Horowitz, "New Knowledge of the Earth's Atmosphere

FOOTNOTES

from the Aeronomy Satellite," paper presented at the Fifth International Space Science Symposium, Florence, Italy, May 1964 (also GSFC Report No. X–651–64–114, May 1964).

[125] For *Telstar II* results, see *Electronics*, XXXVI (May 10, 1963), 29; *Bell Laboratories Record*, XLI (April 1963), 181; *Aviation Week and Space Technology*, LXXXI (May 6, 1963), 30.

[126] For *Tiros VII* results, see Sigmund Fritz, "Pictures from Meteorological Satellites and Their Interpretation," *Space Science Reviews*, III (November 1964), 541–580; W. R. Bandeen, B. J. Connath, and R. A. Hanel, "Experimental Confirmation from the *Tiros VII* Meteorological Satellite of the Theoretically Calculated Radiance of the Earth within the 15-micron Band of Carbon Dioxide," *Journal of the Atmospheric Sciences*, XX (November 1963), 609–614; W. Nordberg, W. R. Bandeen, G. Warnecke, and V. Kunde, "Stratospheric Temperature Patterns Based on Radiometric Measurements from *Tiros VII* Satellite," GSFC Report No. X–651–64–115 (May 1964); *Sky and Telescope*, XXVI (August 1963), 76.

[127] For *Syncom II* results, see G. E. Mueller and E. R. Spangler, *Communications Satellites*, New York: John Wiley and Sons, Inc., 1964; R. M. Bentley and A. T. Owens, "Syncom Satellite Program," *Journal of Spacecraft*, I (July–August 1964), 395–399; "*Syncom II* Satellite," *1964 IEEE International Convention Record* (March 23–26, 1964), pp. 71–153.

[128] For *Explorer XVIII* results, see T. L. Cline, G. H. Ludwig, and F. B. McDonald, "Detection of Interplanetary 3- to 12-MeV Electrons," GSFC Report X–611–64–362, November 1964; E. Ehrlich, "NASA Particles and Fields Spacecraft," AIAA Paper 64–337, First AIAA Annual Meeting, Washington, D.C., June-July 1964; "Initial Results from the First Interplanetary Monitoring Platform *(IMP 1),*" *International Geophysical Bulletin*, No. 84, June 1964; "Interim Status Report, Interplanetary Monitoring Platform, *IMP 1*, Explorer 18," GSFC Report X–672–64–33, February 1964; F. B. McDonald and G. H. Ludwig, "Measurement of Low-Energy Primary Cosmic-Ray Protons on *IMP–1* Satellite," *Physical Review Letters*, XIII (December 1964), 783–785; N. F. Ness and J. M. Wilcox, "Extension of the Photospheric Magnetic Field into Interplanetary Space," GSFC Report X–612–65–79, February 1965.

[129] On *Tiros VIII*, see M. Tepper and D. S. Johnson, "Toward Operational Weather Satellites," *Astronautics and Aeronautics*, III (June 1965), 16–26.

Chapter 8

[130] *Launch Vehicles of the National Launch Vehicle Program* (NASA SP–10), November 1962; *Goddard Projects Summary: Satellites and Sounding Rockets*, Goddard Space Flight Center; William S. Beller, "New Delta May Prove Most Economical," *Missiles and Rockets* (August 16, 1965), pp. 24–29.

[131] See Arnold W. Frutkin, *International Cooperation in Space*, Englewood Cliffs, N.J.: Prentice-Hall, 1965, pp. 51–59.

[132] For further information on sounding rockets, see *Goddard Space Flight Center Contributions to the COSPAR Meeting, May 1962*, GSFC Technical Note D–1669, 1962; *United States Space Science Program: Report to COSPAR, Sixth Meeting, Warsaw, Poland, June, 1963*, Washington: National Academy of Sciences–National Research Council, 1963; *United States Space Science Program: Report to COSPAR, Seventh Meeting, Florence, Italy, May, 1964*, Washington: National Academy of Sciences-National Research Council, 1964.

[133] *NASA Authorization for Fiscal Year 1961 —Part I*. 86th Congress, 2d Session—Senate. Testimony of Homer E. Newell, Jr., Deputy Director, Office of Space Flight Programs, NASA, pp. 23–24.

Chapter 9

[134] Harry J. Goett, "Scientific Exploration of Space," paper presented before the Franklin Institute, Philadelphia, Pa., March 8, 1962.

[135] James E. Webb, Address before American Institute of Aeronautics and Astronautics, New York Lecture, October 21, 1963.

[136] *Sperryscope*, Sperry Rand Corporation, October 1966.

APPENDIXES

Man, being the servant and interpreter of Nature, can do and understand so much and so much only as he has observed in fact or in thought of the course of nature: beyond this he neither knows anything nor can do anything.
—Francis Bacon

Appendix A

Introduction to the United States Space Sciences Program*

Excerpts from the report of the National Academy of Sciences dated March 12, 1959, to the Committee on Space Research

... A space sciences program is being developed by the U.S. National Aeronautics and Space Administration on as broad a basis as possible. In the planning and programing, advantage is being taken of the advice of the National Academy of Sciences' Space Science Board and also of specialists and experts in the scientific community. In the conduct of satellite and space probe experiments broad participation of the scientific community and industry, along with government, is planned, and steps are being taken to secure such participation. The developing program uses and will increase the momentum in space research developed during the International Geophysical Year. . . .

Although the program planning is still in its preliminary stages, it is hoped that in each of the next 2 years between 75 and 100 sounding rockets may be launched and on the order of one or two satellite or space probes every two months.

In the rocket sounding program, emphasis will be placed upon experiments relating to atmospheric structure, electric and magnetic fields, astronomy, energetic particles, and the ionosphere.

The satellite program will emphasize atmospheres, ionospheres, astronomy, energetic particles, electric and magnetic fields, and gravitation.

Space probes will investigate energetic particles, fields, and ionospheres.

Although the approximate magnitude and emphasis of the program has been described, much remains uncertain regarding the special vehicles to be used, their orbits or trajectories, their specific schedules, launching sites, tracking and telemetering support, and special technology support.

I. Atmospheres

Objectives

To determine and understand the origin, evolution, nature, spatial distribution, and dynamical behavior of the atmospheres of the earth, moon, sun, and planets; and their

---

* Report on the second meeting of the Committee on Space Research, held at The Hague, March 12–14, 1959.

relations to the medium of interplanetary space; to investigate atmospheric phenomena associated with interactions between photons, energetic particles, fields, and matter; to understand the relations between the earth's upper atmosphere and its surface meteorology; to evaluate atmospheric effects on space flight.

Program

(a) Long range

Long-range plans for achieving the above objectives include: (1) instrumented satellite stations around other planets; (2) rocket probes deep into the atmospheres of other planets including soft landings onto the surface with automatic and eventually manned recording stations; (3) probes deep into the solar atmosphere; (4) special probes for measuring the density and nature of gas and dust particles in interplanetary space and within comets; and (5) extensive theoretical studies to understand the basic natural phenomena taking place within the atmospheres.

(b) Immediate

Short-range plans include extensive and intensive studies of the structure and composition of the earth's atmosphere by direct measurements with sounding rockets and with satellites. Diurnal, latitudinal, and temporal variations in these parameters will be studied and will be correlated with energy and momentum balances in the earth's upper atmosphere. Models of the earth's atmosphere will be formulated for (1) providing basic data needed in understanding ionospheric, auroral, and other phenomena; and (2) providing guidance in the study of the atmospheres of other planets.

Short-range plans for studies up to about 50 miles include scores of synoptic rocket flights and several cloud cover satellites to establish the relationships between surface meteorology and the structure and dynamics of the upper atmosphere.

II. Ionospheres

Objectives

To determine and understand the source, nature, spatial distribution, and dynamical behavior of the ionized regions of the solar system, including the ionospheres of the earth, moon, and planets; to investigate ionospheric phenomena resulting from interactions between photons, particles, ions, and magnetic, electrostatic, and electromagnetic fields; to understand the relationship between solar activity and the terrestrial and other planetary ionospheres, magnetic fields, and upper atmospheric current systems; to evaluate ionospheric effects on space flight, including communications.

Program

(a) Long range

The long-range program will exploit present techniques for determining the terrestrial ionospheric structure, its propagation characteristics, and its influence on space flight by observation from below, within, and above. New techniques for evaluating the least known parameters will be developed. All of the applicable methods will then be used for the study of other planetary ionospheres. Eventually, propagation sounding stations may be established on the surface of the moon. All the data will then be applied to

APPENDIX A

understand the interrelations between solar activity, magnetic fields, the aurora, the Great Radiation Belt, and other phenomena.

(b) Immediate

The immediate program is concerned with obtaining electron density profiles at altitudes above the F_2 layer by inclusion of proven propagation experiments in space probes. Concurrently latitude and temporal variations of this parameter will be obtained by use of a polar orbiting satellite beacon. Topside sounders in satellite will be used for synoptic studies of electron density in the outer ionosphere. This technique promises less ambiguity than that obtainable from satellite beacons. Present knowledge of electromagnetic propagation will be extended by inclusion of very low frequency receivers in polar-orbiting satellites. Ion spectrum studies will be extended to lower mass numbers and higher altitudes by inclusion of rf mass spectrometers in space probes and satellites. Direct measurements using devices such as antenna probes, ion probes, and electric field meters will be made in rockets and satellites, to better define ionospheric structure and to study the interaction between the ionosphere and space vehicles.

III. Energetic Particles

Objectives

To determine and understand the origin, nature, motion, spatial distribution, and temporal variation of particles having energies appreciably greater than thermal; to understand their relation to the origin of the universe; to understand interactions between such particles, fields, photons, and matter; to evaluate possible hazards to life and other effects of energetic particles and photons in space.

Program

(a) Long range

Measurements using deep space probes will be made from the close proximity of the sun to the limits of the solar system. Extensive measurements in the vicinity of the planets, especially the earth, will be made to determine the interactions of the energetic particles with the atmospheres and fields of these bodies. These measurements will require satellite orbits around the earth, the moon, and other planets. The establishment of an observatory on the surface of the moon or on some other planet might be desirable, depending on the data previously acquired by artificial satellites.

(b) Immediate

In the near future the measurement of energetic particles will be pursued with satellites and rockets in the vicinity of the earth and with interplanetary probes. These measurements will be aimed at determining the interactions of these particles with the earth's atmosphere and field, their interactions with interplanetary fields, the types and energies of these particles, their spatial distribution, and the origin of the energetic particles.

The immediate program includes specifically measurements of the cosmic ray intensity in interplanetary space; of time and latitude cosmic ray intensity variations; of the composition and spatial extent of the Great Radiation Belt; of the cosmic ray energy and charge spectrum; and of the nature of the particles producing auroras.

IV. Electric and Magnetic Fields

Objectives

To determine and understand the origin, nature, method of propagation, spatial distribution, and temporal variation of magnetic and electric fields throughout the universe; to understand interactions between these fields and matter in space, and the influence of existing fields on solar and planetary atmospheres; to use these fields in the investigation of the internal constitution of astronomical bodies; to evaluate their effects and interactions on space flights.

Program

(a) Long range

Results from satellites, probes, and rockets to be flown in 1959 will be an important factor in determining the long-range program for studying the earth's fields. One can, however, anticipate that an important item will be establishing an earth satellite observatory which will include instruments for measuring particle flux and solar radiations as well as magnetic and electric field instruments such that direct correlations can be made between the various phenomena. Also rocket soundings into the ionosphere will continue to study details of ionospheric currents more thoroughly.

Attempts will be made to measure the fields of the moon, Mars, and Venus from probes making those approaches and eventually from packages landing and serving as observatories.

Probes will be launched toward the sun to obtain solar field measurements as close to the sun as feasible.

Theoretical analyses and correlations between electric and magnetic field phenomena and other phenomena will be an integral part of the program.

(b) Immediate

The short-range magnetic field program includes the use of sounding rockets, satellites, and space probes to carry magnetometers for investigation of the existence of ring currents above the ionosphere during magnetic storms, ionospheric currents, information on radiation belt currents, for measuring electric currents and the form of the earth's field at great distances, interplanetary fields, and the moon's magnetic field, and to study the complete spectra of field variations and for comprehensive field mapping.

It is also anticipated that simple magnetometers which can detect only the existence of a perceptible field will be placed in several rockets and space vehicles as secondary experiments. The short-range electric field program includes the use of electric field meters and Langmuir probes to explore satellite charging and ion sheath characteristics.

V. Gravitational Fields

Objectives

To determine and understand the origin, nature, method of propagation, spatial distribution, temporal variation, and effect of gravitational fields throughout the universe; to determine and understand the external form and internal constitution of the earth, planets, and stars; to determine and understand the relations between gravitational and electromagnetic fields; to evaluate effects of gravitational fields of different magnitudes, including weightlessness, on space flights.

APPENDIX A

Program

(a) Long range

For the study of the fundamental nature of the gravitational fields, two avenues are opened by the ability to launch satellites and space probes. The first of these is the ability to try experiments on a scale of hundreds and thousands of kilometers by probing the fields of planetary masses with bodies capable of being accurately observed. This is significant because gravitational fields, except that of the earth, are almost unmeasurable in laboratory-scale experiments. In the second place, the periods of artificial satellites are so much shorter than those of the moon and other natural satellites that in a few years a number of revolutions corresponding to thousands of years for natural satellites may be observed.

By the first avenue, it is possible to seek the links which must exist between the theory of the electromagnetic field and gravitational field. It is planned in particular to test the equality of gravitational and inertial masses by experiments in space which are a repetition of the experiment of Galileo in the Leaning Tower of Pisa. An attempt will be made to devise experiments which will reveal the velocity of propagation of gravitation, if any.

It is planned to determine the masses of the inner planets by direct observation of probes passing near them or possibly around them. These probes will at the same time help to determine the value of the astronomical unit.

It is planned to test the hypothesis that gravitational attraction depends on the average density of matter in the universe and that it therefore is slowly weakening as the universe expands. For this purpose, it is planned to compare an atomic clock on the ground with a gravitational clock of some kind. A proposal for a gravitational clock consisting, in effect, of a high satellite with a very well measured orbit is being studied.

It is planned to employ moon probes to obtain improved values for the overall mass of the moon and for the moments of inertia about its three principal axes. It is planned to attempt to determine the strength of the materials in the moon's interior from this information.

It is planned to measure the mass of Venus and of Mercury in order to test Bullen's ideas about the nature of the cores of the planets.

Using the second avenue, it is planned to observe the motions of close satellites of the earth over a long period and to make precise comparison with theory, searching for systematic trends in the inclination and the eccentricity, which might shed light on the history of the solar system.

(b) Immediate

(1) Studies are now being made on existing satellites with the object of determining the low harmonics of the earth's field from tracking data.

(2) It is also planned to put into orbit a special geodetic satellite which will be capable of refining the observations on the harmonics, and of determining intercontinental distances with high precision. It should be possible to carry the study of the form of the geoid much further than has been possible to date.

The information developed in (1) and (2) above will be applied to the question of the basic hypothesis of geodesy. This hypothesis, as formulated by some theorists, is in essence that the low harmonics of the earth's gravitational field have amplitudes of a meter or so. The hypothesis is not universally accepted; other theorists consider that the amplitudes are on the order of scores of meters. A decision between these two hypotheses is important because Heiskanen, in particular, proposes extensive work revolving around the Stokes' Theorem—work which is only warranted if the basic hypothesis is satisfied. This information will also be used in an attempt to evaluate hypotheses of convection in the mantle. These hypotheses seem to go with the ideas of the first-mentioned school of theorists, and it may be possible to decide between these hypotheses and the alternative contraction hypothesis on the basis of our information.

(3) It is planned to put in orbit a satellite carrying a very precise clock in order to test the theory of Einstein which predicts a change in the clock's speed depending upon the strength of the earth's gravitational field.

VI. Astronomy

Objectives

To determine the spatial distributions of matter and energy over the entire universe, and to understand their cosmological origins, evolutions, and destinies; to observe from above the earth's atmosphere the spectral distributions of energy radiated from objects in the solar system, in this and other galaxies, and in the intervening space, with emphasis on observations that are prevented or comprised by the absorption, background emission, and differential refraction of the earth's atmosphere; to determine and understand the geology of the planets; to determine the effects of meteors, radiations, and other astronomical influences on space flights.

Program

(a) Long range

The first phase of the long-range program will be the development of an orbiting and stabilized platform. With such a platform, it will be possible to orient a wide range of telescopic instruments so as to make detailed observations of specific quantities of interest at selected locations on the celestial sphere. Command control for the platform will be incorporated so that redirection of the instrumentation will be possible. The obvious advantages of observations made beyond the earth's atmosphere will be available to us with such an orbiting observatory. There will still remain some observational difficulties because of the backscattered light of the sun and the Doppler shifts resulting from the high velocity of the satellite.

(b) Immediate

The immediate program will continue and extend to the Southern sky the survey of the newly discovered nebulosities in the far ultraviolet by means of rockets. These measurements are being undertaken to determine the nature and sources of these emissions. Concurrently stellar photometry measurements will be made in the near and far ultraviolet spectrum region to extend magnitude systems to ultraviolet. Emphasis is being given to extending observations into the previously unexplored far infrared and high energy gamma-ray spectral regions by means of scanning satellite and rockets. Apart from their intrinsic value, these surveys are essential as ground work for the satellite observatory program.

Studies of the solar ultraviolet and x-ray spectra will be extended to include long term variations, line profiles, distribution across the disk, and the spectra of the coronal x-ray flux. These studies will be carried out in a series of rocket firings and with satellite-borne pointing devices.

Deep space probes will be used to determine the nature of the interplanetary medium.

Satellites will be used to map the emissions of the high atmosphere which arise from charged particle interactions and photochemical reactions.

APPENDIX A

VII. Biosciences

Objectives

To determine the effects on living terrestrial organisms of conditions in the earth's atmosphere, in space and in other planetary atmospheres, and of flight through these regions; to investigate the existence of life throughout the solar system, and to study such life forms in detail; to develop information necessary to achieve and maintain healthful artificial environments for terrestrial organisms, including man, throughout the solar system.

Appendix B
Goddard Space Flight Center Satellite and Space Probe Projects

GODDARD SPACE FLIGHT CENTER satellite launchings for 1959 through 1963 are given in the following table. The listings include the name of the satellites, their international designation (the international designation changed after 1962), the NASA designation, the project manager, and project scientist. The tabulation also gives the date of launch, the date on which the satellite became silent, the launch vehicle, and launch site. The period of the satellite is given in minutes, unless otherwise designated, and the perigee and apogee are given in statute miles. Orbital elements change over time. Any inconsistencies between text and appendixes derive from the date of measurement. The following abbreviations have been used.

Affiliations:

AFCRL	Air Force Cambridge Research Laboratories
ARC	Ames Research Center
BTL	Bell Telephone Laboratories
CRPL	Central Radio Propagation Laboratory
DRTE	Defence Research Telecommunications Establishment
DSIR	Department of Scientific and Industrial Research
ETR	Eastern Test Range
GSFC	Goddard Space Flight Center
JPL	Jet Propulsion Laboratory
MIT	Massachusetts Institute of Technology
NRC	National Research Council
NRL	Naval Research Laboratory
TRW/STL	Thompson Ramo Wooldridge/Space Technology Laboratories
WTR	Western Test Range

Scientific disciplines:

- R Aeronomy
- E Energetic Particles and Fields
- I Ionospheric Physics
- A Astronomy
- P Planetary Atmospheres
- S Solar Physics

Goddard Space Flight Center

Designation	Objectives	Launch and orbit data					Project manager and project scientist
		Launch date/ silent date	Vehicle and launch site	Period, min.	Statute miles		
					Perigee	Apogee	
Explorer VI 1959 Delta 1 S-2	To measure three specific radiation levels of earth's radiation belts; test scanning equipment for earth's cloud cover; map earth's magnetic field; measure micrometeoroids; study behavior of radiowaves.	Aug. 7, 1959 Oct. 6, 1959	Thor-Able ETR	12.5 hours	156	27,357	Dr. John C. Lindsay Dr. John C. Lindsay
Vanguard III 1959 Eta 1	To measure the earth's magnetic field, X-radiation from the sun, and several aspects of the space environment through which the satellite travels.	Sept. 18, 1959 Dec. 11, 1959	Vanguard ETR	130	319	2,329	
Explorer VII 1959 Iota 1	Variety of experiments, including solar ultraviolet, X-ray cosmic-ray, earth radiation, and micrometeoroid experiments.	Oct. 13, 1959 Aug. 24, 1961	Juno II ETR	101.33	342	680	H. E. LaGow

APPENDIX B

Satellite and Space Probe Projects

Experiment data				Remarks
Instrumentation summary	Experiment and discipline	Experimenter	Affiliation	
Equipment to measure radiation levels; TV-type scanner; micro-meteoroid detector; two types of magnetometers and devices for space communication experiments.	Triple coincidence telescopes—A Scintillation counter—E Ionization chamber Geiger counter—E Spin-coil magnetometer—E Fluxgate magnetometer—E Aspect sensor Image-scanning television system Micrometeoroid detector —P	J. A. Simpson C. Y. Fan P. Meyer T. A. Farley Allan Rosen C. P. Sonett J. Winckler E. J. Smith D. L. Judge P. J. Coleman	U. of Chicago TRW/STL U. of Minnesota TRW/STL TRW/STL TRW/STL TRW/STL AFCRL TRW/STL	Orbit achieved. All experiments performed. First complete televised cloud-cover picture was obtained. Detected large ring of electrical current circling earth; first detailed study of region now known as the Van Allen radiation belt. Weight: 142 lb Power: Solar
Proton precession magnetometer, ionization chambers for solar X-rays, micrometeorite detectors and thermistors.	Proton magnetometer—E Ionization chambers—E Environmental measurements	J. P. Heppner H. Friedman H. E. LaGow	GSFC NRL GSFC	Orbit achieved. Provided comprehensive survey of earth's magnetic field over area covered; surveyed location of lower edge of Van Allen radiation belt. Accurate count of micrometeoroid impacts. Power: Solar
Sensors for measurements of earth-sun heat balance; Lyman-alpha and X-ray solar radiation detectors; micrometeoroid detectors, Geiger-Mueller tubes for cosmic ray count; ionization chamber for heavy cosmic rays.	Thermal radiation balance Solar X-ray and Lyman-alpha—S Heavy cosmic radiation—E Radiation and solar-proton observation—E Ground-based ionosphere observation—I	V. Suomi H. Friedman R. W. Kreplin T. Chubb G. Groetzinger P. Schwed M. Pomerantz J. Van Allen G. Ludwig H. Whelpley G. Swenson C. Little C. Reid O. Villard, Jr. W. Ross W. Dyke	U. of Wisconsin NRL Martin Co. Bartol Research St. U. of Iowa U. of Illinois Nat. Bu. of Standards U. of Alaska Stanford U. Penn. State U. Linfield Res. Inst.	Orbit achieved. Provided significant geophysical information on radiation and magnetic storms; demonstrated method of controlling internal temperatures; first micrometeoroid penetration of a sensor in flight. Weight: 91.5 lb Power: Solar

Goddard Space Flight Center

Designation	Objectives	Launch and orbit data					Project manager and project scientist
		Launch date/ silent date	Vehicle and launch site	Period, min.	Statute miles		
					Perigee	Apogee	
Explorer VII —Cont.							
Pioneer V 1960 Alpha 1	To investigate interplanetary space between orbits of earth and Venus, test extreme long-range communications, study methods for measuring astronomical distances.	Mar. 11, 1960 Jun. 26, 1960	Thor-Able ETR	311.6 days	Perihelion 74.9 million from sun	Aphelion 92.3 million from sun	Dr. John C. Lindsay Dr. John C. Lindsay
Tiros I 1960 Beta 2 A-1	To test experimental television techniques leading to eventual worldwide meteorological information system.	Apr. 1, 1960 June 17, 1960	Thor-Able ETR	99.1	428.7	465.9	W. G. Stroud H. I. Butler S. Fritz (U.S. Weather Bureau)
Echo I 1960 Iota 1 A-11	To place 100-foot inflatable sphere into orbit; measure reflective characteristics of sphere and propagation; study effects of space environment.	Aug. 12, 1960 Passive satellite	Thor-Delta ETR	110.3	945	1,049	R. J. Mackey
Explorer VIII 1960 Xi S-30	To investigate the ionosphere by direct measurement of positive ion and electron composition; collect data on the frequency, momentum,	Nov. 3, 1960 Dec. 28, 1960	Juno II ETR	112.7	258	1,423	Robert E. Bourdeau Robert E. Bourdeau

APPENDIX B

Satellite and Space Probe Projects (Cont.)

Instrumentation summary	Experiment and discipline	Experimenter	Affiliation	Remarks
	Micrometeoroid penetration—P	H. LaGow	GSFC	
High-intensity radiation counter, ionization chamber Geiger-Mueller tube to measure plasmas, cosmic radiation, and charged solar particles. Magnetometer and micrometeoroid measurements.	Triple coincidence proportional counter cosmic-ray telescope—E Search-coil magnetometer and photoelectric cell aspect indicator—E Ionization chamber and G-M tube—E Micrometeoroid counter—P	J. Simpson D. Judge J. Winckler E. Manring	U. of Chicago TRW/STL U. of Minnesota AFCRL	Highly successful exploration of interplanetary space between orbits of earth and Venus; established communication record of 22.5 million miles on June 26, 1960; made measurements of solar flare effects, particle energies and distribution, and magnetic-field phenomena in interplanetary space. Weight: 94.8 lb Power: Solar
One wide and one narrow angle camera, each with tape recorder for remote operation. Picture data can be stored on tape or transmitted directly to ground stations.	TV camera systems (2)			Provided first global cloud-cover photographs (22,952 total) from near-circular orbit. Weight: 370 lb Power: Solar
Two tracking beacons 107.94 Mc and 107.97 Mc.	Communications		JPL BTL NRL	Demonstrated use of radio reflector for global communications; numerous successful transmissions. Visible to the naked eye. Orbit characteristics perturbed by solar pressure due to high area-to-mass ratio. Still in orbit. Weight: 124 lb (including inflation powder) Power: Passive
RF-impedance probe using a 20-foot dipole sensor; single-grid ion trap; four multiple-grid ion traps; Langmuir probe experiment, rotating shutter electric field meter; photomultiplier and micrometeoroid microphone; thermistors for reading internal	RF impedance—I Ion traps—I Langmuir probe—I Rotating-shutter electric field meter—I Micrometeoroid photomultiplier—I	J. Cain R. Bourdeau G. Serbu E. Whipple J. Donley R. Bourdeau G. Serbu E. Whipple J. Donley J. Donley M. Alexander C. McCracken	GSFC GSFC GSFC GSFC GSFC	The micrometeoroid influx rate was measured. Weight: 90.14 lb Power: Battery

Goddard Space Flight Center

Designation	Objectives	Launch and orbit data					Project manager and project scientist
		Launch date/ silent date	Vehicle and launch site	Period, min.	Statute miles		
					Perigee	Apogee	
Explorer VIII —Cont.	and energy of micrometeoroid impacts; establish the altitude of the base of the exosphere.						
Tiros II 1960 Pi 1 A-2	To test experimental television techniques and infrared equipment leading to eventual worldwide meteorological information system.	Nov. 23, 1960 July 12, 1961	Delta ETR	98.2	406	431	R. A. Stamfl
Explorer IX 1961 Delta 1 S-56a (A project of the Langley Research Center with GSFC participation)	To study performance, structural integrity, and environmental conditions of Scout research vehicle and guidance controls system. Inject inflatable sphere into earth orbit to determine density of atmosphere.	Feb. 16, 1961 April 9, 1964 Passive satellite	Scout Wallops Island	118.3	395	1,605	
Explorer X 1961 Kappa 1 P-14	To gather definite information on earth and interplanetary magnetic fields and the way these fields affect and are affected by solar plasma.	Mar. 25, 1961 Mar. 27, 1961	Thor-Delta ETR	112 hours	100	186,000	J. P. Heppner J. P. Heppner

APPENDIX B

Satellite and Space Probe Projects (Cont.)

Experiment data				Remarks
Instrumentation summary	Experiment and discipline	Experimenter	Affiliation	
and surface temperatures of the spacecraft, and despin mechanisms to reduce spin from 450 to 30 rpm.	Micrometeoroid microphone—I	O. Berg M. Alexander C. McCracken	GSFC	
Included one wide-angle and one narrow-angle camera, each with tape recorder for remote operation; infrared sensors to map radiation in various spectral bands; attitude sensors; experimental magnetic orientation control.	Two TV camera systems Widefield radiometer Scanning radiometer	W. Nordberg R. Hanel	GSFC GSFC	Orbit achieved. Narrow-angle camera and IR instrumentation sent good data. Transmitted 36,156 pictures. Weight: 277 lb Power: Solar
Radio beacon on balloon and in fourth stage.				Vehicle functioned as planned. Balloon and fourth stage achieved orbit. Transmitter on balloon failed to function properly requiring optical tracking of balloon. Weight: 80 lb Power: Passive
Included rubidium vapor magnetometer, two fluxgate magnetometers, a plasma probe, and an optical aspect sensor.	Rubidium-vapor magnetometer and fluxgate magnetometers—E Plasma probe—E Spacecraft attitude	J. P. Heppner T. L. Skillman C. S. Scearce H. Bridge F. Scherb B. Rossi J. Albus	GSFC MIT GSFC	Probe transmitted valuable data continuously for 52 hours as planned. Demonstrated the existence of a geomagnetic cavity in the solar wind and the existence of solar proton streams transporting solar interplanetary magnetic fields past the earth's orbit. Weight: 79 lb Power: Battery

Goddard Space Flight Center

Designation	Objectives	Launch and orbit data					Project manager and project scientist
		Launch date/ silent date	Vehicle and launch site	Period, min.	Statute miles		
					Perigee	Apogee	
Explorer XI 1961 Nu 1 S-15	To orbit a gamma-ray astronomy telescope satellite to detect high-energy gamma rays from cosmic sources and map their distribution in the sky.	Apr. 27, 1961 Dec. 6, 1961	Juno II ETR	108.1	304	1,113.2	Dr. J. Kupperian, Jr. Dr. J. Kupperian, Jr.
Tiros III 1961 Rho 1 A-3	To develop satellite weather observation system; obtain photos of earth's cloud cover for weather analysis; determine amount of solar energy absorbed, reflected and emitted by the earth.	July 12, 1961 Feb. 1962	Delta ETR	100.4	461.02	506.44	Robert Rados
Explorer XII Energetic Particles Explorer 1961 Upsilon 1 S-3	To investigate solar wind, interplanetary magnetic fields, distant portions of earth's magnetic field, energetic particles in interplanetary space and in the Van Allen belts.	Aug. 15, 1961 Dec. 6, 1961	Thor-Delta ETR	26.45 hours	180	47,800	Paul Butler Dr. F. B. McDonald
Explorer XIII 1961 Chi 1 (A project of the Langley Research Center with GSFC	To test performance of the vehicle and guidance; to investigate nature and effects on	Aug. 25, 1961 Aug. 28, 1961	Scout Wallops Island	97.5	74	722	C. T. D'Aiutolo

APPENDIX B

Satellite and Space Probe Projects (Cont.)

Experiment data				Remarks
Instrumentation summary	Experiment and discipline	Experimenter	Affiliation	
Gamma-ray telescope consisting of a plastic scintillator, crystal layers, and a Cerenkov detector; sun and earth sensors; micrometeoroid shields; temperature sensor; damping mechanism.	Gamma-ray telescope—E	W. Kraushaar G. Clark	MIT	Orbit achieved. Detected first gamma rays from space. Directional flux obtained. Disproved one part of "steady-state" evolution theory. Weight: 82 lb Power: Solar
Two wide-angle cameras, two tape recorders and electronic clocks, infrared sensors, five transmitters, attitude sensors, magnetic attitude coil.	Omnidirectional radiometer Widefield radiometer Scanning radiometer Two TV cameras	V. Suomi R. Hanel W. Nordberg	U. of Wisconsin GSFC GSFC	Orbit achieved. Cameras and IR instrumentation transmitted good data. Transmitted 35,033 pictures. Weight: 285 lb Power: Solar
Ten particle detection systems for measurement of protons and electrons and three orthogonally mounted fluxgate sensors for correlation with the magnetic fields, optical aspect sensor, and one transmitter. PFM telemetry transmitting continuously.	Proton analyzer—E Magnetometer—E Cosmic ray—E Ion-electron—E Solar cell	M. Bader L. Cahill B. O'Brien F. B. McDonald L. Davis G. Longanecker	ARC U. of New Hampshire St. U. of Iowa GSFC GSFC GSFC	Orbit achieved. All instrumentation operated normally. Ceased transmitting on Dec. 6, 1961, after sending 2,568 hours of real-time data. Provided significant geophysical data on radiation and magnetic fields. Weight: 83 lb Power: Solar
Micrometeoroid impact detectors; transmitters.	Cadmium sulfide photoconductor—A Wire grid	M. W. Alexander L. Secretan	GSFC	Orbit was lower than planned. Reentered Aug. 28, 1961. Weight: 187 lb including 50-lb 4th stage and 12-lb transition section. Power: Solar

Goddard Space Flight Center

Designation	Objectives	Launch and orbit data					Project manager and project scientist
		Launch date/ silent date	Vehicle and launch site	Period, min.	Statute miles		
					Perigee	Apogee	
participation) S-55a	space flight of micrometeoroids.						
P-21 Electron Density Profile Probe P-21	To measure electron densities and to investigate radio propagation at 12.3 and 73.6 Mc under daytime conditions.	Oct. 19, 1961 Oct. 19, 1961	Scout Wallops Island				John E. Jackson Dr. S. J. Bauer
Tiros IV 1962 Beta 1 A-9	To develop principles of a weather satellite system; obtain cloud and radiation data for use in meteorology.	Feb. 8, 1962 June 19, 1962	Delta ETR	100.4	471	525	Robert Rados
OSO I 1962 Zeta 1 OSO-1	To measure solar electromagnetic radiation in the ultraviolet, X-ray and gamma-ray regions; to investigate effect of dust particles on surfaces of spacecraft.	Mar. 7, 1962 Aug. 6, 1963	Delta ETR	96.15	343.5	369	Dr. John C. Lindsay Dr. John C. Lindsay
P-21a Electron Density Profile Probe	To measure electron density profile, ion density, and intensity	Mar. 20, 1962 Mar. 20, 1962	Scout Wallops Island				John E. Jackson Dr. S. J. Bauer

APPENDIX B

Satellite and Space Probe Projects (Cont.)

	Experiment data			Remarks
Instrumentation summary	Experiment and discipline	Experimenter	Affiliation	
Continuous-wave propagation experiment for the ascent portion of the trajectory, and an RF-probe technique for the descent.	RF probe—I CW propagation—I	H. Whale G. H. Spaid J. E. Jackson	GSFC GSFC GSFC	Probe achieved altitude of 4,261 miles and transmitted good data. Electron density was obtained to about 1,500 miles, the first time such measurements had been taken at this altitude. Weight: 94 lb Power: Battery
Two TV camera systems with clocks and recorders for remote pictures, infrared sensors, heat budget sensors, magnetic orientation control horizon sensor, north indicator.	Omnidirectional radiometer Widefield radiometer Scanning radiometer Two TV camera systems	V. Suomi R. Hanel W. Nordberg	U. of Wisconsin GSFC GSFC	Orbit achieved. All systems operated properly. Tegea Kinoptic lens used on one camera, Elgeet lens on the other. Supported Project Mercury. Weight: 285 lb Power: Solar
Devices to conduct 13 different experiments for study of solar electromagnetic radiations; investigate dust particles in space and thermoradiation characteristics of spacecraft surface materials.	X-ray spectrometer—S 0.510 MeV gamma-ray monitoring; 20–100 keV X-ray monitoring; 1–8A X-ray monitoring—S. Dust particle—E Solar radiation and solar ultraviolet—A Solar gamma rays, high-energy distribution—A Solar gamma rays, low-energy distribution—A Solar gamma rays, high-energy distribution—A Neutron monitor—E Lower Van Allen belt—E Emissivity stability of surfaces in a vacuum environment—E	W. Behring W. Neupert K. Frost W. White M. Alexander C. McCracken W. White K. Hallam W. White K. Frost J. R. Winckler L. Peterson M. Savedoff G. Fazio W. Hess S. Bloom G. Robinson	GSFC GSFC GSFC GSFC GSFC U. of Minnesota U. of Rochester U. of California U. of California ARC	Orbit achieved. Experiments transmitted as programed. Weight: 458 lb Power: Solar
A continuous-wave propagation experiment to determine electron density and associated param-	CW propagation—I RF probe—I Ion traps—I	S. Bauer H. White R. Bourdeau E. Whipple J. Donley	GSFC GSFC GSFC	Probe achieved altitude of 3,910 miles. Afforded nighttime observations. Determined that characteristics of the iono-

VENTURE INTO SPACE

Goddard Space Flight Center

Designation	Objectives	Launch and orbit data					Project manager and project scientist
		Launch date/ silent date	Vehicle and launch site	Period, min.	Statute miles		
					Perigee	Apogee	
P-21a Electron Density Profile Probe— Con.	of ions in the atmosphere.						
Ariel I International Satellite 1962 Omicron 1 (UK-1) S-51	To study the relationships between ionosphere and cosmic rays.	Apr. 26, 1962 Nov. 9, 1964	Delta ETR	100.9	242.1	754.2	R. C. Baumann Robert E. Bourdeau
Tiros V 1962 Alpha Alpha 1 A-50	To develop principles of a weather satellite system; obtain cloud-cover data for use in meteorology.	June 19, 1962 May 4, 1963	Delta ETR	100.5	367	604	Robert Rados
Telstar I (A project of AT&T) 1962 Alpha Epsilon 1 A-40	Joint AT&T-NASA investigation of wideband communications.	July 10, 1962 Feb. 21, 1963	Delta ETR	157.8	592.6	3,503.2	C. P. Smith, Jr.

APPENDIX B

Satellite and Space Probe Projects (Cont.)

Experiment data				Remarks
Instrumentation summary	Experiment and discipline	Experimenter	Affiliation	
eters of ionosphere. A swept-frequency probe for direct measurements of electron density and a positive ion experiment to determine ion concentration under nighttime conditions.		G. Serbu		sphere differ drastically from daytime state when the temperature of the ionosphere is much cooler. (See P-21) Weight: 94 lb Power: Battery
Electron density sensor, electron temperature gauge, solar aspect sensor, cosmic-ray detector, ion mass sphere, Lyman-alpha gauges, tape recorder, X-ray sensors.	Electron density sensor—I	J. Sayers	U. of Birmingham (U.K.)	Orbit achieved. First international satellite. Contained six British experiments launched by American Delta vehicle. All experiments except Lyman-alpha transmitted as programed. Lyman-alpha gauge failed during launch, ion mass sphere, Sept. 1962; X-ray emission, Oct. 1962; cosmic-ray detector, Dec. 1962, and electron density sensor, Mar. 1963. Tracking and data acquisition stopped on request of the project on June 30, 1964. Restarted on Aug. 25, 1964, for a 2-month period. Good data were acquired from electron temperature gauge.
	Electron temperature gauge—I	R. L. F. Boyd	U. College, London (U.K.)	
	Cosmic-ray detector—E	H. Elliot	Imperial College, London (U.K.)	
	Ion mass sphere—I	R. L. F. Boyd	U. College, London (U.K.)	
	Lyman-alpha gauge—I	R. L. F. Boyd	U. College, London (U.K.)	
	X-ray emission—I	R. L. F. Boyd	U. College, London (U.K.)	
Two TV camera systems with tape recorders for recording remote picture areas, magnetic orientation control, horizon sensor, north indicator.	Two TV camera systems			Launched at a higher inclination (58°) than previous Tiros satellites, to provide greater coverage. Time of launch chosen to include normal hurricane season for South Atlantic. One TV system transmitted good data for 10½ months. Weight: 285 lb Power: Solar
The system provided TV, radio, telephone and data transmission via a satellite repeater system.	Included electron detector for range 0.25-1 MeV; proton detectors in the following energy ranges: 2.5-25.0 MeV, ranges greater than 50 MeV.	W. Brown	BTL	Orbit achieved. Television and voice transmissions were made with complete success. BTL provided spacecraft and ground stations facilities. Government was reimbursed for cost incurred.

167

Goddard Space Flight Center

Designation	Objectives	Launch and orbit data					Project manager and project scientist
		Launch date/ silent date	Vehicle and launch site	Period, min.	Statute miles		
					Perigee	Apogee	
Telstar I—Cont.							
Tiros VI 1962 Alpha Psi 1 A-51	To study cloud cover and earth heat balance; measurement of radiation in selected spectral regions as part of a program to develop meteorological satellite systems.	Sept. 18, 1962 Oct. 11, 1963	Delta ETR	98.73	425	442	Robert Rados
Alouette I Swept Frequency Topside Sounder (Canada) 1962 Beta Alpha 1 S-27	To measure the electron density distribution in the ionosphere between the satellite height (620 miles) and the F2 peak (approx. 180 miles) and to study for a period of one year the variations of electron density distribution with time of day and with latitude under varying magnetic and auroral conditions with particular emphasis on high-latitude effects. To obtain galactic-noise measurements, study	Sept. 29, 1962	Thor-Agena WTR	105.4	620	638	John E. Jackson

APPENDIX B

Satellite and Space Probe Projects (Cont.)

Experiment data				Remarks
Instrumentation summary	Experiment and discipline	Experimenter	Affiliation	
				Conducted more than 300 technical tests and over 400 demonstrations; 50 TV programs—5 in color. Weight: 175 lb Power: Solar
Two TV camera systems (78° and 104° lens), clocks and tape recorders for remote operation, infrared and attitude sensors, magnetic-attitude coil.	Two TV camera systems			Inclination 58.3°; velocity at perigee 16,822; apogee, 16,756. Medium-angle camera failed Dec. 1, 1962, after taking 1,074 pictures. TV camera provided good data for 13 months after launch. Weight: 300 lb Power: Solar
The satellite was spin-stabilized and contained a swept-frequency pulse sounder covering the frequency range 1.6 to 11.5 Mc. Sounder data were transmitted via a 2-watt FM telemetry system. Data from the other experiments and housekeeping data were transmitted through a ¼-watt PM-telemetry system. There were two sets of sounder antennas, the longest set measuring 150 ft. tip to tip. Data were acquired on command and in real time only.	Topside sounder—I	E. S. Warren G. L. B. Nelms G. E. Lockwood E. L. Hagg L. E. Petrie D. B. Muldrew	DRTE	The Alouette satellite was a project of the Canadian Defence Research Board. This international project was a part of NASA's topside sounder program and was the first NASA-launched satellite from the WTR. Alouette had the distinction of being the first spacecraft designed and built by any country other than the U.S. and the U.S.S.R. Weight: 320 lb Power: Solar
		R. W. Knecht T. E. Van Zandt W. Calvert	CRPL NBS	
		J. W. King	DSIR England	
		S. J. Bauer L. Blumle R. Fitzenreiter J. E. Jackson	GSFC	
	Energetic particle counters—E	D. C. Rose I. B. McDiarmid	NRC Canada	
	VLF receiver (whistler)—I	J. S. Belrose	DRTE	
	Cosmic noise—A	T. R. Hartz	DRTE	

169

Designation	Objectives	Launch and orbit data					Project manager and project scientist
		Launch date/ silent date	Vehicle and launch site	Period, min.	Statute miles		
					Perigee	Apogee	
Alouette I— Cont.	the flux of energetic particles, and investigate whistlers.						
Explorer XIV Energetic Particles Satellite 1962 Beta Gamma 1 EPE–B S–3a	To correlate energetic particles activity with observations of the earth's magnetic fields; to monitor the existence of transient magnetic fields associated with plasma streams.	Oct. 2, 1962 Feb. 1964	Delta ETR	36.58 hours	174	61,090	Paul G. Marcotte Dr. F. B. McDonald
Explorer XV 1962 Beta Lambda 1 EPE–C S–3b	To study artificial radiation belt created by nuclear explosion.	Oct. 27, 1962 Feb. 9, 1963	Delta ETR	5 hours	195	10,950	Dr. John W. Townsend Dr. Wilmot Hess
Relay I 1962 Beta Upsilon 1 A–15	To investigate wideband communications between ground stations by means of low-altitude orbiting spacecraft. Communications signal to be evaluated will be an assortment of TV signals, multichannel telephony,	Dec. 13, 1962	Delta ETR	185.09	819.64	4,612.18	Wendell Sunderlin Dr. R. Waddel

APPENDIX B

Satellite and Space Probe Projects (Cont.)

Experiment data				Remarks
Instrumentation summary	Experiment and discipline	Experimenter	Affiliation	
A low-energy (0.1 to 20 keV) proton analyzer; a three-core magnetometer; one omnidirectional and three directional electron-proton detectors; a cosmic-ray package; an ion-electron scintillation detector; and devices to determine the effects of radiation on solar cells and the effects of space on electrolytic timers.	Proton analyzer—E Magnetic field (magnetometer)—E Trapped-particle radiation—E Cosmic-ray, ion-electron detector, solar-cell, and electrolytic timer—E	M. Bader L. Cahill J. A. Van Allen B. J. O'Brien F. B. McDonald L. R. Davis U. Desai	ARC U. of New Hampshire State U. of Iowa GSFC	Velocity at apogee 1,507 mph; perigee 23,734 mph. Inclination to equator 33°. Weight: 89.25 lb Power: Solar
Similar to Explorer XII	Electron energy distribution—I Omnidirectional detector—I Angular detector—E Directional detector—I Ion-electron detector—E Magnetic field—E Solar cell damage—I	W. Brown U. Desai C. McIlwain W. Brown C. McIlwain L. Davis L. Cahill H. K. Gummel	BTL GSFC U. of California BTL U. of California GSFC U. of New Hampshire BTL	Good data received on artificial radiation belt. Weight: 100 lb Power: Solar
The spacecraft contained an active communications repeater to receive and retransmit communications between the U.S. and Europe, U.S. and South America, U.S. and Japan, and Europe and South America; and an experiment to assess radiation damage to solar cells, and to measure proton and electron energy.	Determine radiation damage to solar cells and semiconductor diodes—E Measure proton energy (2.5–25.0 MeV)—E Measure electron energy (1.25–2.0 MeV)—E Measure integral omnidirectional proton flux energy (35.0–300.0 MeV)—E Measure directional electron energy (0.5–1.2 MeV)—E Measure directional proton energy (15.0–60.0 MeV)—E	R. Waddel W. Brown W. Brown C. McIlwain C. McIlwain C. McIlwain	GSFC BTL BTL U. of California U. of California U. of California	Orbit achieved. TV, telephone, teletype, facsimile, and digital-data transmissions were made with very satisfactory results. Conducted more than 2,000 technical tests and 172 successful demonstrations. Weight: 172 lb Power: Solar

VENTURE INTO SPACE

Goddard Space Flight Center

Designation	Objectives	Launch and orbit data					Project manager and project scientist
		Launch date/ silent date	Vehicle and launch site	Period, min.	Statute miles		
					Perigee	Apogee	
Relay I— Con.	and other communications. To measure the effects of the space environment on the system; to include radiation damage to solar cells and radiation flux density. To provide tests and demonstrations of low-altitude communications satellite.						
Syncom I 1963 4A A-25	To provide experience in using communications satellites in a 24-hour orbit. To flight-test a new, simple approach to satellite attitude and period control. To develop transportable ground facilities to be used in conjunction with communications satellites. To develop capability of launching satellites into 24-hour orbit using existing vehicles, plus apogee kick techniques and to test components' life at 24-hour-orbit altitude.	Feb. 14, 1963 Feb. 14, 1963	Delta ETR	24 hours	Near-synchronous orbit	22,300	R. J. Darcey

172

APPENDIX B

Satellite and Space Probe Projects (Cont.)

Experiment data				Remarks
Instrumentation summary	Experiment and discipline	Experimenter	Affiliation	
	Measure directional proton energy (1.0–8.0 MeV)—E	C. McIlwain	U. of California	
The 24-hour communications satellite consists of a spin-stabilized active repeater in a near-synchronous low-inclination orbit. The spacecraft is in the form of a cylinder 28 inches in diameter and 15 inches high. The repeater consists of a 7200-Mc receiver and an 1800-Mc transmitter with an output of 2 watts. In addition, the spacecraft contains a vernier velocity control system for orientation of spin axis and adjustment of the orbit.				Twenty seconds after firing apogee rocket, all satellite transmissions stopped. The satellite was sighted on Feb. 28, 1963, and later dates. It was traveling in a near-synchronous orbit eastward at about 2.8° per day. Weight: 78 lb Power: Solar

173

Goddard Space Flight Center

Designation	Objectives	Launch and orbit data					Project manager and project scientist
		Launch date/ silent date	Vehicle and launch site	Period, min.	Statute miles		
					Perigee	Apogee	
Explorer XVII Atmosphere Explorer 1963 9A S-6	To measure the density, composition, pressure, and temperature of the earth's atmosphere from 135 to 540 nautical miles and to determine the variations of these parameters with time of day, latitude, and in part, season.	Apr. 3, 1963 July 10, 1963	Delta ETR	96.4	158.1	598.5	N. W. Spencer
Telstar II 1963 13A (A project of AT&T) A-41	Joint AT&T–NASA investigation of wideband communications.	May 7, 1963	Delta ETR	221	575	6,559	C. P. Smith, Jr.
Tiros VII 1963 24A A-52	To launch into orbit a satellite capable of viewing cloud cover and the earth's surface and atmosphere by means of television cameras and radiation sensors. To acquire and process collected data from satellite and to control its attitude by magnetic means.	June 19, 1963	Delta ETR	97.4	385.02	401.14	Robert Rados
Syncom II 1963 31A A-26	To provide experience in using communications satellites in a 24-hour orbit.	July 26, 1963	Delta ETR	24 hours	22,300 near-synchronous orbit		R. J. Darcey

APPENDIX B

Satellite and Space Probe Projects (Cont.)

Experiment data				Remarks
Instrumentation summary	Experiment and discipline	Experimenter	Affiliation	
Primary detectors employed (two each) are: Double focusing magnetic sector mass spectrometer, hot-cathode total-pressure ionization gauges, and cold-cathode total-pressure ionization gauges. The remaining satellite instrumentation converts the outputs from six detectors to radio signals.	Two mass spectrometers —P Four vacuum (pressure) gauges—P Two electrostatic probes —I	C. Reber R. Horowitz G. Newton N. Spencer L. Brace	GSFC GSFC GSFC GSFC	Confirmed that the earth is surrounded by a belt of neutral helium at an altitude of from 150 to 600 miles. Weight: 405 lb Power: Silver-zinc batteries
The system provides for TV, radio, telephone, and data transmission via a satellite repeater system.	Included electron detector for energy range 0.75 to 2 MeV			Weight: 175 lb Power: Solar
Two vidicon camera systems with tape recorder for recording remote picture area, five-channel medium-resolution radiometer, electron temperature probe, and magnetic attitude coil.	Omnidirectional radiometer—P Scanning radiometer Electron temperature experiment—R Two TV camera systems	V. Suomi A. McCulloch N. Spencer	U. of Wisconsin GSFC GSFC	TV coverage extended to 65° N and 65° S latitudes. Launch date selected to provide maximum northern hemisphere coverage during 1963 hurricane season. Electron temperature probe malfunction 26 days after launch. First Tiros to have two operational camera systems and fully functioning IR subsystem 15 months after launch. Weight: 297 lb Power: Solar Inclination: 58° to equator
The 24-hour communications satellite consists of a spin-stabilized active repeater in a near-synchronous low-inclination				Orbit and attitude control of the spin-stabilized synchronous satellite achieved. Data telephone and facsimile transmission were excellent.

VENTURE INTO SPACE

Goddard Space Flight Center

Designation	Objectives	Launch and orbit data					Project manager and project scientist
		Launch date/ silent date	Vehicle and launch site	Period, min.	Statute miles		
					Perigee	Apogee	
Syncom II— Con.	To flight-test a new, simple approach to satellite attitude and period control. To develop transportable ground facilities to be used in conjunction with communications satellites. To develop capability of launching satellites into 24-hour orbit using existing vehicles, plus apogee kick techniques and to test components' life at 24-hour orbit altitude.						
Explorer XVIII Interplanetary Monitoring Platform 1963 46A IMP-A	To study in detail the radiation environment of cislunar space and to monitor this region over a significant portion of a solar cycle. To study the quiescent properties of the interplanetary magnetic field and its dynamical relationships with particle fluxes from the sun. To develop a solar flare prediction capa-	Nov. 27, 1963 May 1965	Delta ETR	93 hours	122	121,605	Paul Butler Dr. F. B. McDonald

176

APPENDIX B

Satellite and Space Probe Projects (Cont.)

Experiment data				Remarks
Instrumentation summary	Experiment and discipline	Experimenter	Affiliation	
orbit. The spacecraft is in the form of a cylinder 28 inches in diameter and 15 inches high. The repeater consists of a 7,200-Mc receiver and an 1,800-Mc transmitter with an output of 2 watts. In addition, the spacecraft contains a vernier velocity-control system for orientation of spin axis and adjustment of the orbit.				Television video signals also were successfully transmitted, even though the satellite was not designed for this capability. Weight: 70 lb Power: Solar
To carry 10 experiments; essentially a combination of the successful GSFC Explorer X and XII satellites. It is spin-stabilized and powered by solar cells. The system is designed so that data can be received from apogee by the GSFC Minitrack stations.	Plasma: measure thermal ions and electrons 0.10 eV—I	G. P. Serbu R. Bourdeau	GSFC	All experiments and equipment operated satisfactorily except for thermal ion experiment which gave only 10 percent good data. Continued to provide significant data. First accurate measure of the interplanetary magnetic field, and the shock front. First satellite to survive a severe earth shadow of 7 hr 55 min. Electronics equipment estimated to have cooled to below −60° C. Weight: 137.5 lb Power: 38 watts solar
	Magnetic field experiment (fluxgate magnetometer)—E	N. F. Ness	GSFC	
	Measure solar and galactic protons and alpha particles—E	J. A. Simpson	U. of Chicago	
	Measure total ionization produced per unit time in a unit volume of standard density air—E	K. A. Anderson	U. of California	
	Measure flux of low-energy interplanetary plasma—E	H. S. Bridge	MIT	
	Measure solar and galactic protons, electrons, alpha particles, heavy primaries, and isotropy of solar proton events and of cosmic-ray modulation—E	F. McDonald G. Ludwig	GSFC	
	Magnetic field (rubidium-vapor magnetometer)—E	N. F. Ness	GSFC	

VENTURE INTO SPACE

Goddard Space Flight Center

Designation	Objectives	Launch and orbit data					Project manager and project scientist
		Launch date/ silent date	Vehicle and launch site	Period, min.	Statute miles		
					Perigee	Apogee	
Explorer XVIII— Con.	bility for Apollo. To extend the knowledge of solar-terrestrial relationships. To further the development of simple, inexpensive, spin-stabilized spacecraft for interplanetary investigations.						
Tiros VIII 1963 54A A-53	To launch into orbit a satellite capable of viewing cloud cover and the earth's atmosphere by means of television cameras. To acquire and process collected data from satellite and to control its attitude by magnetic means.	Dec. 21, 1963	Delta ETR	99.35	435.01	468.30	Robert Rados

APPENDIX B

Satellite and Space Probe Projects (Cont.)

Experiment data				Remarks
Instrumentation summary	Experiment and discipline	Experimenter	Affiliation	
	Solar-wind proton concentrations—E	John Wolfe	ARC	
One standard Tiros vidicon with a wide-angle lens camera system, and one automatic picture transmission camera system; magnetic attitude coil.	One standard Tiros TV system One APT camera system	C. Hunter	GSFC	This satellite proved for the first time the feasibility of APT (automatic picture transmission), an inexpensive direct facsimile readout. Weight: 265 lb Power: Solar

Appendix C
NASA Sounding Rocket Flights

Notes

Numbering System

1. Aerobee-100
2. Arcon
3. Nike-Asp
4. Aerobee-150, 150A
5. Iris
6. Aerobee-300
7. Argo E-5
8. Argo D-4
9. Skylark
10. Nike-Cajun
11. Argo D-8
12. Special Projects
14. Nike-Apache
15. Arcas
16. Astrobee-1500
17. Aerobee-350
18. Nike-Tomahawk

Identifying Letters

The letters which follow each rocket number identify (1) the instrumenting agency, and (2) the experiment according to the following list:

Agency		Experiment	
G	Goddard	A	Aeronomy
N	Other NASA Centers	M	Meteorology
U	College or University	E	Energetic Particles and Fields
D	DOD	I	Ionospheric Physics
A	Other Government Agency	S	Solar Physics
C	Industrial Corporations	G	Galactic Astronomy
I	International	R	Radio Astronomy
		B	Biological
		P	Special Projects
		T	Test and Support

Firing Sites

ARG	Chamical, Argentina	IND	Thumba, India
ASC	Ascension Island	Italy	Sardinia, Italy
AUS	Woomera, Australia	NOR	Andöya, Norway
BRZ	Natal, Brazil	NZ	Karikari, New Zealand
EGL	Eglin Air Force Base, Florida	PB	Point Barrow, Alaska
FC	Fort Churchill, Canada	PMR	Pacific Missile Range

PAK	Karachi, Pakistan	WI	Wallops Island, Virginia
SWE	Kronogård, Sweden	WS	White Sands Missile Range, New Mexico
SUR	Coronie, Surinam		

Abbreviations

AFCRL	Air Force Cambridge Research Laboratories, Bedford, Mass.	U. Pitt.	University of Pittsburgh, Pittsburgh, Pa.
Ames	NASA, Ames Research Center, Moffett Field, Calif.	U. Wisc.	University of Wisconsin, Madison, Wis.
AS&E	American Science and Engineering, Inc., Cambridge, Mass.	Varian	Varian Associates, Palo Alto, Calif.
		Harvard	Harvard College, Cambridge, Mass.
BRL	Ballistics Research Laboratories, Aberdeen, Md.	JHU	Johns Hopkins University, Baltimore, Md.
BuStds	National Bureau of Standards, Boulder, Colo.	JPL	Jet Propulsion Laboratory, Pasadena, Calif.
CRPL	Central Radio Propagation Laboratories, National Bureau of Standards, Boulder, Colo.	LaRC	NASA, Langley Research Center, Hampton, Va.
AIL	Airborne Instruments Laboratory, New York	LeRC	NASA, Lewis Research Center, Cleveland, Ohio
DRTE	Canadian Defence Research Telecommunications Establishment, Ottawa, Canada	Lockheed	Lockheed Missiles and Space Division, Palo Alto, Calif.
		U. Minn.	University of Minnesota, Minneapolis, Minn.
GCA	Geophysics Corporation of America, Bedford, Mass.	NYU	New York University, New York, N.Y.
NRL	Naval Research Laboratory, Washington, D.C.	Penn State	Penn State University, University Park, Pa.
U. Colo.	University of Colorado, Boulder, Colo.	Princeton	Princeton University, Princeton, N.J.
U. Ill.	University of Illinois, Urbana, Ill.		
U. Mich.	University of Michigan, Ann Arbor, Mich.	Rice	Rice University, Houston, Tex.
UNH	University of New Hampshire, Durham, N.H.	SCAS	Southwest Center for Advanced Studies, Dallas, Tex.

APPENDIX C

NASA Sounding Rocket Flights

NASA No.	Firing			Aeronomy			Results[a]
	Date	Site	Performance[a]	Experimenter	NASA scientist and location	Experiment	
	1960						
4.09 GA	Apr. 29	WI	S	Horowitz, GSFC	Horowitz, GSFC	Atmospheric Composition	S
10.03 GA	June 16	WI	P	Nordberg, GSFC	Nordberg, GSFC	Grenade	X
10.04 GA	July 9	WI	S	Nordberg, GSFC	Nordberg, GSFC	Grenade	S
10.01 GA	14	WI	S	Nordberg, GSFC	Nordberg, GSFC	Grenade	X
4.14 GA	Nov. 15	WI	S	Taylor, GSFC	Taylor, GSFC	Atmospheric Composition	S
10.06 GA	Dec. 14	WI	S	Nordberg, GSFC	Nordberg, GSFC	Grenade	S
	1961						
10.07 GA	Feb. 14	WI	S	Nordberg, GSFC	Nordberg, GSFC	Grenade	S
10.08 GA	17	WI	P	Nordberg, GSFC	Nordberg, GSFC	Grenade	S
10.33 GA	Apr. 5	WI	S	Nordberg, GSFC	Nordberg, GSFC	Grenade	P
10.34 GA	27	WI	X	Smith, GSFC	Smith, GSFC	Grenade	X
10.02 GA	May 5	WI	S	Smith, GSFC	Smith, GSFC	Grenade	S
10.28 GA	6	WI	S	Smith, GSFC	Smith, GSFC	Grenade	S
10.29 GA	9	WI	S	Smith, GSFC	Smith, GSFC	Grenade	P
10.30 GA	July 13	WI	S	Smith, GSFC	Smith, GSFC	Grenade	S
10.31 GA	14	WI	S	Smith, GSFC	Smith, GSFC	Grenade	S
10.32 GA	20	WI	S	Smith, GSFC	Smith, GSFC	Grenade	S
10.35 GA	21	WI	S	Smith, GSFC	Smith, GSFC	Grenade	X

[a] S—Successful
P—Partial success } Subject to interpretation
X—Unsuccessful

NASA Sounding Rocket Flights (Cont.)

NASA No.	Firing Date	Firing Site	Firing Performance[a]	Experimenter	NASA scientist and location	Experiment	Results[a]
	1961			Aeronomy—Continued			
10.36 GA	Sept. 16	WI	P	Smith, GSFC	Smith, GSFC	Grenade	P
10.37 GA	17	WI	S	Smith, GSFC	Smith, GSFC	Grenade	X
1.08 GA	23	FC	S	Varian Associates	Martin, GSFC	Atmospheric Structure	S
1.09 GA	30	FC	S	Varian Associates	Martin, GSFC	Atmospheric Structure	S
8.23 GA	Oct. 10	WI	S	Taylor, GSFC	Taylor, GSFC	Ionosphere	S
1.10 GA	15	FC	S	Varian Associates	Martin, GSFC	Atmospheric Structure	S
1.07 GA	17	FC	S	Varian Associates	Martin, GSFC	Atmospheric Structure	S
1.11 GA	Nov. 2	FC	S	Varian Associates	Martin, GSFC	Atmospheric Structure	S
1.12 GA	5	FC	S	Varian Associates	Martin, GSFC	Atmospheric Structure	S
10.64 GA	Dec. 21	WI	S	U. Mich.	Spencer, GSFC	Atmospheric Structure	S
	1962						
10.38 GA	Mar. 2	WI	S	Smith, GSFC	Smith, GSFC	Grenade	S
10.39 GA	2	WI	S	Smith, GSFC	Smith, GSFC	Grenade	S
4.18 GA	19	WI	X	U. Mich.	Spencer, GSFC	Atmospheric Structure	X
10.40 GA	23	WI	S	Smith, GSFC	Smith, GSFC	Grenade	S
10.41 CA	28	WI	S	Smith, GSFC	Smith, GSFC	Grenade	S
10.42 GA	Apr. 17	WI	S	Smith, GSFC	Smith, GSFC	Grenade	S
5.04 GA	May 3	WI	P	Taylor, GSFC	Taylor, GSFC	Atmospheric Structure	S
10.43 GA	June 7	WI	S	Smith, GSFC	Smith, GSFC	Grenade	S
10.44 GA	8	WI	S	Smith, GSFC	Smith, GSFC	Grenade	S
10.55 GA	Nov. 16	FC	X	Smith, GSFC	Smith, GSFC	Grenade	X
6.06 GA	20	WI	S	U. Mich.	Brace, GSFC	Thermosphere Probe	S

APPENDIX C

10.45 GA	Dec. 1	WI	S	Smith, GSFC	Smith, GSFC	Grenade	S
10.68 GA	1	FC	S	Smith, GSFC	Smith, GSFC	Grenade	X
10.46 GA	4	WI	S	Smith, GSFC	Smith, GSFC	Grenade	X
10.67 GA	4	FC	X	Smith, GSFC	Smith, GSFC	Grenade	S
10.47 GA	6	WI	S	Smith, GSFC	Smith, GSFC	Grenade	S
10.66 GA	6	FC	S	Smith, GSFC	Smith, GSFC	Grenade	S
	1963						
10.48 GA	Feb. 20	WI	S	Smith, GSFC	Smith, GSFC	Grenade	S
10.58 GA	20	FC	S	Smith, GSFC	Smith, GSFC	Grenade	S
10.53 GA	28	WI	S	Smith, GSFC	Smith, GSFC	Grenade	S
10.59 GA	28	FC	S	Smith, GSFC	Smith, GSFC	Grenade	S
10.54 GA	Mar. 9	WI	S	Smith, GSFC	Smith, GSFC	Grenade	S
10.60 GA	9	FC	S	Smith, GSFC	Smith, GSFC	Grenade	S
6.07 GA	Apr. 18	WI	S	U. Mich.	Brace, GSFC	Thermosphere Probe	S
10.55 GA	Dec. 7	WI	S	Smith, GSFC	Smith, GSFC	Grenade	S
	1961						
10.72 NA	Nov. 18	WI	S	LaRC	Hord, LaRC	Airglow	S
	1962						
10.79 NA	Apr. 5	WI	S	LeRC	Potter, LeRC	Ozone	S
1.13 NA	Sept. 6	WI	S	JPL	Dubin, HQ	UV Airglow	S
1.14 NA	Nov. 20	WI	X	JPL	Dubin, HQ	UV Airglow	X
	1963						
10.80 NA	Jan. 17	WI	S	LeRC	Potter, LeRC	Ozone	S
10.92 NA	Sept. 25	WI	S	LaRC	LaRC	Chemical Release	S

[a] S—Successful
P—Partial success } Subject to interpretation.
X—Unsuccessful

NASA Sounding Rocket Flights (Cont.)

Aeronomy—Continued

NASA No.	Firing Date	Firing Site	Firing Performance[a]	Experimenter	NASA scientist and location	Experiment	Results[a]
	1963						
10.93 NA	Sept. 25	WI	S	LaRC	LaRC	Chemical Release	S
14.102 NA	Oct. 9	WI	S	LeRC	Potter, LeRC	Chemical Release	S
14.103 NA	Oct. 10	WI	S	LeRC	Potter, LeRC	Chemical Release	S
4.85 NA	Nov. 18	WI	S	JPL	Dubin, HQ	Airglow	S
	1960						
10.09 UA	Nov. 2	WI	S	U. Mich.	Dubin, HQ	Atmospheric Composition	X
10.10 UA	16	WI	S	U. Mich.	Dubin, HQ	Atmospheric Composition	X
	1961						
10.50 UA	June 6	WI	S	U. Mich.	Dubin, HQ	Atmospheric Structure	S
10.56 UA	9	WI	S	U. Mich.	Dubin, HQ	Atmospheric Composition	X
10.57 UA	July 26	WI	S	U. Mich.	Dubin, HQ	Atmospheric Composition	X
	1962						
10.90 UA	Feb. 20	WI	S	U. Mich.	Dubin, HQ	Atmospheric Composition	X
10.91 UA	May 18	WI	S	U. Mich.	Dubin, HQ	Atmospheric Composition	S
14.19 UA	June 6	WI	S	U. Mich.	Spencer, GSFC	Atmospheric Structure	S
14.20 UA	Dec. 1	WI	S	U. Mich.	Spencer, GSFC	Atmospheric Structure	S
4.74 UA	13	WI	X	JHU	Dubin, HQ	Airglow	X

APPENDIX C

	1963						
4.73 UA	Jan. 29	WI	X	JHU	Dubin, HQ	Airglow	X
14.08 UA	Mar. 28	WI	S	U. Mich.	Dubin, HQ	Atmospheric Composition	S
14.09 UA	28	WI	S	U. Mich.	Dubin, HQ	Atmospheric Composition	X
4.98 UA	May 7	WI	S	JHU	Dubin, HQ	Airglow	S
4.75 UA	July 20	FC	X	JHU	Dubin, HQ	Airglow	X
10.75 UA	Aug. 2	WI	S	U. Mich.	Holtz, HQ	Atmospheric Density	S
4.76 UA	Nov. 12	WI	S	JHU	Dubin, HQ	Airglow	S
14.10 UA	26	WI	S	U. Mich.	Dubin, HQ	Atmospheric Composition	S
10.131 UA	26	WI	S	U. Mich.	Dubin, HQ	Atmospheric Density	S
14.21 UA	Dec. 7	WI	S	U. Mich.	Smith, GSFC	Atmospheric Structure	S
	1963						
14.140 DA	May 18	EGL	S	AFCRL	Dubin, HQ	Sodium Vapor	S
14.141 DA	18	EGL	S	AFCRL	Dubin, HQ	Sodium Vapor	S
10.130 DA	22	EGL	S	AFCRL	Dubin, HQ	Sodium Vapor	S
	1959						
3.13 CA	Aug. 17	WI	S	GCA	Dubin, HQ	Sodium Vapor	S
3.14 CA	19	WI	X	GCA	Dubin, HQ	Sodium Vapor	X
3.15 CA	Nov. 18	WI	S	GCA	Dubin, HQ	Sodium Vapor	S
3.16 CA	19	WI	S	GCA	Dubin, HQ	Sodium Vapor	X
3.17 CA	20	WI	S	GCA	Dubin, HQ	Sodium Vapor	X
	1960						
3.23 CA	May 24	WI	X	GCA	Dubin, HQ	Sodium Vapor	X
3.24 CA	25	WI	S	GCA	Dubin, HQ	Sodium Vapor	S
10.05 CA	Sept. 20	WI	S	Nordberg, GSFC	Nordberg, GSFC	Grenade	X

[a] S—Successful
P—Partial success } Subject to interpretation.
X—Unsuccessful

NASA Sounding Rocket Flights (Cont.)

NASA No.	Firing Date	Firing Site	Performance[a]	Experimenter	NASA scientist and location	Experiment	Results[a]
				Aeronomy—Continued			
	1960						
8.04 CA	Nov. 10	WI	S	Lockheed	Dubin, HQ	Ionosphere	P
10.11 CA	Dec. 9	WI	X	GCA	Dubin, HQ	Sodium Vapor	X
10.12 CA	9	WI	S	GCA	Dubin, HQ	Sodium Vapor	S
8.05 CA	10	WI	S	GCA	Dubin, HQ	Sodium Vapor	S
	1961						
3.05 CA	Apr. 19	WI	S	GCA	Dubin, HQ	Sodium Vapor	S
3.06 CA	21	WI	S	GCA	Dubin, HQ	Sodium Vapor	S
3.07 CA	21	WI	X	GCA	Dubin, HQ	Sodium Vapor	X
3.08 CA	21	WI	S	GCA	Dubin, HQ	Sodium Vapor	S
8.06 CA	Sept. 13	WI	S	GCA	Smith, GSFC	Sodium Vapor	S
8.22 CA	13	WI	S	GCA	Smith, GSFC	Sodium Vapor	S
3.09 CA	16	WI	X	GCA	Smith, GSFC	Sodium Vapor	X
3.18 CA	16	WI	S	GCA	Smith, GSFC	Sodium Vapor	S
3.19 CA	17	WI	S	GCA	Smith, GSFC	Sodium Vapor	S
	1962						
10.100 CA	Mar. 1	WI	S	GCA	Smith, GSFC	Sodium Vapor	S
10.101 CA	2	WI	S	GCA	Smith, GSFC	Sodium Vapor	S
10.102 CA	23	WI	S	GCA	Smith, GSFC	Sodium Vapor	S
10.103 CA	27	WI	S	GCA	Smith, GSFC	Sodium Vapor	S
3.20 CA	Apr. 17	WI	S	GCA	Smith, GSFC	Sodium Vapor	S

Vehicle	Date	Site	Result	Agency	Experimenter	Experiment	Result
3.21 CA	June 7	WI	S	GCA	Smith, GSFC	Sodium Vapor	S
3.22 CA	June 7	WI	X	GCA	Smith, GSFC	Sodium Vapor	X
14.30 CA	Aug. 23	WI	P	Lockheed	Depew, GSFC	Atmospheric Structure	X
14.16 CA	Nov. 7	WI	S	GCA	Smith, GSFC	Sodium Vapor	S
14.17 CA	Nov. 30	WI	S	GCA	Smith, GSFC	Sodium Vapor	S
14.18 CA	Dec. 5	WI	S	GCA	Smith, GSFC	Sodium Vapor	P
1963							
3.11 CA	Feb. 18	WI	X	GCA	Smith, GSFC	Sodium Vapor	X
14.35 CA	Feb. 20	WI	S	GCA	Smith, GSFC	Sodium Vapor	S
14.39 CA	Feb. 21	WI	S	GCA	Smith, GSFC	Sodium Vapor	S
14.110 CA	May 8	WI	S	Lockheed	Bourdeau, GSFC	Massenfilter	X
14.13 CA	May 22	FC	S	GCA	Dubin, HQ	Sodium Vapor	S
14.14 CA	May 22	FC	S	GCA	Dubin, HQ	Sodium Vapor	S
14.15 CA	May 23	FC	S	GCA	Dubin, HQ	Sodium Vapor	S
14.40 CA	May 24	WI	S	GCA	Dubin, HQ	Sodium Vapor	S
14.41 CA	May 24	WI	S	GCA	Dubin, HQ	Sodium Vapor	X
14.42 CA	May 25	WI	S	GCA	Dubin, HQ	Sodium Vapor	S
10.77 IA	May 16	PAK	S	Pakistan	Dubin, HQ	Sodium Vapor	X
14.137 IA	May 20	Italy	S	Italy	Dubin, HQ	Sodium Vapor	S
14.138 IA	May 21	Italy	S	Italy	Dubin, HQ	Sodium Vapor	S
14.139 IA	May 21	Italy	S	Italy	Dubin, HQ	Sodium Vapor	S
14.128 IA	Nov. 21	IND	S	India	Dubin, HQ	Sodium Vapor	P
1962							
Rehbar 1[b]	June 7	PAK	S	Pakistan	Dubin, HQ	Sodium Vapor	X
Rehbar 2[b]	June 11	PAK	S	Pakistan	Dubin, HQ	Sodium Vapor	X

[a] S—Successful
P—Partial success } Subject to interpretation.
X—Unsuccessful
[b] Nike-Cajun

NASA Sounding Rocket Flights (Cont.)

NASA No.	Firing			Experimenter	NASA scientist and location	Experiment	Results[a]
	Date	Site	Performance[a]				
	1960			Energetic Particles and Fields			
10.17 GE	June 6	FC	S	Fichtel, GSFC	Fichtel, GSFC	SBE	S
8.07 GE	30	WI	X	Heppner, GSFC	Heppner, GSFC	Magnetic Field	S
10.18 GE	July 22	FC	X	Fichtel, GSFC	Fichtel, GSFC	SBE	S
10.19 GE	Sept. 3	FC	S	Fichtel, GSFC	Fichtel, GSFC	SBE	S
10.20 GE	3	FC	S	Fichtel, GSFC	Fichtel, GSFC	SBE	S
11.01 GE	19	PMR	S	Naugle, GSFC	Naugle, GSFC	NERV 1	S
10.21 GE	27	FC	S	Fichtel, GSFC	Fichtel, GSFC	SBE	S
10.22 GE	Nov. 11	FC	S	Fichtel, GSFC	Fichtel, GSFC	SBE	P
10.23 GE	11	FC	S	Fichtel, GSFC	Fichtel, GSFC	SBE	S
10.24 GE	12	FC	S	Fichtel, GSFC	Fichtel, GSFC	SBE	S
10.15 GE	12	FC	S	Fichtel, GSFC	Fichtel, GSFC	SBE	S
10.16 GE	13	FC	S	Fichtel, GSFC	Fichtel, GSFC	SBE	S
10.13 GE	16	FC	S	Fichtel, GSFC	Fichtel, GSFC	SBE	S
10.14 GE	17	FC	S	Fichtel, GSFC	Fichtel, GSFC	SBE	S
10.26 GE	18	FC	S	Fichtel, GSFC	Fichtel, GSFC	SBE	S
10.27 GE	18	FC	S	Fichtel, GSFC	Fichtel, GSFC	SBE	S
8.08 GE	Dec. 12	WI	S	Heppner, GSFC	Heppner, GSFC	Magnetic Fields	S
	1961						
10.76 GE	Dec. 10	FC	S	Ogilvie-Fichtel, GSFC	Ogilvie-Fichtel, GSFC	Cosmic Ray	S

APPENDIX C

ID	Date	Launch	Experimenter	Status[a]	Experimenter	Experiment	Status[a]
4.91 GE	1963 Sept. 4	FC	Fichtel, GSFC	S	Fichtel, GSFC	Heavy Cosmic Rays	S
4.16 UE	1960 Aug. 23	WI	NYU	S	Meredith, GSFC	Cosmic Ray	S
14.03 UE	1961 July 14	WI	UNH	S	Heppner, GSFC	Magnetic Field	S
14.04 UE	14	WI	UNH	S	Heppner, GSFC	Magnetic Field	S
14.05 UE	20	WI	UNH	S	Heppner, GSFC	Magnetic Field	S
11.06 UE	1963 Feb. 12	PMR	U. Minn.	S	Cline, GSFC	Electron Spect.	S
14.06 UE	Sept. 9	WI	UNH	S	Schardt, HQ	Electrojet	S
Ionospheric Physics							
4.08 GI	1959 Sept. 11	FC	Jackson, GSFC	S	Jackson, GSFC	Ionosphere	S
4.07 GI	14	FC	Jackson, GSFC	S	Jackson, GSFC	Ionosphere	S
1.01 GI	1960 Nov. 23	FC	Whipple, GSFC	S	Whipple, GSFC	Ionosphere	S
1.02 GI	27	FC	Whipple, GSFC	S	Whipple, GSFC	Ionosphere	S

[a] S—Successful
P—Partial success } Subject to interpretation.
X—Unsuccessful

NASA Sounding Rocket Flights (Cont.)

NASA No.	Firing Date	Firing Site	Firing Performance[a]	Experimenter	NASA scientist and location	Experiment	Results[a]
				Ionospheric Physics—Continued			
	1961						
8.10 GI	Apr. 27	WI	S	Jackson, GSFC	Jackson, GSFC	Ionosphere	P
8.09 GI	June 13	WI	S	Jackson, GSFC	Jackson, GSFC	Ionosphere	P
10.74 GI	Dec. 21	WI	S	Kane, GSFC	Kane, GSFC	Ionosphere	S
	1962						
10.110 GI	Apr. 26	WI	S	Serbu, GSFC	Serbu, GSFC	Electron Temperature	S
8.21 GI	May 3	WI	S	Serbu, GSFC	Serbu, GSFC	ELF Electron Trap	S
10.112 GI	16	WI	S	Serbu, GSFC	Serbu, GSFC	Electron Temperature	S
10.111 GI	17	WI	S	Serbu, GSFC	Serbu, GSFC	Electron Temperature	S
14.12 GI	June 15	WI	S	Kane, GSFC	Kane, GSFC	Ionosphere	S
K62–1[b]	Aug. 7	SWE	S	Sweden	Witt, Sweden	Air Sample	S
K62–3[b]	11	SWE	S	Sweden	Smith, GSFC	Air Sample	S
K62–4[b]	11	SWE	S	Sweden	Smith, GSFC	Air Sample	P
K62–5[b]	31	SWE	S	Sweden	Smith, GSFC	Air Sample	X
14.31 GI	Oct. 16	WI	S	Bauer, GSFC	Bauer, GSFC	Ionosphere	S
14.32 GI	Dec. 1	WI	S	Bauer, GSFC	Bauer, GSFC	Ionosphere	S
	1963						
14.107 GI	Mar. 8	WI	S	Whipple, GSFC	Whipple, GSFC	Ionosphere	P
14.108 GI	Apr. 9	WI	S	Kane, GSFC	Kane, GSFC	D-Region	S
4.44 GI	23	WI	S	Bauer, GSFC	Bauer, GSFC	Electron Density	S
8.14 GI	July 2	WI	S	Bauer, GSFC	Bauer, GSFC	Ionosphere	S

APPENDIX C

6.08 GI		20	WI	S	Brace, GSFC	Brace, GSFC	Thermosphere Probe	S
K63-1[b]	July	27	SWE	S	Sweden	Smith, GSFC	Grenade	S
K63-2[b]		29	SWE	S	Sweden	Smith, GSFC	Grenade	S
K63-3[b]	Aug.	1	SWE	S	Sweden	Smith, GSFC	Grenade	S
K63-4[b]		7	SWE	S	Sweden	Smith, GSFC	Heavy Cosmic Rays	S
4.65 GI	Sept.	25	WI	S	GSFC	Serbu, GSFC	Ionosphere	S
						Hirao, Japan		
4.64 GI		28	WI	S	GSFC	Serbu, GSFC	Ionosphere	S
						Hirao, Japan		
8.18 GI		29	WI	S	Bauer, GSFC	Bauer, GSFC	Ionosphere	
14.37 GI	Dec.	13	WI	P	Whipple, GSFC	Whipple, GSFC	Ionosphere	
	1960							
6.01 UI	Mar.	16	FC	S	U. Mich.	Bourdeau, GSFC	Ionosphere	S
3.10 UI		17	FC	X	U. Mich.	Bourdeau, GSFC	Ionosphere	X
6.02 UI	June	15	FC	S	U. Mich.	Bourdeau, GSFC	Ionosphere	S
6.03 UI	Aug.	3	WI	S	U. Mich.	Bourdeau, GSFC	Ionosphere	S
	1961							
6.04 UI	Mar.	26	WI	S	U. Mich.	Bourdeau, GSFC	Ionosphere	S
6.05 UI	Dec.	22	WI	S	U. Mich.	Wright, GSFC	Ionosphere	S
	1963							
4.58 UI	Apr.	3	WI	S	Stanford	Bourdeau, GSFC	Ionosphere	S
4.59 UI	July	10	WI	S	Stanford	Bourdeau, GSFC	Ionosphere	S

[a] S—Successful
P—Partial success } Subject to interpretation
X—Unsuccessful
[b] Nike-Cajun

NASA Sounding Rocket Flights (Cont.)

NASA No.	Firing Date	Firing Site	Firing Performance[a]	Experimenter	NASA scientist and location	Experiment	Results[a]
				Ionospheric Physics—Continued			
14.36 DI	*1963* Oct. 7	FC	S	BRL	Bourdeau, GSFC	Ionosphere	P
8.15 AI	*1961* June 24	WI	S	CRPL/AIL	Jackson, GSFC	Ionosphere	S
8.17 AI	Oct. 14	WI	S	Jackson, GSFC	Jackson, GSFC	Ionosphere	S
8.16 AI	*1962* Feb. 7	WI	S	Jackson, GSFC	Jackson, GSFC	Ionosphere	X
3.12 CI	*1960* Aug. 22	WI	X	GCA	Bourdeau, GSFC	Langmuir Probe	X
10.25 CI	Dec. 8	WI	S	GCA	Bourdeau, GSFC	Langmuir Probe	S
10.51 CI	*1961* Aug. 18	WI	S	GCA	Bourdeau, GSFC	Langmuir Probe	S
10.52 CI	Oct. 27	WI	S	GCA	Bourdeau, GSFC	Langmuir Probe	S
10.99 CI	*1962* Nov. 7	WI	S	GCA	Bourdeau, GSFC	Ionosphere	S

APPENDIX C

ID	Date	Site	Result	Agency	PI	Experiment	Result
10.108 CI	30	WI	S	GCA	Bourdeau, GSFC	Ionosphere	S
10.109 CI	Dec. 5	WI	S	GCA	Bourdeau, GSFC	Ionosphere	S
	1963						
14.86 CI	Feb. 27	WI	S	GCA	Bourdeau, GSFC	Ionosphere	S
14.87 CI	Mar. 28	WI	P	GCA	Bourdeau, GSFC	Ionosphere	S
14.88 CI	July 14	FC	P	GCA	Bourdeau, GSFC	Ionosphere	P
14.89 CI	20	FC	X	GCA	Bourdeau, GSFC	Eclipse Ionosphere	X
14.90 CI	20	FC	X	GCA	Bourdeau, GSFC	Eclipse Ionosphere	X
14.91 CI	20	FC	S	GCA	Bourdeau, GSFC	Eclipse Ionosphere	S
14.92 CI	20	FC	S	GCA	Bourdeau, GSFC	Eclipse Ionosphere	S
14.93 CI	20	FC	S	GCA	Bourdeau, GSFC	Eclipse Ionosphere	S
14.94 CI	20	FC	S	GCA	Bourdeau, GSFC	Eclipse Ionosphere	S
	1959						
4.02 II	Sep. 17	FC	S	DRTE	Jackson, GSFC	Ionosphere	S
4.03 II	20	FC	P	DRTE	Jackson, GSFC	Ionosphere	X
	1961						
8.13 II	June 15	WI	S	DRTE	Jackson, GSFC	Antenna Test	S
	1962						
4.79 II	Nov. 16	WI	X	Australia	Cartwright, Australia	Ionosphere	X
4.80 II	Dec. 11	WI	X	Australia	Cartwright, Australia	Ionosphere	X
Ferdinand III[b]	11	NOR	S	Norway	Kane, GSFC	Ionosphere	S
Ferdinand II[b]	14	NOR	S	Norway	Kane, GSFC	NASA T/M only	S

[a] S—Successful
P—Partial success } Subject to interpretation.
X—Unsuccessful
[b] Nike-Cajun

195

NASA Sounding Rocket Flights (Cont.)

NASA No.	Firing			Experimenter	NASA scientist and location	Experiment	Results[a]
	Date	Site	Performance[a]				
	1963			Ionospheric Physics—Continued			
4.96 II	Apr. 12	WI	S	Australia	Cartwright, Australia	VLF	S
4.97 II	May 9	WI	S	Australia	Cartwright, Australia	VLF	S
Ferdinand V[b]	Sept. 8	NOR	S	Norway	Kane, GSFC	Ionosphere	X
Ferdinand IV[c]	Sept. 11	NOR	S	Norway	Kane, GSFC	Ionosphere	S
4.93 II	Oct. 17	WI	S	France	Shea, GSFC	Ionosphere	S
4.94 II	Oct. 31	WI	S	France	Shea, GSFC	Ionosphere	S
				Solar Physics			
	1960						
3.01 GS	Mar. 1	WI	S	Hallam, GSFC	Hallam, GSFC	Solar Study	X
3.02 GS	Mar. 3	WI	S	Hallam, GSFC	Hallam, GSFC	Solar Study	X
3.03 GS	Apr. 27	WI	X	Hallam, GSFC	Hallam, GSFC	Solar Study	X
3.04 GS	May 25	WI	X	Hallam, GSFC	Hallam, GSFC	Solar Study	X
	1961						
4.25 GS	Sept. 30	WI	S	Behring, GSFC	Behring, GSFC	Solar Studies	S
	1963						
4.77 GS	July 20	WI	S	Hallam, GSFC	Hallam-Wolff, GSFC	Solar Studies	X
4.78 GS	Oct. 1	WI	S	Hallam, GSFC	Hallam, GSFC	Solar Studies	P

APPENDIX C

ID	Date	Loc		PI	PI	Subject	Status
4.33 GS	15	WI		Muney, GSFC	Muney, GSFC	Solar Studies	S
	1962						
4.23 US	July 24	WI	S	U. Colo.	Lindsay, GSFC	Sunfollower	P
4.21 US	Nov. 27	WI	S	Harvard	Lindsay, GSFC	Solar	X
	1963						
4.22 US	Sept. 6	WI	S	Harvard	Lindsay, GSFC	Solar Studies	S
	1963						
4.61 AS	June 20	WI	S	NRL	Packer, NRL	Coronagraph	P
4.62 AS	28	WI	S	NRL	Packer, NRL	Coronagraph	P

Galactic Astronomy

ID	Date	Loc		PI	PI	Subject	Status
	1960						
4.40 GG	Apr. 27	WI	P	Kupperian, GSFC	Kupperian, GSFC	Stellar Fluxes	P
4.05 GG	May 27	WI	S	Boggess, GSFC	Boggess, GSFC	Stellar Fluxes	P
4.06 GG	June 24	WI	S	Boggess, GSFC	Boggess, GSFC	Stellar Fluxes	S
4.11 GG	Nov. 22	WI	S	Stecher, GSFC	Stecher, GSFC	Stellar Spectra	S
	1961						
4.34 GG	Mar. 31	WI	P	Boggess, GSFC	Boggess, GSFC	Stellar Fluxes	P
9.01 GG	Sept. 18	AUS	S	Boggess, GSFC	Boggess, GSFC	Stellar Photo	S
9.02 GG	Oct. 4	AUS	S	Boggess, GSFC	Boggess, GSFC	Stellar Photo	S

[a] S—Successful
P—Partial success } Subject to interpretation.
X—Unsuccessful

[b] Nike-Apache

NASA Sounding Rocket Flights (Cont.)

NASA No.	Firing Date	Firing Site	Performance[a]	Experimenter	NASA scientist and location	Experiment	Results[a]
				Galactic Astronomy—Continued			
	1961						
9.03 GG	Nov. 1	AUS	S	Boggess, GSFC	Boggess, GSFC	Stellar Photo	P
9.04 GG	20	AUS	S	Boggess, GSFC	Boggess, GSFC	Stellar Photo	S
	1962						
4.35 GG	Feb. 7	WI	X	Stecher, GSFC	Stecher, GSFC	Stellar Spectra	X
4.36 GG	Sept. 22	WI	S	Stecher, GSFC	Stecher, GSFC	Stellar Photo	S
	1963						
4.30 GG	Mar. 28	WI	S	Boggess, GSFC	Boggess, GSFC	Stellar Spectra	S
4.37 GG	July 19	WI	S	Stecher, GSFC	Stecher, GSFC	Stellar Spectra	S
4.29 GG	23	WI	S	Stecher, GSFC	Stecher, GSFC	Stellar Spectra	S
4.31 GG	Oct. 10	WI	X	Boggess, GSFC	Boggess, GSFC	Stellar Spectra	X
	1962						
4.54 UG	Oct. 30	WI	S	U. Wisc.	Kupperian, GSFC	Stellar Studies	S
	1962						
4.69 CG	Sept. 30	WI	S	Lockheed	Dubin, HQ	Night Sky Mapping	S

APPENDIX C

	Date	Launch	Success[a]	Lockheed	Biological — Depew, GSFC	Stellar Spectra	Success[a]
4.70 CG	*1963* Mar. 16	WI	S				S
	1961						
11.04 GB	Nov. 15	Pt. A	S	Ames	Smith, HQ	BIOS 1	X
11.05 GB	18	Pt. A	P	Ames	Smith, HQ	BIOS 1	X
					Special Projects		
	1960						
1.03 GP	Sept. 15	FC	S	Baumann, GSFC	Baumann, GSFC	AMPP	S
1.05 GP	24	FC	S	Baumann, GSFC	Baumann, GSFC	AMPP	P
4.43 GP	Oct. 5	FC	S	NRL	Baumann, GSFC	AMPP	S
	1961						
1.04 GP	May 17	FC	S	Baumann, GSFC	Baumann, GSFC	AMPP	P
1.06 GP	19	FC	S	Baumann, GSFC	Baumann, GSFC	AMPP	S
	1961						
4.38 NP	Feb. 5	WI	S	LeRC	Gold, LeRC	Hydrogen Zerog	P
4.39 NP	Apr. 21	WI	S	LeRC	Gold, LeRC	Hydrogen Zerog	S
4.42 NP	Aug. 12	WI	S	LeRC	Plohr, LeRC	Hydrogen Zerog	P
4.40 NP	Oct. 18	WI	S	LeRC	Regetz, LeRC	Hydrogen Zerog	S

[a] S—Successful
P—Partial success } Subject to interpretation.
X—Unsuccessful

NASA Sounding Rocket Flights (Cont.)

NASA No.	Firing Date	Firing Site	Firing Performance[a]	Experimenter	NASA scientist and location	Experiment	Results[a]
	1962			Special Projects—Continued			
4.41 NP	Feb. 17	WI	S	LeRC	Dillon, LeRC	Hydrogen Zerog	S
4.46 NP	May 8	WI	P	JPL	Brown, JPL	Radar	X
4.26 NP	June 20	WI	S	LeRC	Flagge, LeRC	Hydrogen Zerog	P
4.47 NP	July 10	WI	S	JPL	Brown, JPL	Radar	X
4.27 NP	Nov. 18	WI	S	LeRC	Corpas, LeRC	Hydrogen Zerog	S
	1963						
4.66 NP	May 14	WI	S	LaRC	Kinard, LaRC	Paraglider	X
4.28 NP	June 19	WI	S	LeRC	Corpas, LeRC	Hydrogen Zerog	P
4.32 NP	Sept. 11	WI	S	LeRC	Corpas, LeRC	Hydrogen Zerog	S
	1962						
4.71 UP	June 29	WI	S	JHU	Depew, GSFC	Airglow	S
4.72 UP	29	WI	S	JHU	Depew, GSFC	Airglow	S
				Test and Support			
	1959						
2.01 GT	May 14	WI	X	Medrow, GSFC	Medrow, GSFC	Rocket Test	S
2.02 GT	15	WI	X	Medrow, GSFC	Medrow, GSFC	Rocket Test	S

APPENDIX C

ID	Date	Site	Result	Experimenter	Sponsor	Test	Result
2.03 GT	Aug. 15	WI	X	Medrow, GSFC	Medrow, GSFC	Rocket Test	X
2.04 GT	Aug. 7	WI	X	Medrow, GSFC	Medrow, GSFC	Rocket Test	X
2.05 GT	Aug. 7	WI	X	Medrow, GSFC	Medrow, GSFC	Rocket Test	X
2.06 GT	Aug. 7	WI	X	Medrow, GSFC	Medrow, GSFC	Rocket Test	S
8.01 GT	Dec. 22	WI	S	GSFC/NRL/DRTE	Winkler, GSFC	X248 Vibration Test	S
1960							
8.02 GT	Jan. 26	WI	S	GSFC/NRL/DRTE	Winkler, GSFC	X248 Vibration Test	S
4.01 GT	Feb. 16	WI	X	Medrow, GSFC	Medrow, GSFC	Rocket Test	X
4.12 GT	Mar. 25	WI	S	Medrow, GSFC	Medrow, GSFC	Rocket Test	S
4.10 GT	Apr. 23	WI	S	Medrow, GSFC	Medrow, GSFC	Rocket Test	S
5.01 GT	July 22	WI	S	Sorgnit, GSFC	Sorgnit, GSFC	Rocket Test	S
3.28 GT	Aug. 9	WI	S	Sorgnit, GSFC	Sorgnit, GSFC	Rocket Test	S
5.02 GT	Oct. 18	WI	S	Sorgnit, GSFC	Sorgnit, GSFC	Rocket Test	S
3.29 GT	Nov. 3	WI	S	Sorgnit, GSFC	Sorgnit, GSFC	Rocket Test	S
1961							
3.36 GT	Jan. 17	WI	S	Sorgnit, GSFC	Sorgnit, GSFC	Rocket Test	S
5.03 GT	Jan. 19	WI	X	Sorgnit, GSFC	Sorgnit, GSFC	Rocket Test	P
10.49 GT	Mar. 15	WI	S	Sorgnit, GSFC	Sorgnit, GSFC	Cajun Fin Test	S
4.19 GT	Apr. 14	WI	S	Russell, GSFC	Russell, GSFC	Attitude Control	S
12.01 GT	May 2	WI	S	U. Mich.	Spencer, GSFC	Cone Test	P
14.01 GT	May 25	WI	S	Sorgnit, GSFC	Sorgnit, GSFC	Rocket Test	S
4.20 GT	June 26	WI	S	Russell, GSFC	Russell, GSFC	Attitude Control	P
14.02 GT	Aug. 16	WI	S	Sorgnit, GSFC	Sorgnit, GSFC	Rocket Test	S
1962							
4.68 GT	Jan. 13	WI	S	Russell, GSFC	Russell, GSFC	Attitude Control	S
10.69 GT	Mar. 1	WI	X	Donn, GSFC	Donn, GSFC	Water Launch	S

[a] S—Successful
P—Partial success } Subject to interpretation
X—Unsuccessful

NASA Sounding Rocket Flights (Cont.)

NASA No.	Firing			Experimenter	NASA scientist and location	Experiment	Results[a]
	Date	Site	Performance[a]				
	1962			Test and Support—Continued			
10.70 GT	Mar. 2	WI	S	Donn, GSFC	Donn, GSFC	Water Launch	S
4.48 GT	May 25	WI	S	Pressly, GSFC	Pressly, GSFC	Sea Recovery	S
4.60 GT	Aug. 8	WI	P	Russell, GSFC	Russell, GSFC	Attitude Control	P
	1963						
16.01 GT	Apr. 8	WI	X	Sorgnit, GSFC	Sorgnit, GSFC	ACS Test	X
4.87 GT	June 17	WI	S	Russell, GSFC	Russell, GSFC	Attitude Control	S
14.111 GT	Oct. 31	WI	S	Williams, GSFC	Williams, GSFC	Vibration Test	S

[a] S—Successful
P—Partial success } Subject to interpretation.
X—Unsuccessful

Appendix D
A Chronology of Events Related to the Goddard Space Flight Center

THE FOLLOWING CHRONOLOGY is parallel and supplementary to the text. Events considered important for the achievements and early history of Goddard Space Flight Center and its missions have been included.

1915

April 15: The Secretary of War called the first meeting of the National Advisory Committee for Aeronautics (NACA) in his office. Brig. Gen. George P. Scriven, Chief Signal Officer, was elected temporary Chairman, and Dr. Charles D. Walcott, Secretary of the Smithsonian Institution, was elected first Chairman of the important NACA Executive Committee.

1918

November 6–7: Robert H. Goddard fired several rocket devices before representatives of the Signal Corps, Air Service, Army Ordnance, and others at Aberdeen Proving Ground, Md.

1919

May 26: Date of Dr. Robert H. Goddard's progress report to the Smithsonian Institution entitled "A Method of Reaching Extreme Altitudes." It was published by the Smithsonian in January 1920.

1923

November 1: Robert H. Goddard successfully operated a liquid oxygen and gasoline rocket motor on a testing frame, both fuel components being supplied by pumps installed on the rocket.

1926

March 16: Robert H. Goddard launched the world's first liquid-fueled rocket at Auburn, Mass., which traveled 184 feet in 2½ seconds. This event was the "Kitty Hawk" of rocketry.

1929

July 17: A liquid-fueled, 11-foot rocket, fired by Robert Goddard at Auburn, Mass., carried a small camera, thermometer, and a barometer which were recovered intact after the flight. Much "moon rocket" publicity was made of this flight.

1930

December 30: Robert H. Goddard fired 11-foot liquid-fueled rocket to a height of 2,000 feet and a speed of almost 500 mph near Roswell, N.Mex.

1932

April 19: First flight of Goddard rocket with gyroscopically controlled vanes for automatically stabilized flight, near Roswell, N.Mex.

1935

March 28: Robert Goddard launched the first rocket equipped with gyroscopic controls, which attained a height of 4,800 feet, a horizontal distance of 13,000 feet, and a speed of 550 mph, near Roswell, N.Mex.

1936

March 16: Robert H. Goddard's classic report on "Liquid Propellant Rocket Development," reviewing his liquid-fuel rocket research and flight testing since 1919, was published by the Smithsonian Institution.

1940

May 28: Robert H. Goddard offered all his research data, patents, and facilities for use by the military services at a meeting arranged by Harry Guggenheim with representatives of Army Ordnance, Army Air Corps, and Navy Bureau of Aeronautics. Nothing resulted from this except an expression of possible use of rockets in jet-assisted takeoffs of aircraft.

1943

During September: Rocket Development Branch was created in Army Ordnance to direct and coordinate development of rockets.

1945

May 8: At time of Germany's surrender, more than 20,000 V-weapons (V–1's and V–2's) had been fired. Although figures vary, best estimate is that 1,115 V–2 ballistic rockets had been fired against England and 1,675 against continental targets. Great disparity between production figures and operational missions was caused by series production and development testing being performed concurrently, there being as many as 12 major modifications in basic design features.

August 10: Dr. Robert H. Goddard, American rocket pioneer, died.

December 17: Rocket Sonde Research Branch was constituted in Naval Research Laboratory to conduct scientific exploration of the upper atmosphere.

1946

March 22: First American rocket to escape earth's atmosphere, the JPL-Ordnance Wac, reached 50-mile height after launch from WSPG.

May 17: Original design and development of Aerobee sounding rocket began when contract was given to Aerojet Engineering Corp.

June 6: Joint Army-Navy Research and Development Board was created for purpose of coordinating all activities of joint interest in fields of aeronautics, atomic energy, electronics, geographical exploration, geophysical sciences, and guided missiles.

September 25: First successful firing of Applied Physics Laboratory Aerobee research rocket at White Sands Proving Ground, N.Mex.

During September: After completing studies, Project RAND reported that earth satellites were technically feasible.

November 14: First complete Aerobee rocket was fired to a height of 190,000 feet from White Sands Proving Ground, N.Mex.

1948

October 19: Photographs of the earth's surface, taken from altitudes between 60 and 70 miles by cameras installed in rockets, were released by the Navy.

1949

May 11: President Harry S Truman signed a bill providing a 5,000-mile guided-missile test range, subsequently established at Cape Canaveral, Fla.

1950

June 13: Department of Defense assigned range responsibilities to the armed services: Army: White Sands (N.Mex.) Proving Ground and nearby Holloman AFB at Alamogordo; Navy: Point Mugu, Calif.; Air Force: Long-Range Proving Ground at Banana River, Fla. (later called Cape Canaveral).

1954

March 17: President Dwight D. Eisenhower signed Executive Order 10521 on the "Administration of Scientific Research by Federal Agencies," which gave the National Science Foundation major responsibility in pure scientific research.

APPENDIX D

During April: Bell Laboratories announced invention of the silicon solar battery.

August 26: The Supplemental Appropriations Act, 1955, appropriated $2 million to the National Science Foundation to support the U.S. IGY program sponsored and coordinated by the National Academy of Sciences.

During October: NRL Aerobee fired at White Sands took photographs at 100-mile altitude, first picture taken of complete hurricane, off the Texas gulf coast.

1955

During March: The Navy proposed a program for the launch of an elementary uninstrumented satellite in 2 or 3 years. This program, jointly developed by the Office of Naval Research and the Army, was known as Project Orbiter. It called for the use of the Redstone booster and Loki rockets (small solid-propellant rockets).

April 26: Moscow Radio reported U.S.S.R. planned to explore the moon with a tank remotely controlled by radio, foresaw trips by man in 1 to 2 years and reported formation of scientific team to devise satellite able to circle earth.

July 20: President Eisenhower endorsed IGY earth satellite proposal, the White House announced: "The President has approved plans by this country for going ahead with the launching of small, unmanned, earth-circling satellites as part of the U.S. participation in the International Geophysical Year which takes place between July 1957 and December 1958." Scientific responsibility was assumed by the National Academy of Sciences, fiscal responsibility by the National Science Foundation, and responsibility for logistic and technical support by the Department of Defense.

July 30: U.S.S.R. announced that it planned to launch an earth satellite in connection with IGY.

September 9: Project Vanguard was born when the Department of Defense wrote a letter to the Secretary of the Navy authorizing him to proceed with the Naval Research Laboratory proposal for launch of at least one U.S. satellite in the IGY, which was to end in December 1958.

October 2: National Academy of Sciences' IGY committee established Technical Panel for the Earth Satellite Program, with Richard W. Porter as Chairman, to plan the scientific aspects of the program, including the selection of experiments, the establishment of optical tracking stations, and the handling of international and interdisciplinary relations.

During November: Naval Research Laboratory transmitted transcontinental communications from Washington, D.C., to San Diego, Calif., by reflecting teletype messages off the moon.

1956

Spring: A plan was developed at NRL for seven test vehicles and six satellite-launching vehicles in Project Vanguard.

September 10–15: Scientists from 40 nations, including the U.S. and U.S.S.R., at a meeting in Barcelona of the Special Committee for the IGY (CSAGI), approved resolutions calling for, among other things, countries having satellite programs to use tracking and telemetering radio systems compatible with those announced at the current CSAGI meeting, and to release technical information on tracking equipment and scheduling and planning information essential to preparation for and execution of optical and radio observations.

December 8: NRL Test Vehicle 0 (TV-0), a Viking rocket carrying no Vanguard components, was successfully fired in a test of range facilities, telemetry, and instrumentation.

1957

April 11: U.S.-IGY scientific satellite equipment, including a radio transmitter and instruments for measuring temperature, pressure, cosmic rays, and meteoric dust encounters, was tested above earth for the first time, as a rocket containing this equipment was fired by the Navy to a 126-mile altitude.

During April: Upper Atmosphere Rocket Research Panel was renamed the Rocket and Satellite Research Panel. Its chairman was James A. Van Allen of the State University of Iowa.

May 1: NRL TV-1, with a Viking first stage, launched a Vanguard third stage in a successful test of the control system and of the

1957 Continued

third-stage separation, spin-up, ignition, and propulsion.

October 4: *Sputnik I*, the first manmade earth satellite, was launched by the U.S.S.R. It remained in orbit until January 4, 1958.

October 14: The American Rocket Society presented to President Dwight D. Eisenhower a program for outer space research which proposed establishment of an Astronautical Research and Development Agency similar to NACA and AEC, with responsibility for all space projects except those directly related to military defense.

October 23: The launch of NRL TV-2 was the first successful launch of the complete Vanguard configuration—a successful test of the first-stage engine, control system, and vehicle structure; second and third stages were dummies. A 109-mile altitude was reached at 4,250 mph.

During October: Project Vanguard worldwide tracking system became operational.

November 21: The National Advisory Committee for Aeronautics (NACA) authorized establishment of a special committee on space technology, headed by H. Guyford Stever.

December 4: The American Rocket Society's proposal for an Astronautical Research and Development Agency, which had been presented to President Eisenhower on October 14, 1957, was announced.

December 6: An attempt to launch NRL TV-3, the first test of the complete Vanguard vehicle and control system, failed when the first engine lost thrust after 2 seconds and the vehicle burned on the pad. This was the first Vanguard vehicle with three live stages and orbit capability.

1958

January 4: The American Rocket Society and the Rocket and Satellite Research Panel issued a summary of their proposals for a National Space Establishment. Preferably independent of DOD, but in any event not under one of the military services, this establishment would be responsible for the "broad cultural, scientific, and commercial objectives" of outer space research.

January 9: In his State of the Union message, President Eisenhower reported: "In recognition of the need for single control in some of our most advanced development projects, the Secretary of Defense has already decided to concentrate into one organization all antimissile and satellite technology undertaken within the Department of Defense."

January 16: NACA adopted a resolution recommending that the national space program could be most effectively implemented by the cooperative effort of DOD, the National Academy of Sciences, the National Science Foundation, and NACA, together with universities, research institutions, and industrial companies of the Nation. Military development and operation of space vehicles would be the responsibility of DOD, and research and scientific space operations the responsibility of NACA.

January 31: *Explorer I*, the first U.S. satellite, was launched by an Army Ballistic Missile Agency-Jet Propulsion Laboratory team on a modified Jupiter C, with the U.S.-IGY scientific experiment of James A. Van Allen which would discover the radiation belt region around the earth.

February 5: NRL TV-3 backup was a repeat of the TV-3 launch attempt on Dec. 6, 1957. It failed when a control malfunctioned after 57 seconds of flight; the vehicle broke up at about 20,000 feet.

March 17: The second U.S.-IGY satellite, *Vanguard I*, was launched into orbit with a life expectancy of perhaps 1,000 years. A highly successful scientific satellite, its data proved the earth to be slightly pear shaped. Operating on solar-powered batteries, it transmitted for more than 6 years.

March 26: The third U.S.-IGY satellite, *Explorer III*, another joint ABMA-JPL project, was successfully launched by an Army Juno II rocket. It yielded valuable data on the radiation belt region, micrometeoroid impacts, and temperatures before reentering on June 27.

April 2: In a message to Congress, President Eisenhower proposed the establishment of a national aeronautics and space agency into which NACA would be absorbed. This agency was to have responsibility for civilian space science and aeronautical research. It would conduct research in

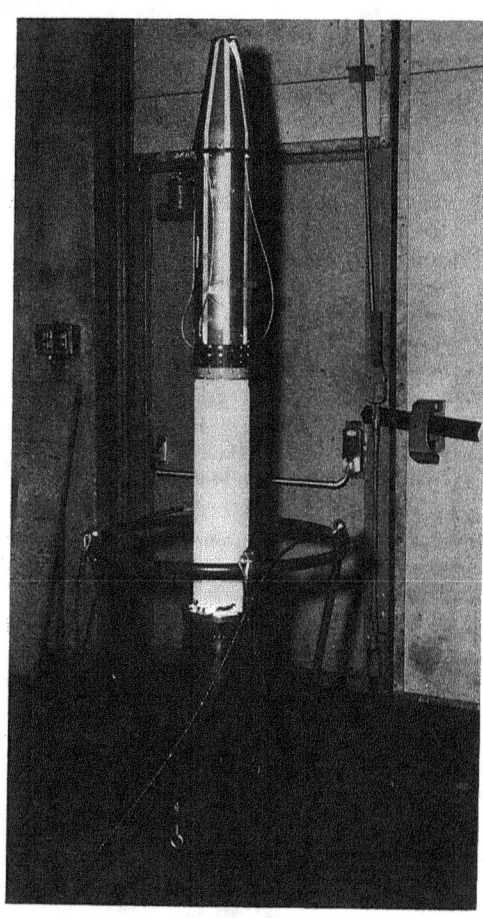

Explorer I, the first U.S. satellite, launched January 31, 1958.

these fields in its own facilities or by contract and would also perform military research required by the military departments. Interim projects pertaining to the civilian program which were under the direction of the Advanced Research Projects Agency (ARPA) would be transferred to this civilian space agency.

April 14: A proposal for a National Aeronautics and Space Agency, drafted by the Bureau of the Budget, was submitted to the Congress by the President.

April 28: Project Vanguard's TV-5 failed to put its 21.5-pound spacecraft into orbit when the control system release failed and the third stage was not ignited.

May 1: Scientific findings from the two Explorer satellites (I and III) disclosed an unexpected band of high-intensity radiation extending from 600 miles above the earth to possibly 8,000 miles.

—Responsibility for the Project Vanguard portion of the U.S.-IGY scientific satellite program was transferred within DOD from the Navy to ARPA.

May 27: The first Vanguard satellite launch vehicle (NRL SLV-1) generally was successful in its launch with exception of premature second-stage burnout, which prevented achievement of satisfactory orbit.

June 26: The NRL Vanguard SLV-2 launch failed when the second stage cut off prematurely because of low chamber pressure, terminating the flight.

July 26: Explorer IV, under the project direction of the DOD's ARPA, was launched for the purpose of studying radiation belts detected by *Explorer I* and *Explorer III* and for measurement of artificial radiation created by DOD nuclear experiments.

July 29: President Eisenhower signed the National Aeronautics and Space Act of 1958.

Summer 1958: At Langley, Edmond C. Buckley unofficially formed a network study group for the embryonic manned satellite program (later to become Project Mercury). Key technical personnel were: George B. Graves, Jr. (network arrangement); Robert L. Kenimer (tracking); James H. Schrader (telemetry and capsule communication); William J. Boyer (ground communication); Eugene L. Davis, Jr. (computing); Howard C. Kyle (control center).

August 1: U.S. Senator J. Glenn Beall, Maryland, announced that the new "outer space agency" would establish its laboratory and plant in Maryland. The location at Greenbelt, Md., was considered as "ideal" for the new agency. The Greenbelt laboratory was to employ 650 technicians, mostly electronic engineers and some chemists, the announcement stated. Construction on the plant was expected to start immediately in view of the fact that legislation authorizing appropri-

1958 Continued

ations of $47,800,000 for construction of the "space projects center" (S4208) had passed the Senate and was expected to clear the House of Representatives shortly.

August 14: Public Law 85-657 was approved, authorizing appropriations to NASA for construction and other purposes and specifically for a "space projects center in the vicinity of Washington, D.C.": a space projects building; research projects laboratory; posts and appurtenances; utilities; equipment and instrumentation, $3,750,000.

August 19: T. Keith Glennan was sworn in as Administrator and Hugh L. Dryden as Deputy Administrator. Forty days later, October 1, 1958, NASA was declared ready to function.

August 21: NACA held its final meeting and invited T. Keith Glennan, newly appointed Administrator of NASA, to attend.

September 24: First meeting of the senior staff of the National Aeronautics and Space Administration (NASA) was held, with T. Keith Glennan, Administrator, and Hugh L. Dryden, Deputy Administrator.

September 26: The third Vanguard satellite launch vehicle (SLV-3) reached an altitude of 265 miles, and was believed to have made one orbit and to have been destroyed 9,200 miles downrange over Central Africa on reentry into the atmosphere.

October 1: First official day of NASA. By executive order of the President, DOD responsibilities for the remaining U.S.-IGY satellite and space probe projects were transferred to NASA; included were Project Vanguard and the four lunar probes and three satellite IGY projects which had previously been assigned by ARPA to the Air Force Ballistic Missile Division and the Army Ballistic Missile Agency (ABMA). Also transferred were a number of engine development research programs.

October 21: Three weeks after NASA officially began operating, prospective contractors were invited to a briefing at NASA Hq. on development of a 1.5-million-pound-thrust F-1 engine for Saturn V.

November 7: A bidders' conference was held by NASA on a manned-satellite capsule for Project Mercury.

November 17: Senator Lyndon B. Johnson presented U.S. proposal for the international control of outer space before the United Nations in New York.

December 3: The President transferred the functions and facilities of the Jet Propulsion Laboratory (JPL) of the California Institute of Technology, Pasadena, Calif., from the Army to NASA.

December 3: NASA and the Army reached an agreement whereby the Army Ordnance Missile Command (AOMC), Huntsville, Ala., would be responsive to NASA requirements.

December 18: Project Score, the fifth U.S.-IGY satellite—under the project direction of ARPA—was launched at 12:45 a.m. from AMR by a Juno II rocket.

December 19: President Eisenhower's Christmas message was beamed from the Score satellite in orbit—the first voice transmitted from a satellite to earth.

December 31: IGY had been scheduled to end on this date; but in October 1958 the International Council of Scientific Unions, meeting in Washington, approved extension of operation through 1959 (IGC-59) and also approved establishment of the Committee on Space Research (COSPAR) to continue international cooperation in the scientific exploration of space.

1959

January 8: NASA requested eight Redstone launch vehicles from the Army to be used in Project Mercury development.

January 15: In General Notice No. 1 signed by T. E. Jenkins, Administrative Officer, it was announced that "four divisions (Construction Division, Space Sciences Division, Theoretical Division, Vanguard Division) have been designated as comprising the Beltsville Space Center of the National Aeronautics and Space Administration. . . ."

February 17: Vanguard II was successfully launched. It was a 21.5-pound satellite with infrared sensors for cloud-cover measurement.

February 28: Discoverer I, a 1,450-pound USAF satellite, was successfully launched into first near-polar orbit by a Thor-Hustler

APPENDIX D

booster from PMR; stabilization difficulties hampered tracking acquisition.

March 3: Pioneer IV, the fourth U.S.-IGY space probe, a joint AMBA-JPL project under direction of NASA, was launched by a Juno II rocket from AMR and achieved an earth-moon trajectory, passing within 37,000 miles of the moon before going into a permanent solar orbit. Radio contact was maintained to a record distance of 406,620 miles.

March 11: NASA granted $350,000 to the National Academy of Sciences-National Research Council for a program of research appointments in theoretical and experimental physics to stimulate basic research in the space sciences.

March 24: NASA announced that Wallops Island, Va., had made 3,300 rocket firings since 1945.

April 10: The first construction contract for the Beltsville Space Center was awarded.

April 13: Vanguard's SLV-5 launch failed when the second stage did not operate properly and the vehicle tumbled. The 23.3-pound payload included a 13-inch ball with a magnetometer attached for mapping the earth's magnetic field and a 30-inch inflatable sphere to measure atmospheric drag.

—*Discoverer II* was successfully placed into a polar orbit by a Thor-Agena A booster but capsule ejection malfunctioned, causing it to impact in the vicinity of the Spitsbergen Islands (Arctic Ocean) on Apr. 14 instead of in the vicinity of Hawaii. It was the first vehicle known to have been placed in a polar orbit and was the first attempt to recover an object from orbit.

April 15: A NASA-DOD joint working group discussed procedures for search and recovery aspects of Project Mercury involving Army, Navy, and Air Force units.

April 20: NASA announced acceptance of proposals by the Canadian Defence Research Telecommunications Establishment for continuing joint rocket and satellite ionospheric experiments of a nonmilitary nature.

April 24: Construction began at Goddard Space Flight Center on Buildings 1 and 2, the Space Projects Building, and Research Projects Laboratory, respectively.

April 28: NASA announced the signing of a $24-million contract with Douglas Aircraft Company, Inc., for a three-stage Thor-Vanguard launching rocket called "Delta."

During April: The Tiros meteorological satellite program was transferred from DOD to NASA.

May 1: NASA Administrator Glennan announced that the Beltsville Space Center was renamed "Goddard Space Flight Center," in commemoration of Dr. Robert H. Goddard, the American pioneer in rocket research. John W. Townsend, Jr., was named Assistant Director for Space Science and Satellite Applications; John T. Mengel was named Assistant Director for Tracking and Data Systems; and Dr. Robert R. Gilruth was named Assistant Director for Manned Satellites. The three Assistant Directors reported to the Director of Flight Development, NASA Hq. The announcement also stated that Dr. Michael J. Vaccaro would head the Office of Business Administration.

June 1: The Smithsonian Optical Tracking Station at Woomera, Australia, successfully photographed *Vanguard I* at the apogee of its orbit, nearly 2,500 miles from the earth. This feat has been compared with taking a picture of a golf ball 600 miles away.

July 7: A four-stage rocket with an Air Research and Development Command payload was fired from Wallops Island to an altitude of 750 miles. This was the first in a series of launchings to measure natural radiation surrounding the earth.

July 20: NASA selected the Western Electric Company to build its worldwide network of tracking and ground instrument stations to be used in Project Mercury.

During August: A conference of the International Telecommunication Union at Geneva, Switzerland, allocated radio frequency bands for space and earth-space use.

August 7: Explorer VI was launched. All experiments performed; it provided the first complete televised cloud-cover pictures. A better map of the Van Allen radiation belt region was obtained. This was the first scientific satellite under the project direction of the Goddard Space Flight Center.

August 14: While *Explorer VI* was passing

209

Above, Building 1 under construction; below, Building 2 under construction.

1959 Continued

over Mexico at an altitude of about 17,000 miles, it successfully transmitted a crude picture of a sunlit, crescent-shaped portion of the North Central Pacific Ocean. The area of the earth photographed was 20,000 square miles.

August 17: The first of the Nike-Asp sounding rockets, which were to provide geophysical information on wind activity between 50 and 150 miles high, was launched successfully from Wallops Island.

September 1: Dr. Harry J. Goett was appointed Director of GSFC. Dr. Goett came from the NASA Ames Research Center, Moffett Field, Calif., where he had been Chief of the Full Scale and Flight Research Division since 1948.

September 16: Goddard's Building 2, the Research Projects Laboratory, was fully occupied.

September 18: Vanguard III was successfully launched. It had a 50-pound payload which measured the earth's magnetic field, solar x-rays, and space environmental conditions. This vehicle was SLV-7, the TV-4 backup vehicle, and had a more powerful third stage than previous Vanguards.

September 21: Construction began on Building 3, the Central Flight Control and Range Operations Building, at GSFC.

October 13: Explorer VII, the seventh and last U.S.-IGY earth satellite, under the direction of NASA with the Army as executive agent, was launched into an earth orbit by a modified Army Juno II booster.

October 28: A 100-foot-diameter inflatable sphere for Project Echo was launched on a suborbital test flight from Wallops Island to an altitude of 250 miles by the first Sergeant-Delta rocket. It was an aluminum-coated Mylar-plastic sphere; when in orbit, such a sphere would be used as a passive electronic reflector.

December 11: The transmitters of *Vanguard III*, launched Sept. 18, became silent after providing tracking signals and scientific data for 85 days. The satellite was expected to remain in orbit 40 years.

December 17: T. Keith Glennan, NASA Administrator, offered the services of the U.S. worldwide tracking network in support of any manned space flight the U.S.S.R. might plan to undertake, in a speech before the Institute of World Affairs in Pasadena, Calif.

December 22: In a Canadian-U.S. cooperative project, NASA launched a four-stage Javelin sounding rocket from Wallops Island to measure the intensity of galactic radio noise.

December 31: Approximately 300 research rockets were launched during the 30-month IGY and IGC-59 periods; 221 of these had been launched during IGY.

1960

February 27: A 100-foot-diameter inflatable sphere for Project Echo was launched on the third suborbital test from Wallops Island.

March 11: Pioneer V space probe was successfully launched by Thor-Able on a historic flight that measured radiation and magnetic fields in space and that communicated over great distances.

March 24: Pioneer V radio signals were received 2,000,000 miles from earth, more than 4 times the distance radio signals had previously been transmitted from a satellite.

April 1: Tiros I, a weather observation satellite, was launched into orbit by a Thor-Able and took pictures of the earth's cloud cover on a global scale from about 450 miles above the surface.

— Fourth suborbital test of the 100-foot-diameter Echo sphere was launched from Wallops Island to an altitude of 235 miles.

April 23: NASA fired the first of five Aerobee-Hi sounding rockets from Wallops Island in a program to measure ultraviolet radiation.

May 8: A 150-MW transmitter on *Pioneer V*, commanded at 5:04 a.m. EDT, worked satisfactorily at 8,001,000 miles from the earth.

May 19: Tiros I spotted a tornado storm system in the vicinity of Wichita Falls, Tex.

May 23: Construction began on Building 4, the Boiler House and Electric Substation, at GSFC; it would house service shops, a central powerplant, a refrigeration plant, and office areas.

May 31: NASA launched a Project Echo 100-foot inflatable sphere to an altitude of 210 miles to test a payload configuration carrying two beacon transmitters.

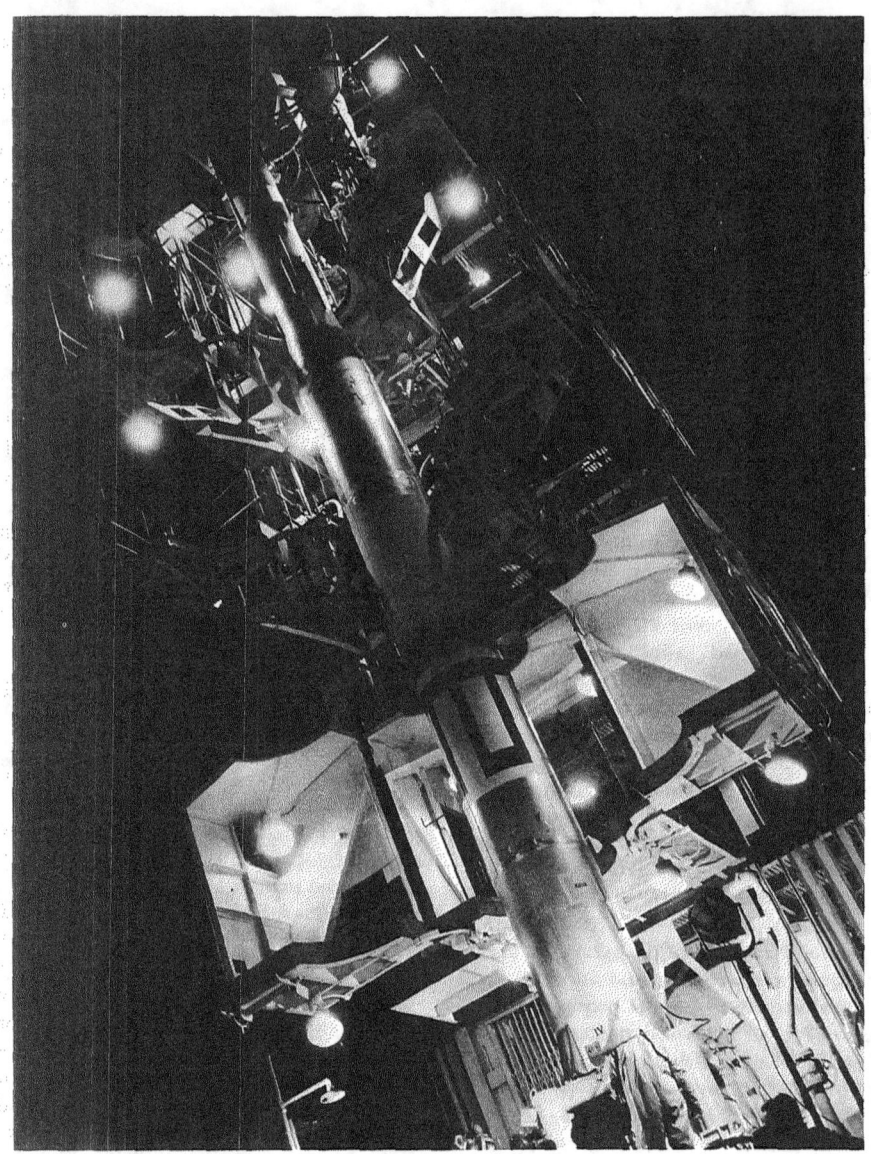

Vanguard vehicle in a gantry.

1960 Continued

June 26: A 6-minute message was received by the Jodrell Bank tracking station, England, from *Pioneer V*, the last communication received from this spacecraft, then 22.5 million miles from the earth, moving at a relative velocity of 21,000 mph.

June 28: The Smithsonian Institution posthumously awarded its highest honor, the Langley Medal, to Robert H. Goddard.
—Building 1, the Space Projects Building at GSFC, was fully occupied.

July 1: The first complete Scout rocket vehicle was launched from Wallops Island, but the

APPENDIX D

fourth-stage separation and firing were not accomplished.

July 21: A Nike-Cajun sounding rocket was fired from Fort Churchill, Manitoba, Canada, containing an instrumental payload to measure data on energetic particles during a period of low solar activity.

August 12: Echo I, the first passive communications satellite, was successfully launched into orbit; it reflected a radio message from President Eisenhower across the nation, thus demonstrating the feasibility of global radio communications via satellites. *Echo I*, visible to skywatchers, provided reflection for numerous long-range radio transmissions by private and Government research agencies.

August 24: Echo I first went into the earth's shadow, with its two tracking beacons still operating. Since going into orbit on August 12, it had bounced back hundreds of telephonic experiments and transmissions.

October 4: The second complete NASA Scout rocket was fired successfully from Wallops Island to its predicted 3,500-mile altitude and 5,800-mile impact range.

October 12: NASA Administrator Glennan announced that communications satellites developed by private companies would be launched by NASA at cost to assist private industry in developing a communications network.

November 3: Explorer VIII was launched. Measurements were taken of the electron density, temperature, ion density and composition, and charge on the satellite in the upper atmosphere.

November 19: Construction began on Building 6, the Space Sciences Laboratory, at GSFC.

November 22: An Aerobee-Hi was fired to a 105-mile altitude from Wallops Island, with four stellar spectrometers developed for an experiment by the University of Rochester's Institute of Optics.

November 23: Tiros II was launched by Thor-Delta booster from AMR—the fourteenth successful U.S. satellite launched in 1960.

November 26: Construction began on Building 5, the Instrument Construction and Installation Laboratory, at GSFC.

— Patent awarded to Stephen Paull, GSFC's Spacecraft Technology Division, for a Variable Frequency Multivibrator Subcarrier Oscillator for Telemeter System.

1961

January 29: The Goddard Institute for Space Studies was established in New York City.

January 31: A contract was awarded for construction of Buildings 7 and 10, the Payload **Testing Facility and Environmental Testing Laboratory** at GSFC.

— Experiments with *Echo I* were discontinued except for occasional checks, having provided innumerable communications since launch on August 12, 1960.

February 5: The orientation of *Tiros II* made it impossible to obtain Northern Hemi-

Architect's drawing of Buildings 7 and 10.

1961 Continued

sphere pictures, and malfunctions made remote picture taking undesirable; use of the satellite's cameras was suspended until orbit precession again made Northern Hemisphere pictures possible.

February 10: A voice message was sent from Washington to Woomera, Australia, by way of the moon. NASA Deputy Administrator Hugh L. Dryden spoke on telephone to Goldstone, Calif., which "bounced" it off the moon to the deep space instrumentation station at Woomera. The operation was held as part of the official opening ceremony of the Deep Space Instrumentation Facility site in Australia.

Antenna at Goldstone, California.

February 15: James E. Webb was sworn in as second NASA Administrator.

February 16: Explorer IX, a 12-foot inflatable sphere of Mylar and aluminum, painted with white "polka dots," was placed in orbit by a four-stage Scout booster from Wallops Island. This was the first satellite launching from Wallops and the first satellite boosted by a solid-fuel rocket.

— France and NASA agreed to establish a joint program to test NASA communications satellites in Projects Relay and Rebound, to be launched by NASA in 1962 and 1963.

February 17: Explorer IX was located in orbit, by visual and photographic means, after failure of its radio beacon delayed orbit confirmation.

February 23: NASA Administrator James E. Webb and Deputy Secretary of Defense Roswell L. Gilpatric signed a letter of understanding confirming the national launch vehicle program—the integrated development and procurement of space boosters by DOD and NASA. Neither DOD nor NASA would initiate the development of a launch vehicle or booster for use in space without the written acknowledgment of the other agency.

— Tiros II completed 3 months in orbit, continuing useful observations beyond the original estimate of useful life.

February 27: A memorandum of understanding between the Federal Communications Commission and NASA delineating and coordinating civil communication space activities was signed. It stated that "earliest practicable realization of a commercially operable communications satellite system is a national objective."

During February: Dr. Joseph W. Siry, Head of GSFC's Theory and Analysis Office, received the Arthur S. Flemming award "For accomplishments in the field of orbit determination and prediction."

March 1: The installation of computer equipment was completed in Building 3 at GSFC.

March 6: Direct-mode pictures by the Tiros II camera were resumed after a month of inoperation. The quality of the pictures showed slight improvement, supporting the theory that foreign matter might have been deposited on the lens and was gradually evaporating.

March 16: The Goddard Space Flight Center was officially dedicated at Greenbelt, Md.; the dedication address was delivered by Dr. Detlev Bronk, President of the National Academy of Sciences. It was the 35th anniversary of Dr. Goddard's successful launching of the world's first liquid-fuel rocket. Mrs. Goddard accepted the Congressional Medal honoring her husband.

March 23: Tiros II had completed 4 months in orbit and continued to provide useful cloud pictures and radiation data. A signal from Tiros II was used on orbit 1763 to

APPENDIX D

trigger dynamite to break ground for new RCA Space Environment Center at Princeton, N.J.

March 25: Explorer X, on a Thor-Delta, launched into a highly elliptical orbit (apogee 186,000 miles, perigee 100 miles) with instruments to transmit data on the nature of the magnetic field and charged particles in the region of space where the earth's magnetic field merges with that of interplanetary space.

March 27: Its instruments recording a magnetic impulse, *Explorer X* became the first satellite to measure the shock wave generated by a solar flare.

March 28: GSFC scientists reported that *Explorer X* had encountered magnetic fields considerably stronger than expected in its elongated orbit which carried it 186,000 miles from the earth (almost halfway to the moon).

March 31: All stations in the Goddard-managed worldwide Mercury tracking network became operational.

April 19: Preliminary data from *Explorer X* indicated that solar wind blows the sun's magnetic field past the orbit of the earth.

April 27: Explorer XI was launched. It detected directional gamma rays from space.

May 1: Tiros operations at Belmar, N.J., were terminated to begin the move of equipment to Wallops Island.

May 18: The first test inflation of a 135-foot rigidized inflatable balloon (*Echo II* series) in a dirigible hangar was conducted by NASA Langley Research Center and G. T. Schjeldahl Co. at Weeksville, N.C.

— The GSFC Institute for Space Studies in New York announced that its first major project, a 2-month seminar on the origin of the solar system, would be held in the fall of 1961.

May 22: Construction began on Building 7, the Payload Testing Facility, at GSFC.

May 24: FCC endorsed the ultimate creation of a commercial satellite system to be owned jointly by international telegraph and telephone companies and announced a meeting for June 5 to explore "plans and procedures looking toward early establishment of an operable commercial communications satellite system."

May 25: In his second State of the Union message President Kennedy set forth an accelerated space program based upon the long-range national goals of landing a man on the moon and returning him safely to earth, "before this decade is out"; early development of the Rover, a nuclear rocket; speed-up of the use of earth satellites for worldwide communications; and providing "at the earliest possible time a satellite system for worldwide weather observation."

— An additional $549 million was requested for NASA over the new administration's March budget requests; $62 million was requested for DOD for starting development of a solid-propellant booster of the Nova class. (Nova would be capable of placing 100,000 pounds on the moon; its first stage would have six 1,500,000-pound-thrust engines.)

June 8: NASA announced accelerated recruiting of qualified scientists and engineers at its field centers to fill anticipated manpower requirements in the expanded space exploration program. During 1960 NASA had interviewed 3,000 persons on 100 college campuses.

— Astronomers of Lick Observatory positioned a 36-inch refractor telescope to intersect the path of *Echo I* at its predicted point of maximum elevation. The prediction of GSFC was confirmed at the exact time and within 10 minutes of arc.

June 9: A NASA press conference revealed that data from *Vanguard III* (November 15 to 17, 1960) and *Explorer VIII* (also during November 1960) indicated that high-velocity clouds of micrometeoroids moved near the earth, perhaps in a meteor stream around the sun. These new data had just been discovered from completed analysis.

June 14: The Argentine Comisión Nacional de Investigaciones Espaciales and NASA signed a memorandum of understanding for a cooperative space science research program using sounding rockets.

— A four-stage Javelin fired to 560-mile altitude from Wallops Island tested the extension of two 75-foot antenna arms on radio command, a test flight in the Canadian-U.S. Alouette satellite program.

June 15: President Kennedy directed the Na-

Dr. Robert Jastrow addresses the Conference on Origins of the Solar System held shortly after the Institute for Space Studies began operation in New York.

1961 Continued

tional Aeronautics and Space Council to undertake a full study of the Nation's communications satellite policy; he stated that leadership in science and technology should be exercised to achieve worldwide communications through the use of satellites at the earliest practicable date. Although no commitments as to an operational system should be made, the Government would "conduct and encourage research and development to advance the state of the art and to give maximum assurance of rapid and continuing scientific and technological progress."

During June: NASA entered a letter contract with RCA for four Tiros satellites.

July 9: The National Science Foundation released a forecast of the Nation's scientific needs for the next decade, which predicted that the United States would need nearly twice as many scientists in 1970 (168,000) as in 1961 (87,000).

July 12: Tiros III weather satellite was successfully launched into a near-circular orbit by the Thor-Delta booster from Cape Canaveral.

July 13 to 14: Two Nike-Cajun rockets launched Univ. of N.H.-GSFC payloads from Wallops Island.

July 19: Tiros II photographed tropical storm Liza in the Pacific Ocean, pinpointing its location for meteorologists.

July 23: NASA Administrator Webb, in congressional testimony, pointed out that the Tiros cloud-cover program was known to the entire world, involved no surveillance, and promised great benefits to all nations. He said data from the Tiros satellites had been made available to all, including the Soviet Union.

July 28: NASA and AT&T signed a cooperative agreement for the development and testing of two, possibly four, active communications satellites during 1962. AT&T would design and build the TSX satellites at its own expense and would reimburse NASA for the cost of the launchings by Thor-Delta vehicles at Cape Canaveral.

July 31: NASA awarded a contract to the Univ. of Michigan to continue to provide research instrumentation for measurement of temperatures and winds at altitudes up to 150 kilometers with Nike-Cajun and other sounding rockets.

August 8: Over 100 foreign weather services were invited by NASA and the U.S. Weather Bureau to participate in the Tiros III experiment for a 9-week period beginning today. The program provided cooperating services with an opportunity to conduct special meteorological observations synchronized with passes of the satellite.

August 11: NASA announced negotiation of a contract with Hughes Aircraft Co. for con-

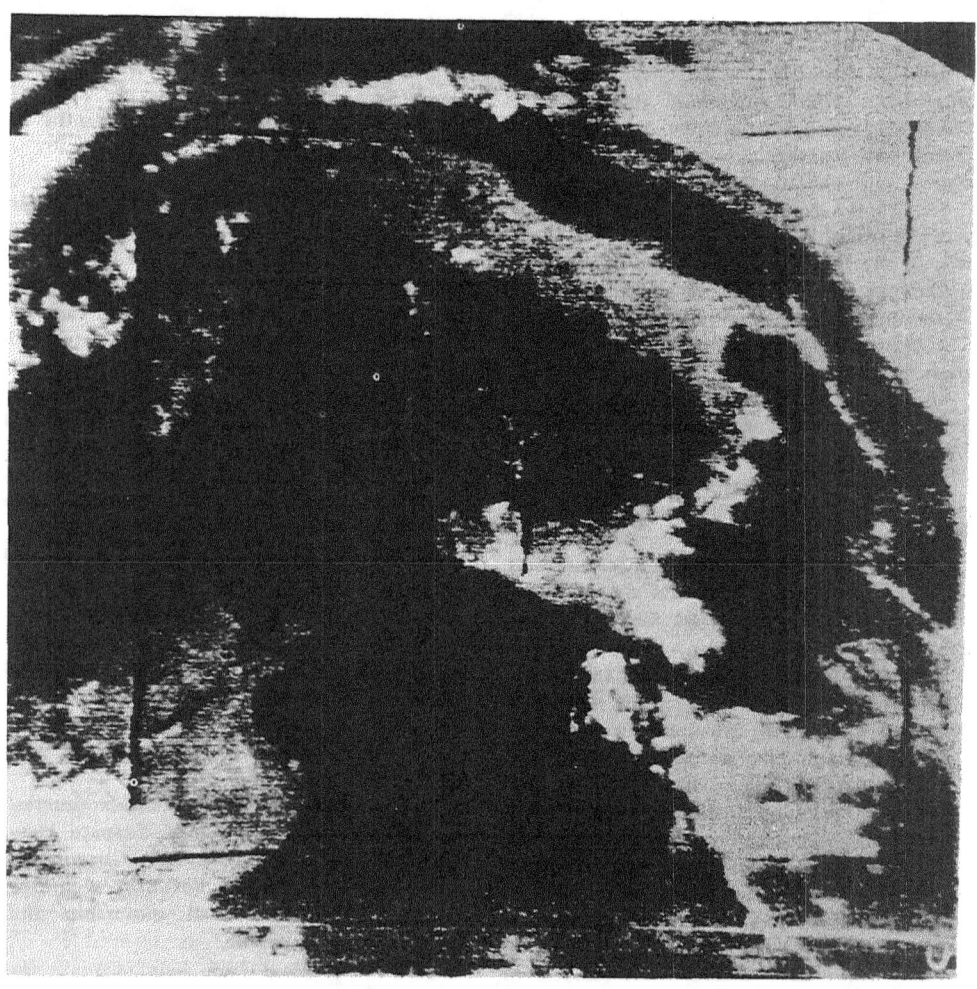

Italy and the Mediterranean as seen from a Tiros satellite.

struction of three experimental synchronous communications satellites (Syncom).

August 12: Echo I completed its first year in orbit, still clearly visible to the naked eye, after traveling 4,480 orbits and 138 million miles. *Echo I* had provided the basis for over 150 communications experiments, recent ones indicating only a 40 percent reduction in transmission reflection, caused by the changed shape. It also provided significant data on atmospheric drag and solar pressure.

— Aerobee 150A, with a liquid-hydrogen experiment, was fired from Wallops Island.

August 15: Explorer XII was placed into a highly eccentric orbit by a Thor-Delta booster from AMR; would provide detailed evaluation of the behavior of energetic particles between a 180- and 47,800-mile altitude. Under GSFC management, this "windmill" satellite carried six experiments developed by Ames Research Center, State Univ. of Iowa, Univ. of N.H., and GSFC. Several days were required to

1961 Continued

confirm the orbit. All instrumentation operated normally.

August 17: NASA announced that *Explorer XII* had successfully completed its first orbit and was sending data on magnetic fields and solar radiation from an apogee near 54,000 miles and a perigee within 170 miles of the earth.

August 21: NASA held a news conference on *Explorer XII*, at which the great amount of continuous coverage of interrelated data in its eccentric orbit was pointed out.

August 25: Explorer XIII was placed into orbit by the Scout rocket from Wallops Island; it was a micrometeoroid counting satellite developed by the Langley Research Center with GSFC participation.

August 29: NASA announced Explorer *XIII* had reentered the atmosphere. Transmitting data on micrometeoroids, the spacecraft was last heard from on August 27 by the Minitrack facility at Antofagasta, Chile.

During August: With the successful launch of *Explorer XII* on August 15, the Delta launch vehicles had successfully launched five satellites in six attempts, the only failure being the first attempt. Delta's high reliability record began with *Echo I* on August 12, 1960, and included *Tiros II* and *Tiros III* and *Explorer X* and *Explorer XII*. Built by prime contractor Douglas Aircraft, the NASA Delta launch vehicle consisted of a Thor first stage (Rocketdyne MB-3 liquid-fuel engine), Aerojet-General second stage (AJ-10-118, an improved Vanguard second stage), and an Allegany Ballistics Laboratory third stage (X-248 rocket is a spin-stabilized version of the Vanguard third stage).

September 13: Two experiments to measure atmospheric winds, temperature, and density in relatively high altitudes were conducted from Wallops Island in two four-stage Argo D-4 rocket launches. Sodium clouds were released at near 120 miles and again at 228 miles in the first launch, and at 118 and 230 miles in the second launch. French scientists participated by using special optical instruments to observe the brilliant orange and yellow clouds which stirred a rash of public inquiries from hundreds of miles.

September 16 to 29: Construction began on Building 8, the Satellite Systems Laboratory, at GSFC.

— A pair of spin-up rockets on *Tiros II* were fired after more than 10 months in orbit.

During September: Congress appropriated funds to the U.S. Weather Bureau for implementation of the National Operational Meteorological Satellite System. To phase in as early as technology warranted and to continue expanding the operational capability through the early Nimbus (advanced weather satellite) launchings by NASA, the system was planned to be fully operational by 1966, when the Nimbus system would become operational. The system would include data acquisition stations in northern latitudes, communications for transmitting the data, and a National Meteorological Center to receive, process, analyze, and disseminate the derived information over domestic and international weather circuits.

October 11: The final report of the House Committee on Science and Astronautics relating to their hearings on "Commercial Applications of Space Communications Systems" was released, having among its conclusions:

(1) Because of the worldwide interest and potential usefulness of a space communications system, the U.S. Government must "retain maximum flexibility regarding the central question of ownership and operation of the system."

(2) NASA would not only evaluate the various commercial proposals but would "conduct all space launches and retain direct control over all launching equipment, facilities, and personnel."

(3) Research and development of military space communications systems should continue to be conducted by DOD, but all research and development in space communications "should be conducted under the general supervision of NASA in accordance with its statutory mandate to plan, direct, and conduct aeronautical and space activities" as well as evaluate the technical merits of proposed systems.

October 13: The Ad Hoc Carrier Committee established by FCC to make an industry

APPENDIX D

proposal on the development and operation of commercial communications satellites recommended a nonprofit corporation be formed, to be owned by companies engaged in international communications, with the U.S. Government having one more representative on the board of directors than any single company. Western Union filed a minority statement proposing a public stock company arrangement to prevent dominance of the corporation by any one company.

— After its second year, *Explorer VII* was still transmitting, although the predicted lifetime of its transmitters had been only one year.

October 14: An Argo D-4 launched from Wallops Island carried a Canadian-U.S. topside sounding satellite payload to 560-mile altitude.

October 19: P-21, the Electron Density Profile Probe, was launched, with good data received. Electron density measurements were obtained to about 1,500 miles.

— Construction began on Building 10, the Environmental Testing Laboratory, at GSFC.

October 23: NASA announced it had ordered 14 additional Delta launch vehicles for Relay, Syncom, Telstar, and Tiros satellites.

October 27: GSFC and the Geophysics Corp. of America launched a Nike-Cajun rocket from Wallops Island with a 60-pound payload that reached a 90-mile altitude in a study of electron density and temperature in the upper level of the atmosphere.

October 29: NASA announced that the first Mercury-Scout launch to verify the readiness of the worldwide Mercury tracking network would take place at AMR.

November 23: Tiros II had completed its first year in orbit, still transmitting cloud-cover photographs of usable quality, although it had been expected to have a useful lifetime of only 3 months. *Tiros II* had completed 5,354 orbits and had transmitted over 36,000 photographs.

November 27: Sen. Robert Kerr announced that he would introduce legislation to authorize private ownership in the U.S. portion of the proposed worldwide communications satellite system. His bill would create the "Satellite Communications Corporation," which the participating firms would buy into.

December 18: NASA announced that the first station in a network of data-gathering stations for use with second-generation (advanced) satellites had been completed near Fairbanks, Alaska. The site for the second of the $5 million installations (each had a high-gain antenna 85 feet in diameter) was announced to be Rosman, N.C., 40 miles southwest of Asheville.

During December: The West German Post Office indicated it would construct near Munich by late 1963 or early 1964, a ground station capable of handling up to 600 phone calls simultaneously for operations with Telstar and Relay satellites.

1962

January 18: GSFC selected Rohr Aircraft Corp. to negotiate the manufacture and erection of three 85-foot-diameter parabolic antenna systems at Pisgah National Forest (Rosman, N.C.); Fairbanks, Alaska; and an undetermined location in eastern Canada.

January 19: At a NASA press conference, scientists described preliminary scientific results obtained by *Explorer XII*, based on a study of 10 percent of the data. It appeared that instead of the two radiation belts (previously called the inner and outer Van Allen belts) there was one magnetosphere extending roughly from 400 miles above the earth to 30,000 to 40,000 miles out.

January 23: GSFC announced the selection of Motorola, Inc., Military Electronics Division, of Scottsdale, Ariz., as contractor for research and development on the Goddard range and range-rate tracking system. Intended for tracking satellites in near-space and cislunar space, the system would measure spacecraft position to within a few feet and velocity to within fractions of a foot per second, by measuring carrier and sidetone modulations.

— Dr. Sigmund Fritz of the U.S. Weather Bureau reported that *Tiros III* had spotted fifty tropical storms during the summer of 1961.

January 25: Explorer X detected a "shadow" on the side of the earth away from the sun;

Rosman tracking facility.

1962 Continued

this shadow was marked by an absence of the solar wind, a belt of plasma moving out from the sun at about 200 miles per second but deflected around the earth by the earth's magnetic field and creating a cone-shaped "shadow" some 100,000 miles across at its larger end. The *Explorer X* findings were reported to the annual meeting of the American Physical Society in New York by Dr. Bruno Rossi of MIT.

— "Satellite Communications Corporation" bills were introduced by Sen. Robert Kerr (S. 2650) and by Rep. George Miller H.R. 9696), which would amend the NASA Act by adding a new section declaring that it is "the policy of the United States to provide leadership in the establishment of a worldwide communications system involving the use of space satellites." The section would create a "Satellite Communications Corporation" which would be privately owned and managed, and which would develop and operate a communications satellite system.

January 31: *Explorer I* had completed its fourth year in orbit and had a life expectancy of several more years.

During January: The Nimbus meteorological satellite underwent a rigorous test program at General Electric's Missile and Space Vehicle Center, Valley Forge, Pa.

— NASA awarded a contract to the Kollsman Instrument Division for a 38-inch-diameter primary mirror in the space telescope to be used in the Orbiting Astronomical Observatory (OAO).

February 8: *Tiros IV* was launched by a three-stage Thor-Delta rocket from Cape Canaveral into a near-circular orbit with an apogee of 525 miles and a perigee of 471 miles. It featured the same basic types of equipment as previous Tiros satellites, including cameras for cloud-cover photography and infrared sensors to measure temperatures at various levels in the atmosphere. The principal innovation was a camera with new type of wide-angle lens covering an area 450 miles on a side, which was expected to provide minimum distortion. The quality of *Tiros IV* pictures was good.

February 16: *Explorer IX* was launched. The

APPENDIX D

balloon and fourth stage orbited. Transmitter on the balloon failed to function properly, so the satellite required optical tracking.

February 25: Soviet scientists claimed to have discovered the third radiation belt around the earth and published such findings 2 years before the findings of *Explorer XII* were made public by NASA on January 19, 1962.

February 28: James S. Albus, an engineer at GSFC, was awarded $1,000 for his invention of a digital solar aspect sensor.

The digital solar aspect sensor.

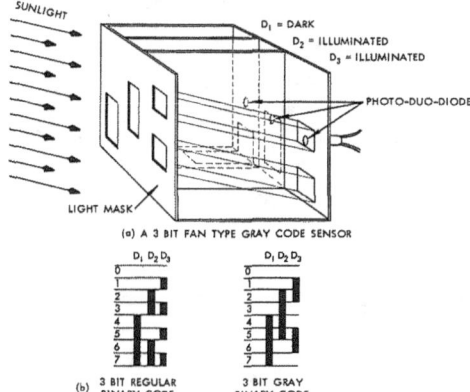

March 7: OSO I (Orbiting Solar Observatory I) was successfully launched into orbit from Cape Canaveral, marking the seventh straight success for the Thor-Delta booster.

March 8: The tracking network that operated during John Glenn's orbital Mercury flight (MA-6) would be sufficient to handle the 18-orbit manned flights to follow, according to Edmond C. Buckley, NASA's Director of Tracking and Data Acquisition, in testimony before a subcommittee of the House Committee on Science and Astronautics.

March 11: NASA announced that *Echo I*, the 100-foot balloon-type passive communications satellite launched on August 12, 1960, had recently become increasingly difficult to see. The sphere now presented only one-half to one-fourth its original size; this was due either to shrinkage or distortion during its 1½ years in orbit.

March 14: A press conference of Smithsonian Astrophysical Observatory and NASA scientists reported that *Explorer IX,* launched on February 16, 1961, had provided new and refined information on the density of the upper atmosphere. *Explorer IX*, a 12-foot aluminum-foil sphere painted with white "polka dots," was expected to have an orbital life of 2 more years. As it spiraled down into denser atmosphere, it was expected to provide much more information on density at altitudes down to 100 miles.

March 16: First anniversary of the dedication of NASA's Goddard Space Flight Center. During this year, seven Goddard satellites were orbited and the Center successfully operated the new 18-station world tracking network for the first manned orbital flight; began expansion of the 13-station scientific satellite tracking and data network; saw some 70 of its sounding rocket payloads launched from Wallops Island; established the Institute for Space Studies in New York; and added three buildings and 700 persons to its staff.

March 27: NASA fired a Nike-Cajun rocket from Wallops Island, releasing a sodium vapor cloud between 25- and 74-mile altitude. Rays of the setting sun colored the sodium cloud red (instead of sodium vapor's normal yellow).

March 28: The Senate Aeronautical and Space Sciences Committee unanimously approved a bill for ownership and operation of the Nation's commercial communications satellites.

March 29: A four-stage NASA Scout rocket carried the P-21A probe payload 3,910 miles into space and 4,370 miles downrange from Wallops Island.

During March: NASA completed work on its first major launching facility on the West Coast, a Thor-Agena pad at Vandenberg AFB, Calif. A used gantry was shipped from Marshall Space Flight Center and installed at a $1 million saving over cost of new construction. This pad would be used for NASA polar-orbit launches, such as for *Echo II*, Nimbus, and POGO (Polar Orbiting Geophysical Observatory).

During March: William G. Stroud, Chief of GSFC's Aeronomy & Meteorology Division, was awarded the Astronautics Engineer

1962 Continued

Achievement Award "For his personal contribution to the technology of meteorological satellites which are now culminating in rapid development of an operational system." Bernard Sisco, Deputy Assistant Director for Administration at GSFC, was awarded the Distinguished Service Award of Prince Georges County Junior Chamber of Commerce, "In recognition of his contributions to the general community welfare during the year."

April 1: Beginning of the third year of successful weather satellite operation by Tiros satellites; *Tiros I*, launched on Apr. 1, 1960, performed beyond all expectations, operated for 78 days, transmitted almost 23,000 cloud photos, of which some 19,000 were useful to meteorologists. *Tiros II*, launched Nov. 23, 1960, had transmitted more than 33,000 photos and one year after launch was still occasionally taking useful photos. *Tiros III*, launched July 12, 1961, took 24,000 cloud photos and was most spectacular as a "hurricane hunter." *Tiros IV*, launched Feb. 8, 1962, had averaged 250 operationally useful photos per day.

April 2: OSO I, launched Mar. 7, 1962, was reported by NASA to be performing well. As of this date, 360 telemetry data tapes had been recorded from 403 orbits. About one year would be required for complete analysis of the data.

April 11: NASA Administrator James E. Webb, testifying before the Senate Commitee on Commerce, supported the President's bill setting up a communications satellite corporation and approved of the Senate amendments, but noted his reservations on the one that would direct the FCC to encourage communications common carriers to build and own their own ground stations.

April 13: NASA Administrator James E. Webb, addressing the National Conference of the American Society for Public Administration in Detroit, said: "No new department or agency in the recent history of the Executive Branch of the Federal Government was created through the transfer of as many units from other departments and agencies as in the case of NASA. Three and one-half years ago, NASA did not exist. Today NASA comprises approximately 20,500 employees, ten major field centers, and an annual budget approaching the $2 billion mark."

April 15: U.S. Weather Bureau began daily international transmissions of cloud maps based on photos taken by *Tiros IV*.

April 17: NASA launched a Nike-Cajun sounding rocket from Wallops Island which detonated 12 grenades at altitudes from 25 to 57 miles.

April 24: The first transmission of TV pictures in space was made via orbiting *Echo I*. Signals were beamed from MIT's Lincoln Laboratory, Camp Parks, Calif., bounced off *Echo I*, and received at Millstone Hill near Westford, Mass.

April 26: Ariel I, the first international satellite (a joint U.K.-U.S. effort), was launched into orbit from Cape Canaveral by a Thor-Delta booster. The 136-pound spacecraft was built by GSFC and carried six British experiments to make integrated measurements in the ionosphere.

— Japan and the U.S. launched their first joint sounding rocket from Wallops Island. NASA provided the Nike-Cajun rocket, launch facilities, data acquisition, and a Langmuir probe to measure electron temperature. Japan furnished other instrumentation. The altitude of the flight was 75.6 miles.

— NASA graduated its first group of Project Mercury tracking personnel from the new course conducted at the Wallops Island training facilities; the seven graduates were personnel from firms having contracts with DOD and NASA.

During April: Tiros IV continued in operation and, to a great extent, provided excellent data. Over 20,000 pictures had been received. A total of 217 nephanalyses had been prepared up to March 26, and 199 had been transmitted over national and international weather circuits. *Tiros II* was turned on late in April by an unknown spurious source. An engineering investigation was run in early May before turning it off again. An analysis of the data indicated that some usable IR data were obtained.

— Robert W. Hutchison, GSFC's Personnel Director, was awarded the Federal Civil

Dr. Kunio Hirao and Toshio Muraoka at the launch of the first U.S.-Japanese sounding rocket experiment in the joint program.

Servant of the Year-State of Maryland, "In recognition of outstanding achievements in contributing to the rapid growth and establishment of Goddard Space Flight Center."

May 2: NASA scientists reported to the COSPAR session that data from *Explorer IX* indicated that the upper atmosphere was heated by sunspot activity.

May 3: Two GSFC scientific sounding rockets were launched from Wallops Island. An Iris research rocket launched with test instrumentation did not achieve its programed altitude and landed 175 statute miles downrange.

— House of Representatives passed the Communications Satellite Act of 1962 by a vote of 354 to 9.

May 17: The second and third joint Japan-U.S. space probes were successfully launched from Wallops Island; the second Nike-Cajun reached a 76-mile altitude and the third, and last—a night shot—reached an 80-mile altitude. The first of a series of 80 rocket probes to determine wind patterns over Cape Canaveral was initiated with the launch of a single-stage Nike to 80,000 feet, where it released a white smoke screen for photographic study.

May 18: Geophysics Corp. of America reported receipt of a Weather Bureau contract to study and explain the formation of vast bands of cloud patterns in the upper atmosphere, a phenomenon first revealed in photographs relayed by *Tiros I.*

May 20: Building 5 at GSFC was completed.

May 22: OSO I, launched March 7, 1962, experienced telemetry failure; it had provided 1,000 hours of data from its solar-pointed experiments.

May 29: NASA announced that *Ariel I,* the U.K.-U.S. ionosphere satellite launched on April 26, was functioning well except for one experiment, the solar ultraviolet detector.

During May: Checkout was completed for the Alaska Data Acquisition Facility near Fairbanks, and the Univ. of Alaska assumed responsibility. Part of the GSFC system, the Alaskan facility was an 85-foot dish; its associated electronics system would be used on tracking and data acquisition of the polar-orbiting Nimbus, EGO (Eccentric Geophysical Observatory), and POGO (Polar-Orbiting Geophysical Observatory) satellites.

June 6: Three sounding rockets were launched from Wallops Island. The first, a Nike-Apache, launched at 7:40 p.m. (EDT) with a 70-pound payload containing a pitot-static

1962 Continued

probe, reached a 78-mile altitude. The second, a Nike-Cajun, launched at 8:05 p.m., consisted of 11 explosive charges and a balloon, released between 25 and 64 miles altitude. The third, a Nike-Asp, was launched at 8:56 p.m. and released sodium vapor clouds to measure atmospheric winds and diffusion, at about 20 miles and extending to a peak altitude of about 100 miles.

June 7: A Nike-Cajun vehicle with an experiment to measure winds and temperatures in the upper atmosphere was launched from Wallops Island. In the night flight, 12 special explosive charges were ejected and detonated at intervals from about 25 up to 58 miles altitude.

— NASA announced selection of Bendix Corp.'s Radio Division, Towson, Md., for a contract to operate five of NASA's worldwide Project Mercury tracking and communications stations.

June 14: Tiros IV was no longer transmitting pictures usable for global weather forecasting, although it was still taking "direct" pictures on command which were suitable for limited U.S. weather analysis.

June 15: A two-stage Nike-Apache sounding rocket was launched from Wallops Island with a 95-pound payload to a 89-mile altitude with a GSFC experiment to measure electron density and electron collision frequency in the ionosphere under undisturbed conditions.

— NASA launched the first two of six tests on the performance of the Canadian Black Brant sounding rocket. The first carried a payload to 58 miles above Wallops Island, the second reached 62 miles.

— In preparation for the 1962 hurricane season, the Weather Bureau arranged to transmit satellite cloud photographs by photofacsimile to warning centers in San Juan, New Orleans, and Miami, where they would be used in forecasting and tracking tropical storms.

June 18: NASA selected Hughes Aircraft Co. for negotiation of a $2.5 million, 6-month study contract on an Advanced Syncom (synchronous communications) satellite. The contract covered satellite subsystems which would require long lead-time developmental and feasibility work. This second-generation Syncom would be a 500-pound, spin-stabilized satellite capable of relaying hundreds of telephone calls or carrying several TV channels. (The first-generation Syncom, for which Hughes was prime contractor, was limited to single telephone channel relay.) The Syncom project was under the technical direction of GSFC.

June 19: Tiros V was launched into orbit by a Thor-Delta booster from Cape Canaveral. A faulty guidance system placed it into an elliptical orbit (apogee, 604 miles; perigee, 367 miles; period, 100.5 minutes) instead of a 400-mile circular orbit.

June 20: An Aerobee 150A was launched from Wallops Island with a 271-pound payload boosted to 97-mile altitude; it carried a camera to study the behavior of liquid hydrogen under conditions of symmetrical heating and zero gravity.

June 24: OSO I began transmitting real-time data on solar observations after 5 weeks of intermittent transmittal.

July 1: OSO I was transmitting continuous signals, and 20 percent of real-time data was being acquired from each 95-minute orbit.

July 3: Ariel I discovery of a new ion belt at an altitude of 450 to 500 miles was announced at the International Conference on the Ionosphere, London, by Prof. James Sayers of Birmingham University.

July 9: Tiros V stopped transmitting pictures from the Tegea-lens, medium-angle camera. The Tegea camera system transmitted 4,701 pictures of which 70 percent were considered of excellent quality. The wide-angle Elgeet-lens camera, which still functioned, had transmitted 5,100 pictures to date, some of which aided in the analysis of Typhoon Joan over the western Pacific.

July 10: Telstar I, the first privately financed satellite, was launched by a Goddard launch team, from AMR on a Delta booster. The satellite was funded by AT&T and launched under a NASA-AT&T agreement of July 27, 1961. Telstar I made the world's first commercial transmission of live TV via satellite and the first transatlantic TV transmission on the same day it was launched. In one test, pictures were telecast from Andover, Me., to Telstar I, then returned and placed

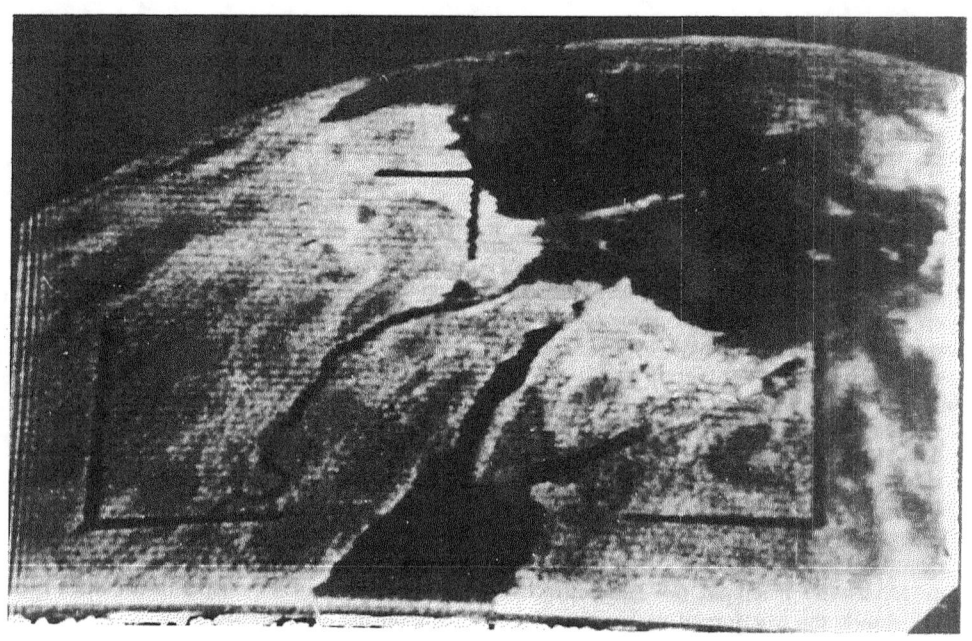

The Nile delta as seen from a Tiros satellite.

A picture transmitted by comsat *Telstar I*.

1962 Continued

on all three major TV networks in the U.S.

July 18: NASA launched a rigidized Echo-type balloon on a Thor booster to 922 miles in an inflation test. Nicknamed "Big Shot," the 135-foot balloon was inflated successfully and was visible for 10 minutes from Cape Canaveral. A movie film capsule parachuted into the sea, northeast of San Salvador, was recovered by three "pararescue" men of the Air Rescue Service. This was the largest man-made object sent into space, the previous record being held by the 100-foot *Echo. I.*

July 20: The Weather Bureau transmitted *Tiros V* photographs to Australia from Suitland, Md., the first time Tiros photographs had been transmitted abroad for current weather analysis by a foreign country. The photographs were of cloud formations west of Australia.

July 23: Telstar I relayed two 20-minute live TV shows, the first formal exchange of programs across the Atlantic via *Telstar I.* The United States Information Agency reported that U.S.S.R. had been invited to participate in the *Telstar I* broadcasts but had never answered the invitation.

July 24: Three major TV networks in the U.S. telecast separate 5-minute newscasts via *Telstar I,* each featuring their respective Paris news correspondents.

July 27: GSFC awarded a contract to IBM's Federal Systems Division for computer support services of Mercury flights, nonrendezvous Gemini flights (orbital flights of two men in one capsule), and the unmanned lunar flights scheduled as part of Project Apollo.

July 31: Former President Dwight D. Eisenhower spoke on the people-to-people benefits to be gained by live international communications in a broadcast televised to the U.S. via *Telstar I* from Stockholm, Sweden.

August 3: It was announced that the Advanced Syncom Satellite, being developed for NASA by the Hughes Aircraft Co., probably would carry four radio signal repeaters and would provide up to 300 two-way telephone channels or one TV channel.

August 7: General Electric announced that the control system for the first Orbiting Astronomical Observatory (OAO) had successfully completed its first simulated space flight.

August 8: NASA launched an Aerobee 150A sounding rocket from Wallops Island. Its 256-pound payload rose to 92-mile altitude and traveled a 60-mile distance downrange.

August 12: Five NASA representatives, led by Ozro M. Covington of GSFC, arrived in Australia to inspect proposed sites for new tracking stations.

August 15: NASA announced that GSFC had awarded three 3-month study contracts on the design of an Advanced OSO, to be launched into polar orbit during 1965. The Advanced OSO would aid development of a method of predicting flares.

August 16: Construction began on Building 11, the Applied Sciences Laboratory, at GSFC.

August 19: NASA launched a Scout vehicle from Wallops Island in an experiment to make direct measurements of radiative heating during atmospheric entry.

August 22: The French government announced its first satellite, weighing 150 pounds, would be launched in March 1966 and would be followed by others three and four times as large. GSFC was to assist in the training of the project staff.

August 27: GSFC announced it was training Italian scientists and engineers for the launching of Italy's first satellite. The 165-pound satellite would be launched by 1965 from a platform in the Indian Ocean off the eastern coast of Africa.

September 5: An agreement establishing the Italy-U.S. cooperative space program, signed in May, was confirmed in Rome by Italian Foreign Minister Attilio Piccioni and U.S. Vice President Lyndon B. Johnson. The Memorandum of Understanding between the Italian Space Commission and NASA provided a three-phase program, expected to culminate in the launching of a scientific satellite into equatorial orbit. Generally, NASA would provide Scout rockets and personnel training; Italians would launch the vehicle with its Italian payload and would be responsible for data acquisition as well as for a towable launch platform located in equatorial waters. Subsequently the satellites in this series were named "San Marco."

APPENDIX D

September 6: ITT announced plans for a NASA Project Relay satellite communication experiment to link North America and South America.

September 15: Signals from *Ariel I* were received again. The satellite had stopped transmitting after radiation from a U.S. high-altitude nuclear test damaged the satellite's solar cells. Although resumed transmission was not continuous, it did demonstrate *Ariel I*'s regained capability to return scientific data from space.

—NASA announced that the sixth Tiros weather satellite would be launched into orbit from Cape Canaveral on Sept. 18, at the earliest. The launch date was moved 2 months ahead to provide backup for *Tiros V* cloud-cover photography during the last half of current hurricane season and to provide weather forecasting support for Astronaut Walter M. Schirra's orbital space flight Sept. 28. The wide-angle TV camera in *Tiros V* continued to operate, but its medium-angle Tegea lens had stopped functioning on July 2 because of "random electrical failure in the camera's system."

September 18: *Tiros VI* was placed in orbit by a three-stage Delta vehicle from Cape Canaveral.

September 22: An Aerobee 150A was launched from Wallops Island; the rocket reached a 177-mile altitude in an experiment to measure the absolute intensity of the spectrum of stars with 50A resolution and to measure ultraviolet fluxes.

September 28: NASA announced plans to launch two Project Echo balloons during October. To be filled with helium while on the ground near the White Sands Missile Range, N. Mex., one balloon would be like *Echo I*, measuring 100 feet in diameter and the other would be an Advanced Echo type measuring 135 feet in diameter.

September 29: *Alouette I*, the Swept Frequency Topside Sounder, was placed in polar orbit by a Thor-Agena B from Vandenberg AFB. It was a Canadian Defence Research Board project.

September 30: NASA launched an Aerobee sounding rocket from Wallops Island. The 259-pound instrumented payload reached a 106-mile altitude in a test to map sources of photons of specific wavelengths in the nighttime sky.

October 2: *Explorer XIV*, an Energetic Particles Satellite, was launched. It was to study trapped corpuscular radiation, solar particles, cosmic radiation, and solar winds.

October 16: A Nike-Apache two-stage sounding rocket carried a 65-pound instrumented payload to 103 miles above Wallops Island.

October 20: An *Echo I*-type balloon launched from the White Sands Missile Range ruptured at a 21-mile altitude and fell back to earth 91 minutes after launch. The 100-foot-diameter balloon was to have reached 24 miles in structural test.

October 21: NASA announced that Swedish and U.S. experimenters were studying samplings of noctilucent clouds obtained in four Nike-Cajun sounding rocket flights during August. Preliminary analysis indicated that samples taken when noctilucent clouds were observed contained significantly more particles than when noctilucent clouds were not visible. Analysis of the origin and structure of the particles might take up to a year. Participants would include scientists from the Univ. of Stockholm Institute of Meteorology, Kiruna (Sweden) Geophysical Observatory, GSFC, and USAF Cambridge Research Laboratories.

October 22: Construction was begun on Building 12, the Tracking and Telemetry Laboratory, at GSFC.

October 27: *Explorer XV* was placed in orbit by a Thor-Delta vehicle launched from Cape Canaveral.

October 29: An Aerobee sounding rocket launched from Wallops Island carried a 230-pound payload to 116 miles. It landed in the Atlantic Ocean 59 miles from the launch site.

—NASA officials said that five experiments aboard *Explorer XV* were working well but that two others had been adversely affected by the satellite's excessive spin rate.

October 31: *Explorer XIV* had transmitted 589 hours of data to GSFC, which had released about 240 hours of data to the various experimenters.

During October: Patents were awarded to the following GSFC employees: Harold J. Peake, Space Technology Division, for a Data Con-

1962 Continued

version Unit; William A. Leavy, Aeronomy and Meteorology Division, for a Switching Mechanism; and Stephen Paull, Spacecraft Technology Division for a V/F Magnetic Multivibrator.

—Robert E. Bourdeau, Head of the Ionospheres Branch, Space Sciences Division, was awarded the NASA Medal for Exceptional Scientific Achievement, for: "Major scientific advances in the study of the ionosphere and significant progress in the understanding of the plasma sheath about satellites."

—Dr. John C. Lindsay, Associate Chief, GSFC's Space Sciences Division, was awarded the NASA Medal for Exceptional Scientific Achievement, "For the achievement of a major scientific advance in the study of the Sun and for significant technological progress in highly precise satellite attitude control."

—Dr. John W. Townsend, Jr., Assistant Director, GSFC's Space Science and Satellite Applications Directorate, was awarded the NASA Medal for Outstanding Leadership, "For outstanding and dynamic leadership in planning, developing, and directing a complex scientific organization whose notable achievements have significantly contributed to the preeminent position of this country in the space sciences, the development of space technology, and the practical application of such research and development."

—The Directorate for Tracking and Data Systems received the NASA Group Achievement Award, "For superior technical and administrative leadership and outstanding results in the operation of the global manned spacecraft tracking network."

November 1: NASA reported that *Explorer XV* radiation satellite was spinning at the rate of 73 rpm instead of a desired 10 rpm because of failure of the despin weights to deploy. Preliminary data indicated most experiments were functioning and that data received were of good quality.

November 5: GSFC announced the award of contracts totaling $12 million for tracking-network modifications in preparation for lengthy manned space flights.

November 7: NASA launched two experimental Nike-Apache rockets into the upper atmosphere within ½ hour of each other, to obtain a comparison of electron density and wind profiles measured at about the same time.

November 8: GSFC announced it would conduct experiments using a laser in tracking the S-66 ionosphere beacon satellite, to be launched into a polar orbit early next year.

November 9: NASA reported Canadian *Alouette I* topside-sounder satellite was performing as expected. Launched Sept. 28, it was considered "a very successful experiment since it is producing not only ionospheric data but also information about the earth's magnetic field. . . . Operation of the satellite continues to be normal. . . ."

November 12: It was reported that TAVE (Thor-Agena Vibration Experiment), flown with the Thor-Agena launching *Alouette I*, measured low-frequency vibrations to the Agena stage and measured spacecraft interfaces during the Thor boost phase.

November 14: In a news conference at MIT, Dr. James A. Van Allen predicted the radiation caused by the U.S. atmospheric nuclear test in July should be "undetectable" by July 1963. Dr. Van Allen reported that signals from *Injun, Telstar I, Explorer XIV,* and *Explorer XV* showed that the electronic stream had disappeared within a few days of the U.S. explosion and that the electrons at a 600-mile altitude were now undetectable. Electrons at a 900-mile altitude were still creating radio-astronomy interference, he acknowledged, but this should be gone by next July.

November 16: A Nike-Cajun sounding rocket was launched from Fort Churchill, Canada, under direction of GSFC. The second stage failed to ignite, so the rocket reached an altitude of only about 9.5 miles.

November 30: Franco-American scientific sounding rocket launchings were coordinated when two U.S. launchings were made from Wallops Island while France launched one from Algeria (and failed to launch one from France). The first U.S. rocket (a Nike-Cajun), fired at 5:57 a.m., carried a Langmuir probe to determine electron density and the temperature of the E layer of the ionosphere (50- to 100-mile altitude); the second (a Nike-Apache), launched at

6:15 a.m., released a sodium vapor cloud to a 106-mile altitude, which spread over 100 miles of the Eastern seaboard.

December 1: The medium-angle camera on *Tiros VI* stopped transmitting pictures during orbit 1,074, but the wide-angle camera was still sending pictures of "excellent quality."

December 4: GSFC launched two Nike-Cajun sounding rockets, one from Wallops Island, and one from Fort Churchill, Canada, for the purpose of comparing data on winds and temperatures in the upper atmosphere.

December 13: Relay I was launched. Its purpose was to investigate wideband communications between ground stations at a low altitude.

December 15: The power supply on *Relay I* remained too low to operate the satellite's instrumentation properly.

December 16: Explorer XVI was launched into orbit by a four-stage Scout vehicle from Wallops Island and it began measuring micrometeoroids in space.

—The *Relay I* 136-Mc beacon was detected by tracking stations at Santiago, Johannesburg, and Woomera, indicating the beacon had spontaneously turned itself on.

December 17: Although the *Relay I* power supply remained low, the Nutley, N.J., ground station was able to obtain about 10 minutes of usable telemetry data.

December 19: U.S. Weather Bureau announced the development of an infrared spectrometer, to be flight-tested in new balloons during the next 6 months. The 100-pound "flying thermometer" was planned for use in Nimbus weather satellites.

December 21: Canada and the U.S. announced a cooperative venture to build a data acquisition station for the Nimbus meteorological satellite program at Ingomish, Nova Scotia.

December 31: Goddard Space Flight Center had over 2,850 people employed or committed for employment.

1963

January 3: Both U.S. communications satellites, *Telstar I* and *Relay I*, came to life. *Telstar*, silent since Nov. 23, respond-

Aerobee sounding rocket fired from Wallops Island, Va.

ed to signals sent by Bell Telephone Laboratories; later in the day, *Relay*, silent since first being orbited Dec. 13, responded twice to television test patterns sent from New Jersey and Maine.

1963 Continued

January 5: Relay I communications satellite made two successful intercontinental television test transmissions between Andover, Me., and Goonhilly Downs, England, one for 23 minutes and the other for 1 hour; teletype tests were also successfully made from Nutley, N.J., to Fucino, Italy. NASA said *Relay I*'s power difficulty had apparently corrected itself, but "project officials have experienced difficulties with *Relay I* responding properly to commands. Tests during the past 3 years were possible by employing special operational procedures and altering command sequences to the satellite. Experiments will continue to evaluate communications and command systems."

January 7: U.K. sent television signals across the Atlantic for first time via *Relay I* communications satellite. Signals sent from Goonhilly Downs to Nutley, N.J., were described as "very good" and "extremely clear"; they were also clearly received at ground station of Italian space communications agency Telespazio in Fucino.

January 8: NASA reported *Relay I* communications satellite's low battery voltage had been result of faulty voltage regulator in one of its twin transponders. Continued tests by RCA and NASA engineers pinpointed the difficulty; the regulator failed to function properly when it became too hot or too cold. Engineers would attempt live television transmission via *Relay I* by sending special command signals to the satellite and concentrating on the remaining good transponder. *Relay I* communications satellite transmitted its first transatlantic television programs, sending British and French viewers clear pictures of ceremonial unveiling of "Mona Lisa" in its visit to Washington and 10 minutes of network program "Today."

January 10: *Explorer XIV* energetic particles satellite developed radio transmission difficulty, not correctable by remote control. Exact cause of difficulty, apparently in one of the binary counters of satellite's encoder system, was not determined.

—French Scientific Research Minister Gas-

The unveiling of "Mona Lisa" at National Gallery, Washington, D.C. (transmitted picture seen in Europe). Left to right, President John F. Kennedy, Madame Malraux, French Minister of Cultural Affairs André Malraux, Mrs. Kennedy, and Vice President Lyndon B. Johnson.

APPENDIX D

ton Palewski told French National Assembly a satellite launching site would be established in Eastern Pyrenees Department near the Spanish border. France's first satellite was scheduled for launching in 1965; other European satellites might also be launched from the site.

January 13: GSFC announced its sodium-vapor cloud experiments during past 2 years had shown wind behavior 44–50 miles above earth became erratic and unpredictable. Below that altitude winds generally follow global pattern, regularly reversing with the seasons. Region between 56- and 68-mile altitude is characterized by "remarkable wind sheers"—within altitude span of less than 3 miles, wind speed was observed to increase swiftly by more than 250 mph and even to reverse direction. Immediately above this band of maximum wind velocity, wind diminishes almost to zero. Above 70 miles, research indicated region of strong but more uniform winds, with velocities of about 200 mph. GSFC experiments, launched on sounding rockets from Wallops Station, did not extend beyond 105-mile altitude.

—NASA announced that it would procure Atlas-Agena B vehicles directly from contractors. NASA already had used seven of the vehicles—five for Ranger and two for Mariner—and was planning to use 20 Atlas-Agena B's over the next 3 years—in Gemini rendezvous flights, OGO, OAO, Ranger, and Mariner R. Prime vehicle contractors were General Dynamics Astronautics for Atlas stage and Lockheed for Agena; USAF had vehicle integration responsibility.

—NASA announced signing of Memorandum of Understanding with India's Department of Atomic Energy providing for cooperative U.S.-India space program. Joint scientific experiments to explore equatorial electrojet and upper-atmosphere winds from geomagnetic equator would be launched from Thumba, India, during 1963.

January 15: Explorer XIV energetic particles satellite transmitted 38 seconds of complete data, and Goddard officials were hopeful the satellite might eventually resume normal operations. *Explorer XIV* developed transmission difficulty Jan. 10, after 100 days of nearly continuous transmission. Project Manager Paul G. Marcotte reported *Explorer XIV* received less than 10 percent degradation from space radiation since its launch Oct. 2.

January 17: *Relay I* satellite transmitted 12-minute Voice of America program as well as AP and UPI news dispatches from Nutley, N.J., to Rio de Janeiro and back. Transmissions were reported perfect, even though ordinary high-frequency radio communication with Rio was not possible because of atmospheric conditions.

January 18: President John F. Kennedy and Dr. Hugh L. Dryden sent teletype and recorded voice messages, respectively, to Italy by way of the *Relay I* satellite.

January 29: *Explorer XIV*, silent since January 10, resumed normal transmission.

—NASA Director of Communications Systems Leonard Jaffe announced NASA would attempt to launch Syncom communications satellite into synchronous orbit with Delta vehicle no earlier than Feb. 6. Syncom launch was postponed "to insure that the [command and control] equipment is completely checked out" aboard USNS *Kingsport*, stationed in Lagos Harbor, and on the launch vehicle at Cape Canaveral.

January 30: The GSFC Spacecraft Systems Branch, Spacecraft Systems and Projects Division, was reorganized and the Spacecraft Projects Office established. The GSFC Constructions Inspection Service was reorganized and retitled the Construction and Renovation Section.

January 31: Representatives of Canadian Defence Research Board and NASA met for preliminary exploration of scientific and technical aspects involved in proposed joint ionospheric research program. Extension of joint Alouette Topside Sounder program would involve design and construction of four satellites in Canada, with first launching proposed for late 1964.

—Ceremonies at Goddard Space Flight Center celebrating the fifth anniversary of *Explorer I* and the GSFC tracking network used for tracking the satellite featured talks by Secretary of State Dean Rusk, NASA Administrator James E. Webb, Astronaut Walter M. Schirra, Jr., Goddard Director

1963 Continued

Harry J. Goett, and Dr. Edward C. Welsh, Executive Secretary of the National Aeronautics and Space Council. Radio transmissions from *Vanguard I*, second U.S. satellite and oldest still transmitting, were piped into Goddard auditorium. Highlighting ceremony was presentation of scrolls of appreciation to ambassadors of 16 nations that have cooperated with U.S. in establishing the tracking networks.

— Contract award was made to the Industrial Engineering Corporation for the construction of an Optical Tracking Observation Building and Ground Plane Test Facilities at Goddard. The construction starting date was Feb. 6, 1963.

During January: In *International Geophysics Bulletin*, NASA proposed contributions to IQSY (1964–65) were outlined. Prominently among them: sounding rockets; ionosphere explorers and monitors; atmospheric structure OSO, EGO, and POGO satellites; IMP, Pioneer, Mariner, and Surveyor probes.

February 1: NASA announced its first contract to study overall systems requirements for Synchronous Meteorological Satellite (SMS) had been awarded to Republic Aviation Corp. Administered by GSFC, contract called for 4-month study to determine "technical systems needed for 24-hour surveillance of the earth's cloud cover, and to identify the major scientific and engineering advances required for the ground stations."

February 4: On effects of artificial radiation on spacecraft solar cells, a joint AEC-DOD-NASA report said: "Improved types of solar cells (employing n-on-p silicon junctions) which are considerably more radiation resistant, are available and were employed on Telstar. With respect to manned missions in space, the shielding provided by normal capsule design effects a considerable reduction in the radiation exposure, and the artificial belt is not regarded as placing any significant restrictions on the conduct of current manned space flights. . . ."

— Sen. Leverett Saltonstall (R. Mass.) introduced in the Senate a bill (S. 656) "to promote public knowledge of progress and achievement in astronautics and related sciences through the designation of a special day (March 16) in honor of Dr. Robert

Astronaut Walter M. Schirra, Jr., addressing the Fifth Anniversary of Space Tracking ceremonies at Goddard Space Flight Center.

APPENDIX D

Hutchings Goddard, the father of modern rockets, missiles, and astronautics. . . ." On March 16, 1926, Dr. Goddard first successfully launched a liquid-fuel rocket.

February 6: Goddard Space Flight Center was host to an optical conference. Approximately 70 persons representing NASA's field centers and installations attended the first intra-agency technical conference on optical communications and tracking.

February 13: Proposal to establish international tracking system using lasers to track the S-66 satellite was made at Third International Congress of Quantum Electronics, Paris, by Richard Barnes of NASA Office of International Programs. Under the proposal, each country would establish and control its own stations, with U.S. furnishing the necessary information on the satellite. Laser system was expected to provide faster and more precise tracking than existing radio and radar systems; used with S-66 satellite, to be launched this spring, it should enable scientists to determine the profile of the ionosphere.

— *Explorer XVI* meteoroid detector satellite recorded 16 punctures by meteoroids during its first 29 days in orbit, NASA reported. Other spacecraft had reported hits by cosmic debris, but this was first time actual punctures were recorded. If *Explorer XVI* continued to report meteoroid data for a full year as expected, it should enable scientists to determine whether meteoroids are hazardous to a spacecraft. The satellite exposed 25 square feet of surface to meteoroid impacts, not large enough to provide good statistical data on larger and rarer particles in space. (On Feb. 5, NASA had announced plans to orbit two meteoroid-detector satellites, each with exposure surface of more than 2,000 square feet.)

February 14: NASA *Syncom I* synchronous-orbit communications satellite was launched into orbit by Thor-Delta vehicle from AMR, entered a highly elliptical orbit. About 5 hours later, apogee-kick motor was fired for about 20 seconds in maneuver designed to place the satellite into near-synchronous, 24-hour orbit 22,300 miles above the earth. At about the time the apogee-kick motor completed its burn, ground stations lost contact with the satellite and could not confirm a synchronous orbit. Attempts to make contact with Syncom were continued.

— A contract was let to the Norair Construction Co. for GSFC's Building No. 16, Development Operations Building. The starting date was Feb. 21, 1963, with a scheduled completion date of Mar. 15, 1964. Partial occupancy in the warehouse portion was estimated for Dec. 1963.

February 15: U.S. worldwide tracking network was not able to locate *Syncom I* communications satellite; radio contact with the satellite was lost Feb. 14, seconds after onboard rocket had fired to transfer *Syncom* from its highly elliptical orbit into near-synchronous orbit.

February 18: Attempted launch of sodium-vapor cloud experiment from NASA Wallops Station was not successful because second stage of Nike-Asp launch vehicle failed to perform properly. A series of rocket grenade and sodium release experiments from Wallops Island and Fort Churchill began on this date and were continued through Mar. 8.

February 19: Dr. Hugh L. Dryden, NASA Deputy Administrator, testified before Communications Subcommittee of Senate Commerce Committee that experiences of both Telstar and Relay communications satellites were being "used continuously to review projects such as Syncom . . . in an attempt to achieve the 24-hour synchronous orbit, as well as all of our other satellite projects. I should like to add, finally, that the experience of Telstar and Relay to date have merely reinforced the opinion which I gave before this committee last year, that considerable research and development have yet to be performed before economic operational systems can be established . . ."

February 20: Following a period of hesitant response by the command decoders on *Telstar I,* due to radiation effects, the spacecraft was inadvertently turned off.

February 23: William Schindler, Goddard manager of the Delta launch vehicle program, was one of 21 engineers and scientists to receive a National Capital Award at the Engineers, Scientists and Architects Day

1963 Continued

awards luncheon. The D.C. Council of Engineering and Architectural Society and the Washington Academy of Sciences honors the men and professions of engineering, science, and architecture each year.

February 25: On communications satellites, Dr. Robert C. Seamans, Jr., NASA Associate Administrator, said: "We re-examined our Communication Satellite program quite carefully in the light of the creation of the Communication Satellite Corporation and the reoriented activities of the DOD following the cancellation of the Advent project. From this programmatic re-examination we have concluded that principal NASA effort should be focused on the research and development problems associated with the synchronous altitude class of communication satellite. We have, therefore, dropped the low altitude multiple passive satellite project, Rebound, and advanced intermediate active satellite projects from hardware development consideration at this time. As a result of these decisions, we reduced our communication satellite program by $35.2 million...."

— 28-nation U.N. Committee on Peaceful Uses of Outer Space approved Indian progress report on plans to sponsor an international rocket base at Quilon for launchings in space above the equatorial regions. Italian delegate reported that the San Marco floating launching facilities would be completed in time for use in the International Quiet Sun Year.

February 28: NASA Director of Meteorological Systems Morris Tepper told House Committee on Science and Astronautics that *Tiros V* and *Tiros VI* (launched in June and September 1962, respectively) were still providing good data. Tiros data "continue to be used by the Weather Bureau for weather analysis and forecasting, storm tracking, hurricane reconnaissance, etc. The Meteorological Soundings project has continued throughout the year as planned. The project at Goddard Space Flight Center, which utilizes the larger meteorological sounding rockets, continues as it has in past years with excellent results. In addition, we have initiated at the Langley Research Center a project which will develop and utilize the smaller meteorological sounding rockets. We expect to have this well underway by the end of the fiscal year...."

— Harvard College Observatory reported that astronomers at Boyden Observatory at Bloemfontein, South Africa, had photographed the *Syncom I* satellite, missing since Feb. 14. The Observatory's photographs indicated *Syncom* probably was in orbit about 22,000 miles high.

— GSFC plans for second-generation OSO satellite—known as Advanced Orbiting Solar Observatory, or Helios—were outlined at Philadelphia technical meeting by Goddard's AOSO Project Manager A. J. Cervenka. AOSO would be designed to have a pointing accuracy of 5 seconds of arc and 70 percent overall systems reliability, Cervenka said.

March 1: At Cape Canaveral, Fla., the team behind NASA's most reliable booster—the Delta—was honored for a success story unique to America's space program. NASA's Group Achievement Award was presented to Goddard's Delta Project Group, which managed the project for NASA. The Delta was used 16 times and was successful the last 15 times.

— U.S. Weather Bureau announced it was purchasing 11 ground stations capable of receiving cloud pictures directly from Nimbus meteorological satellites, to be launched by NASA.

March 2: Boyden Observatory near Bloemfontein, South Africa, had confirmed location of *Syncom I* communications satellite, Harvard University Observatory Director Donald H. Menzel announced. *Syncom I* was tumbling end over end in its orbital path about 19,000 miles high. Boyden's unconfirmed photographs of the satellite, missing since Feb. 14, were reported Feb. 28, and NASA had requested that the findings be confirmed by further observation. "Since then it has cleared and we obtained two good plates showing images in the expected position. With this final confirmation, we have no doubt whatever of the location of the satellite. It behaved approximately as expected."

March 4: U.S. plans for International Year of

Delta Day ceremony, March 1, 1963, at Cape Canaveral. Standing are, left to right, William Schindler, Delta Project Manager, and Dr. Harry J. Goett.

the Quiet Sun (IQSY), 1964-65, were announced by National Academy of Sciences-National Research Council (NAS-NRC), charged by President Kennedy in 1962 to correlate IQSY contribution of Federal agencies. Many IGY observations would be repeated and special experiments made possible by recent scientific advances would be added. IQSY would concentrate more intensively than IGY on the upper atmosphere and space phenomena directly affected by both the large periodic bursts of charged particles and associated magnetic fields escaping from the sun, and the continuous background activity known as "solar wind."

— Dr. Hugh L. Dryden, NASA Deputy Administrator, testifying on NASA's international programs before House Committee on Science and Astronautics, said that the "first substantial fruits of these programs were realized in 1962 and further significant programs were laid down for future years. During 1962, ". . . the first two international satellites, Ariel and Alouette, were successfully placed in orbit, . . . launchings of sounding rockets bearing scientific payloads were carried out in cooperation with eight countries, . . . 37 countries engaged in special projects in support of our weather and communications satellite programs, . . . foreign participation continued to grow in the operation of our global tracking and data acquisition network overseas, . . . and, a new NASA international fellowship program was successfully established in our own universities."

March 5: NASA announced agreement with Australia for establishment of deep space tracking facility about 11 miles southwest of

1963 Continued

Canberra; a manned flight and scientific satellite tracking station at Carnarvon; and a small mobile station at Darwin serving the Syncom communications satellite.

March 7: OSO I solar observatory satellite completed its first year in orbit, exceeding its estimated operating life by 6 months. Eleven of its 13 scientific experiments were still operating and were providing extensive data on behavior and composition of the sun. Preliminary results from *OSO I* would be presented at a symposium Mar. 14.

March 11: U.S.-U.S.S.R. negotiations began in Rome on technical details of a 3-year agreement signed at Geneva in June 1962, for exchange of data to be gained from satellite launchings. Dr. Hugh L. Dryden, Deputy Administrator of NASA, headed U.S. scientific delegation, and Prof. Anatoly A. Blagonravov of the Soviet Academy of Sciences headed the Russian delegation. Joint space research program would include coordination on meteorology, communications studies, and charting of the earth's magnetic field.

— *Relay I* communications satellite was turned off because of severe drain on the onboard power supply, a difficulty similar to that encountered during first week after launch. Power drain was encountered Mar. 9 after *Relay I*'s orbit had been in earth's shadow for 5 weeks and spacecraft temperatures were low.

— NASA and French National Center for Space Studies (CNES) jointly announced signing of Memorandum of Understanding for a cooperative U.S.-France program to investigate propagation of VLF electromagnetic waves. First phase of the program would consist of two electromagnetic-field experiments with French-instrumented payloads to be launched from NASA Wallops Station. Second phase, to be implemented upon mutual consent after Phase I had proved the experiments to be scientifically and technically feasible, would consist of orbiting of scientific satellite, designed and built by France, with a Scout vehicle.

March 13: Relay I communications satellite, its power supply voltage and temperature returned to normal, responded to command signals turning on its telemetry transmitter and encoder. NASA planned to resume normal experimental operations with the satellite Mar. 14. *Relay I* had been turned off because of severe power drain encountered Mar. 9.

March 13 to 15: The Goddard Scientific Satellite Symposium was held at the Interior Department Auditorium in Washington, D.C. The program covered presentations on *Alouette I* (S–27), *Ariel I* (UK–1), *OSO I* (S–16), and *Explorers XII* (S–3), *XIV* (S–3a), and *XV* (S–3b). Data from *Alouette I* showed that ionosphere is usually rough in high latitudes and that electron temperature of ionosphere increases with latitude. This evidence indicated Van Allen radiation belts, which extend to lower altitudes at higher latitudes, possibly are secondary heat source for ionosphere. Where ionospheric and radiation particles collide, ionospheric temperatures rise and F layers of ionosphere spread apart, causing radio waves to scatter. Results from *Ariel I* confirmed the ionospheric temperature relationship with latitude as detected by *Alouette I*. Solar x-ray detectors found solar flares are made up of two phases: (1) heating of sun's corona above sunspot, increasing x-ray flux by factor of 10; and (2) quiet period marked by flux leveling off at accelerated level, followed by streams of electrons pushed into chromosphere, causing x-ray emissions at 500 times greater than normal.

March 14: Dr. John Lindsay and William White, of Goddard Space Flight Center, reported that the *OSO I* satellite had found tentative evidence that solar flares may be preceded by series of microflares whose sequence and pattern may be predictable. *OSO I* reported at least four of these series during a year in orbit.

March 15: Dr. James A. Van Allen said that artificial radition belt caused by U.S. high-altitude nuclear test last July may last for 10 years. At GSFC's Scientific Satellite Symposium, Dr. Van Allen said data from *Injun III* and *Explorer XIV* satellites showed intensity at center of artificial belt had decreased only by a factor of two.

— Data from *Explorer XII* confirmed ex-

APPENDIX D

istence of low-energy proton current ringing earth in east-to-west direction, perpendicular to perpetual north-south spiraling motion along geomagnetic field lines.

— A contract was awarded to the Industrial Engineering Corp. for the construction of a Magnetic Range Control and Test Building and a Magnetic Instrument Test Laboratory at GSFC. The starting date was Apr. 3, 1963, with an estimated completion date of Oct. 1963.

March 19: Goddard Space Flight Center, in cooperation with NBC and RCA, accomplished first known transmission of television in color via *Relay I* communications satellite. Fifteen-minute sequence of movie "Kidnapped" was relayed by *Relay I* from 4,000-mile orbit, and was scheduled to be shown on Walt Disney's TV program on Mar. 24.

March 22: NASA announced *Relay I* communications satellite had achieved all its missions. Performance of *Relay I* included 500 communications tests and demonstrations in 660 orbits between Dec. 13, 1962–Mar. 11, 1964. Although all planned demonstrations were completed, more would be continued while the satellite remained in operation.

— Sixth Annual Robert H. Goddard Memorial Dinner, sponsored by the National Rocket Club, Washington, D.C. In an address, Vice President Lyndon B. Johnson paid tribute to the "father of modern rocketry." He said that those today who "understand the stakes of space" must help "the public to understand these stakes." He urged that communications barriers among scientists, engineers, and politicians be abolished so that public support for public policy can be obtained. "Unless and until this is done," said the Chairman of the National Aeronautics and Space Council, "the technological community cannot justifiably be impatient

Pictures of Italian Premier Fanfani's Chicago trip were transmitted to Europe via *Relay I*. This is a print from the television monitor in New Jersey.

1963 Continued

with those who are chosen to represent and express the public's own will."

—At GSFC Colloquium, Dr. John A. O'Keefe discussed the origin and evolution of the moon, submitting his theory that billions of years ago the moon separated from the still "undifferentiated earth," thereafter was subjected to volcanic eruptions, meteoroid bombardment, eventual cooling, and transformation into a hard cinder-like material. The volcanic dust produced the comparatively smooth lunar maria. If theory is correct, O'Keefe said, the original dust has long since become firm and constitutes "no hazard" for landings of space vehicles. O'Keefe supported his theory with available evidence on tektites.

March 25: Three major U.S. television networks each broadcast 7-minute programs from Paris to New York via *Relay I* communications satellite.

March 26: Dr. Fred S. Singer, Director of National Satellite Weather Center, told House Committee on Science and Astronautics' Subcommittee on Applications and Tracking and Data Acquisition that reports from Tiros weather satellites were being used by Soviet scientists in their weather research. Launching of a weather satellite "is probably an immediate Soviet objective."

March 28: Nike-Apache sounding rocket launched from NASA Wallops Station carried 65-pound instrumented payload to altitude of 100 miles, an experiment to measure electron density profile, electron temperatures, and solar radiation in the ionosphere. Secondary objective of the flight was to check out hardware to be flown from Ft. Churchill, Canada, during solar eclipse in July.

March 29: P-21A was launched from Wallops Island at 2:27 a.m. EST. Preliminary results showed that the ion trap was providing very good data.

During March: Canadian Government authorized four additional satellites for ionospheric research in joint U.S.-Canadian space program. Seven successful sounding rocket experiments were concluded.

April 3: NASA launched *Explorer XVII* (S–6) atmospheric structure satellite from Cape Canaveral, using Thor-Delta launch vehicle (its 16th consecutive success in 17 attempts). Under project management of NASA Goddard Space Flight Center, *Explorer XVII* was first scientific earth satellite to use new GSFC pulse-code-modulation telemetry system, a solid-state system providing output power of 500 milliwatts and capable of supplying 40 separate channels of information in digital form. Useful lifetime of the satellite was estimated at 2 to 3 months.

— Aerobee 150A rocket launched from NASA Wallops Station carried instrumented payload to 147-mile altitude in experiment to flight-test components of equipment for EOGO satellite and to measure propagation of VLF signals through ionosphere. Flight was joint project of Stanford Research Institute and GSFC.

April 7: The six winners of the third annual Federal Women's Award included Eleanor C. Pressly, Head of Vehicles Section, Spacecraft Integration and Sounding Rocket Division, NASA Goddard Space Flight Center. Miss Pressly was cited for her pioneer work in sounding rocket development and her "demonstrated organizational ability in scheduling and coordinating launchings of sounding rocket vehicles in support of upper atmospheric research." She developed the Aerobee Jr. sounding rocket, co-developed Aerobee-Hi 150, and directed improvement of Aerobee-Hi 150 A—all used extensively in IGY.

April 8: Attempt to launch two-stage Astrobee 1500 sounding rocket from NASA Wallops Station failed with first stage of the vehicle failing to perform properly. This was NASA's first attempt to launch the Astrobee; purpose of test was to evaluate the rocket's performance as a NASA test vehicle.

April 9: Televised White House ceremony, with President John F. Kennedy signing bill making Sir Winston Churchill an honorary citizen of the U.S., was transmitted to U.K. and continent via *Relay I* communications satellite. Broadcast was viewed by millions of Britons and Sir Winston himself, and both audio and visual reception were considered perfect.

— In its first few days of operation, *Ex-*

APPENDIX D

plorer *XVII* satellite had obtained data that more than tripled all previous direct measurements of the neutral gases in earth's upper atmosphere, it was announced. New communications system, utilizing special data readout station at GSFC, was providing scientific and technical data from the satellite within minutes of its transmission.

April 17: Five New Jersey newspapermen held first press conference through space, using *Relay I* communications satellite in 25-minute broadcast to Rio de Janeiro, Brazil. Photo of newsmen sent via *Relay I* during conference was of good quality, Rio officials said.

April 18: NASA launched 85-pound scientific payload to 208-mile altitude at exact moment *Explorer XVII* atmospheric structure satellite passed over the Wallops Island, Va., launch site, an unusual "first" in NASA sounding rocket program. Carried on an Aerobee 300A sounding rocket, experiment obtained temperature data on electron and neutral particles and measured ion and neutral particle densities. Data from this experiment would be compared with similar data obtained from *Explorer XVII* as it passed over Wallops Island at 198-mile altitude during its 236th orbit of earth. Preliminary evaluation by GSFC scientists revealed data were of excellent quality. Data from *Explorer XVII* indicated the earth is surrounded by belt of neutral helium atoms, GSFC scientists said at American Geophysical Union meeting. Based on preliminary data received one day after launch, Goddard scientists said *Explorer XVII* atmospheric structure satellite had sent back more than 8 hours of scientific information on physics and chemistry of tenuous gases making up the earth's atmosphere.

— The 3-year milestone in the Tiros success story was officially recognized when the Tiros team received NASA's group achievement award in special ceremonies. Six-out-of-six successful Tiros launches had created an unmatched series of successes for this spacecraft. More than 220,000 cloud-cover pictures had been transmitted back to earth.

— A contract was awarded to Jack Bays, Inc., for the construction of an Anechoic Chamber at Goddard. The starting date

Relay engineers monitor program awarding U.S. citizenship to former Prime Minister Sir Winston Churchill. Segment above shows address by President John F. Kennedy, speaking at the White House ceremony, April 9, 1963.

1963 Continued

was May 10, 1963, with a scheduled completion date of Oct. 1963.

April 20: Preliminary test of instrumentation to be used in joint Italian-U.S. San Marco project was made by two-stage Shotput sounding rocket from Wallops Station; the rocket carried 180-pound instrumented payload to 265-mile altitude. Flight was first in three-phase project being conducted by Italian Commission for Space Research and NASA, to be followed by further tests of San Marco instrumentation with launching of Shotput vehicle from towable platform in Indian Ocean and to be culminated in launching of scientific satellite into equatorial orbit from the platform. Basic objective of San Marco project was to obtain high-altitude measurements of atmospheric and ionospheric characteristics in equatorial region. GSFC assisted in testing the spacecraft.

April 25: Relay I communications satellite was used to transmit electroencephalograms ("brain waves") from Bristol, England, to Minneapolis, Minn., in demonstration experiment conducted in connection with meeting of National Academy of Neurology in Minneapolis.

During April: NASA awarded 4-month study contracts for a synchronous meteorological satellite to Radio Corp. of America and Hughes Aircraft.

May 1: A contract was awarded to the Norair Construction Co. for the construction of GSFC's Building No. 14, Spacecraft Operations Building. The starting date was May 11 with a scheduled completion date of May 15, 1964.

May 3: The Documentation Branch was established in the GSFC Technical Information Division to provide support in writing and publishing documents.

— The Telescopic Systems Section was established in the Astrophysics Branch of the Space Sciences Division at GSFC. The former Detector Section in the Astrophysics Branch was retitled as the Planetary Optics Section.

May 7: Telstar II communications satellite placed in elliptical orbit (6,717-mile apogee, 604-mile perigee, 225.3-mile period, 42.7° inclination to equator). Thor-Delta vehicle launched from Cape Canaveral boosted the satellite into orbit for its 17th straight success, an unmatched record for U.S. satellite-launching vehicles.

May 9: Sen. Margaret Chase Smith (R.-Me.) and NASA Administrator James E. Webb were co-hosts at Senate luncheon for three women accorded national recognition for space age accomplishments—Marcia S. Miner, student at American Univ. and winner of National Rocket Club's 1963 Goddard Memorial Scholarship Award; Dr. Nancy C. Roman, Chief of Astronomy and Solar Physics in NASA Geophysics and Astronomy Program and 1962 winner of Federal Women's Award; and Eleanor C. Pressly, Head of Vehicles Section, Sounding Rocket Branch, in NASA Goddard Space Flight Center's Spacecraft Integration and Sounding Rocket Division, and 1963 winner of Federal Women's Award.

May 16: L. Gordon Cooper completed 22 earth orbits in 34-hour MA-9 space flight. The Goddard-operated Mercury tracking network with 19 stations functioned perfectly, providing "real-time" tracking throughout the entire mission. Final project Mercury flight.

— The launch of two Nike-Cajuns at Wallops Island on this date successfully concluded the series of three cooperative U.S.-Japanese ionospheric experiments. Much useful information was obtained.

May 20: Two-stage sounding rocket instrumented to observe ionosphere was successfully launched to 215-mile altitude by Japanese scientists near Kagoshima, Japan.

May 23: Sodium-vapor experiment to measure high-altitude winds and diffusion rates was launched on Nike-Apache sounding rocket from Wallops Island, Va. Sodium vapor trail, ejected from 27- to 127-mile altitudes, was visible for several hundred miles from launch site.

May 28: GSFC Procurement and Supply Division, Office of Administration, was reorganized and retitled Procurement Division. The former Management Services Division, Office of Administration, was reorganized and retitled Management Services and Supply Division.

APPENDIX D

May 31: As of this date, Goddard Space Flight Center had on board 38 employees in excepted positions, 2,833 employees in Classification Act positions, and 250 Wage Board employees. In addition, 14 military personnel were assigned to the Center.

During May: Six successful sounding rocket projects were carried out. These included aeronomy, ionospheric physics, and test and support experiments.

June 3 to 11: On June 5, Goddard Space Flight Center offered world scientists the design of small rocket payload and ground telemetry station suitable for ionospheric research. GSFC scientists Siegfried J. Bauer and John E. Jackson said payload's "versatility, simplicity and relatively low cost should make it an ideal tool for the investigation of the many problems of the ionosphere by the international scientific community, especially during the IQSY (International Year of the Quiet Sun).

June 7 to 16: Goddard satellites were exhibited at the Paris International Air Show. French President Charles de Gaulle spent some time at the display and expressed deep interest in it. In the United States pavilion at the show was the most complete presentation of present and future space programs ever assembled under one roof. It included a prototype of *OSO I*.

June 14: Goddard Space Flight Center announced series of sounding rocket tests had confirmed association of Sporadic-E disturbances with presence of wind shears in altitude regions measured Nov. 7, Nov. 30, and Dec. 5, 1962. Under NASA contract, Geophysics Corp. of America scientists measured velocity of wind movements (using Nike-Apache rockets with sodium vapor trails) and ionospheric phenomena (using Nike-Cajun with Langmuir probe electrical equipment) at nearly the same time. Experiments confirmed theory of Australian scientist J. D. Whitehead that action of upper atmosphere wind pulls electrons from above and below into thin cloud-like layers, causing Sporadic-E layers that often interfere with radio signals being reflected off higher F layer of ionosphere.

June 19: *Tiros VII* (A-52) meteorological satellite placed in orbit with Thor-Delta launch vehicle launched from Cape Canaveral. On satellite's first orbit, command and data acquisition station at Wallops Island, Va., obtained direct pictures from Camera 2 showing cloud vortex over Newfoundland and set Camera 1 to read out pictures on next orbit. First pictures were transmitted within one hour to Cape Canaveral, Fla.; GSFC; and National Weather Satellite Center, Suitland, Md. In addition to two wide-angle TV cameras, *Tiros VII* carried infrared sensors and electron temperature probe. Orbiting marked 18th straight successful satellite orbiting by Thor-Delta launch vehicles.

June 21: U.S. television audiences witnessed first public appearance of Pope Paul VI via Relay communications satellite.

June 29: The University Building at Adelphi, Maryland, leased by GSFC, was partially occupied by the Space Flight Support Division, Office of Tracking and Data Systems.

July 2: A 50-pound payload of ionosphere measuring instruments was launched with Argo D-4 sounding rocket from Wallops Station, Va., into orbital path of *Alouette I* satellite. Preliminary data indicated measurements were made in upper ionosphere within 2 minutes of soundings taken from *Alouette I*. Payload reached peak altitude of 590 miles. Purpose of experiment was to obtain measurements of ion and electron temperatures and densities; data from payload instruments would be compared with similar data transmitted simultaneously by *Alouette I*.

July 3: With President John F. Kennedy's return to Washington from Europe, NASA communications satellite *Relay I* marked end of its busiest programing period. *Relay I* was "booked solid" during past weeks to cover the President's trip, death of Pope John XXIII, and election of Pope Paul VI. During its 6 months of operation, *Relay I* had been used for 85 public communications demonstrations, including transmission of television, voice, radiophoto, and teletype.

July 9: A 164-pound payload sent to 127-mile altitude with Aerobee 150A sounding rocket from NASA Wallops Station in experiment to obtain nighttime electromagnetic noise and

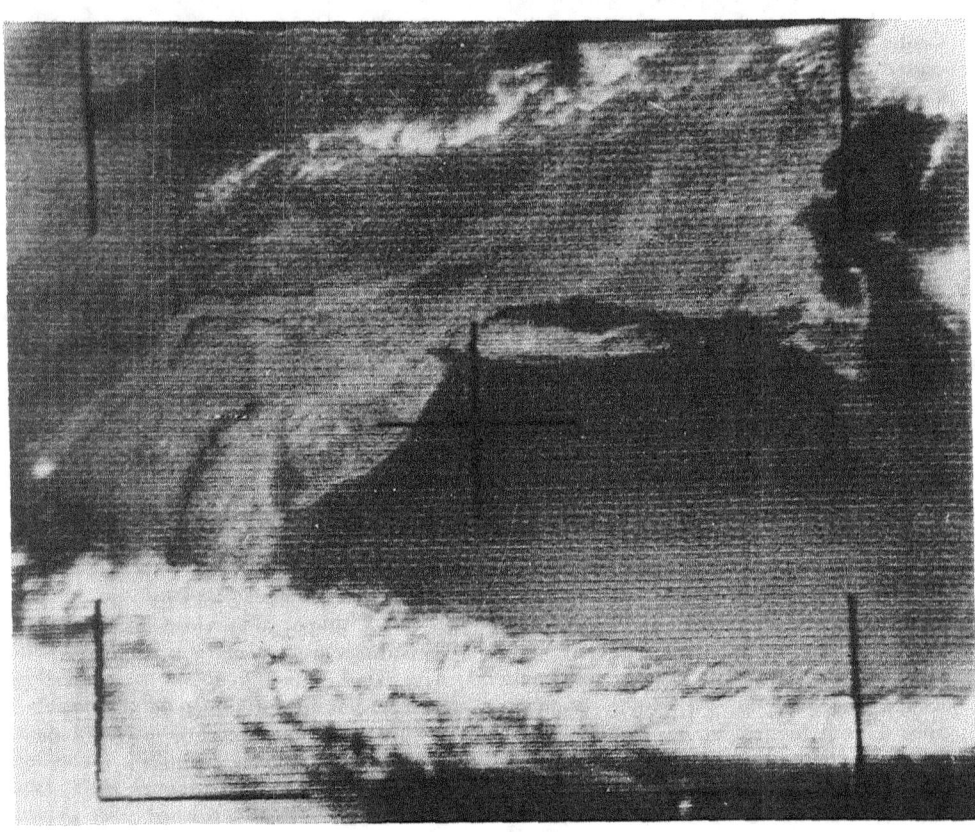

Tiros VII photograph showing cloud-free view of U.S. eastern seaboard from well above Cape Cod to below Chesapeake Bay, June 23, 1963.

1963 Continued

propagation data. Included in payload were three sweeping receivers and a broad-band receiver of the type to be included in EOGO satellite (Eccentric Orbiting Geophysical Observatory) next year. Preliminary telemetry evaluation indicated all experiment objectives were met.

July 19: The former GSFC Communications Branch Spacecraft Systems and Projects Division of the Office of Space Science and Satellite Applications was reorganized and retitled as the Communications Satellite Research Branch. A new branch was established in the Spacecraft Systems and Projects Division and titled Communications Satellite Branch.

July 20: Eclipse of the sun was visible across Canada and northeastern U.S. NASA joined other scientists and astronomers in scientific studies during the eclipse, with emphasis on ionosphere and on sun's corona.

— At Churchill Research Range, USAF OAR facility located at Ft. Churchill, Canada, six Nike-Apache sounding rockets equipped with instruments to measure electron density, electron temperature, and solar radiation in ultraviolet and x-ray regions, were launched for GSFC; Aerobee 150 sounding rocket equipped with spectrophotometric instruments to measure absolute intensity of spectral features in ultraviolet region was launched for Johns Hopkins University; and Canadian Black Brant

APPENDIX D

sounding rocket with instruments to measure variations in D and E layers of ionosphere was launched for USAF Cambridge Research Laboratories. GSFC and AFCRL scientists said preliminary results indicated collected data confirmed previous predictions of composition of the ionosphere.

— At Pleasant Pond, Me., Luc Secretan and Francois V. Dossin of GSFC photographed eclipse with specially made instrument for photographing stars and comets near the sun.

July 26: Syncom II communications satellite was launched into orbit with Thor-Delta launch vehicle from AMR, entering elliptical orbit (140-mile perigee, 22,548-mile apogee). Five hours 33 minutes after launching, apogee kick motor on board fired for 21 seconds, placing *Syncom II* in orbital path ranging from 22,300-mile to 22,548-mile altitude and adjusting its speed to near-synchronous 6,800 mph. Traveling in slightly lower than synchronous orbit and at less than synchronous speed, satellite began drifting eastward at rate of 7.5° per day. Ground signals would attempt to reverse drifting so that satellite would attain synchronous position over Brazil.

— A contract was let to Kalmia Construction Co., Inc., Silver Spring, Md., to alter Goddard's Data Acquisition and Communications Center building for the installation of a microwave antenna. The contract called for a supporting structure, an equipment room, and a cooling tower enclosure for the antenna.

August 2: Sweden successfully launched U.S. Army Nike-Cajun rocket from Kronogård rocket range in test to explore "bright night clouds."

— Schedule of funding for the Goddard Space Flight Center for FY 1964 and previous years was released:

Luc Secretan and Francois V. Dossin with their special instrument that photographed the eclipse of the sun at Pleasant Pond, Me., July 20, 1963.

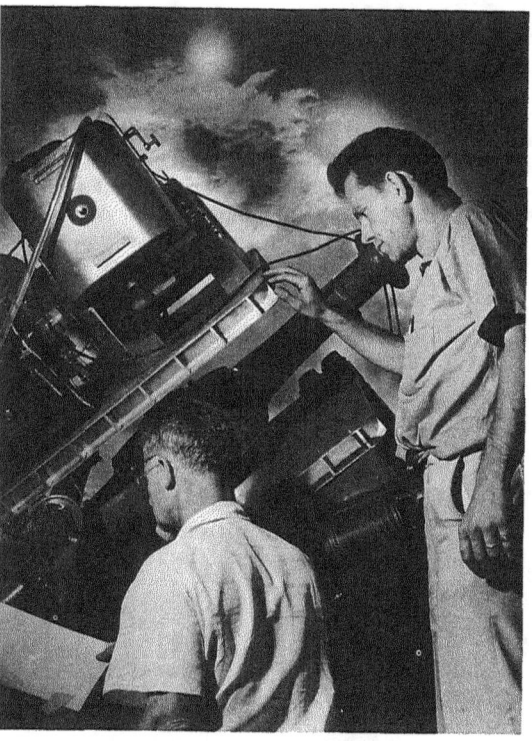

	Inception through FY 1963	FY 1964 Program
	(thousands of dollars)	
Construction of facilities:		
On-site:		
Buildings	$35,123.5	
Equipment	25,614.3	
Total on-site authorized	60,737.8	$20,932.0
Portion Completed and occupied	37,091.0	
Off-site:		
All sites	44,610.2	
Transfer to NASA from Vanguard Project	13,000.0	
Total off-site authorized	57,610.2	111,600.0
Portion completed and operational (includes Vanguard transfer)	40,821.2	
Research, development, and operations:		
Direct allotments		492,286.0

1963 Continued

Anticipated reimbursements	56,350.5
Total R, D, and O	548,636.5
Total GSFC FY 1964 Program	681,168.5

On-board personnel—8/2/63, 3,629 (including 332 summer employees).

— The second San Marco suborbital flight unit was successfully launched from Wallops Island to an altitude of 155 nautical miles and a surface distance of 560 nautical miles. GSFC was assisting in testing this Italian space project.

August 4: First public demonstration of communications exchange via synchronous satellite, when two U.S. wire services and Nigerian newsmen exchanged news stories via *Syncom II* communications satellite, hovering 22,823 miles over Western Africa. Photographs of President Kennedy and Nigerian Governor General Dr. Nnamdi Azikiwe also were exchanged. Transmissions were made from NASA station at Lakehurst, N.J., and USNS *Kingsport* communications ship in Lagos Harbor, Nigeria.

August 5: NASA announced *Syncom II* communications satellite, now drifting westward over Atlantic Ocean at 22,800-mile altitude, would be stopped when it reached desired position at 55° west longitude. At this location *Syncom II* would be lowered into precise synchronous orbit, so it would appear to trace elongated figure-eight pattern along 55° meridian to points 33° north and south to the equator.

August 6: Tracking and data acquisition operations ceased for *OSO I* (Orbiting Solar Observatory), launched March 7, 1962.

August 8: With launching of Nike-Cajun sounding rocket from Kronogård Range, Sweden and U.S. completed series of sounding rocket experiments to study noctilucent clouds near Arctic Circle. Sponsored by NASA and Swedish Committee for Space Research, program included launchings of Arcas rockets with payloads to measure winds and Nike-Cajun rockets with payloads to make direct cloud samplings during 1961 and 1962. Four Nike-Cajun rockets with rocket grenade payloads were successfully launched during summer 1963, these experiments measuring upper atmosphere temperatures, wind pressure and density, and measuring changes in size of artificial cloud particles created by smoke puffs from the payloads. Experimenters were scientists from Institute of Meteorology, University of Stockholm; GSFC had responsibility for U.S. coordination in the project.

August 9: Voice and teletype messages exchanged via *Syncom II* communications satellite between ground station at Paso Robles, Calif., and communications ship,

On August 4 this picture of Nigerian Governor General Nnamdi Azikiwe was transmitted via *Syncom II* satellite.

APPENDIX D

Kingsport, in Lagos Harbor, Nigeria. The test spanned 7,700 miles, greatest surface distance ever spanned between two points on earth via a communications satellite.

— The *Explorer XIV* encoder hung up in a 4-channel (8, 9, 10, and 11) mode of operations. It had completed 310 days of operation with only 15 days of major malfunction.

— The transition of the *Relay I* spacecraft into 100 percent sunlight occurred. Some anomalies were observed but no significant difficulties occurred.

August 11: Tiros VI and VII meteorological satellites observed Hurricane Arlene approximately 600 miles northeast of Bermuda, Typhoon Bess approximately 100 miles west of Japan, and Typhoon Carmen approximately 500 miles east of the Philippine Islands.

August 12: The Program Support Division was established in the Office of Administration. Functions of the Financial Management Analysis Branch were transferred to the Procurement Division. The Business Data Procurement Branch was reconstructed and redesignated as the Business Data Branch. The Reports and Statistics Branch and the Systems Review Branch were established in the Financial Management Division.

August 15: Syncom II communications satellite was successfully maneuvered into synchronous position 55° west longitude, over Brazil and South Atlantic Ocean. The maneuvers were directed by engineers at GSFC, and actual command was executed from ground station at Lakehurst, N.J. *Syncom II* was now stationed about 22,300-mile altitude and traveling at speed of about 6,800 mph, matching earth's rotation speed of 1,040 mph at the equator to keep it on station. It was hovering in figure-eight pattern 33° north and south of equator. NASA Administrator James E. Webb called completion of the positioning maneuvers the culmination of "one of the outstanding feats in the history of space flight."

August 21: The GSFC Sounding Rocket Branch reported that 61 rockets had been fired this year to date. Of this total eight were in the Meteorology Program.

August 23: U.S.-Canada agreement for cooperative testing of communications satellites launched by NASA was announced by NASA and Canada's Department of Transport. Each cooperating national agency would provide a ground station to receive and transmit television and multichannel telephone and telegraphic signals via communications satellites, according to Memorandum of Understanding signed in April and made operative by exchange of notes today.

— *Syncom II* communications satellite relayed its first live telephone conversations, a transmission between President John F. Kennedy and Nigerian Prime Minister Sir Abubaker Tafawa Balewa and other messages between U.S., Nigerian, and U.N. officials. Arranged by USIA, the demonstration program originated from the White House and Voice of America studios in Washington and from ground station aboard USNS *Kingsport* in Lagos Harbor, Nigeria.

September 3: The Alaska Data Acquisition Facility near Anchorage began limited operations by interrogating *Tiros VI* for two orbits. Performance was satisfactory except for some interference on 235 Mc from the command transmission.

— GSFC announced Belgian astrophysicist Dr. Francois V. Dossin, working at GSFC on National Academy of Sciences fellowship, discovered faint comet about 5° from sun during July 20 solar eclipse. Dr. Dossin made seven camera-plate exposures of comet from Pleasant Pond, Me., during 60 seconds of total eclipse. He used blue-green filter to bring out the light of carbon molecules in the comet. Microscopic examination of developed plates showed a diffuse image emitting the light of molecular carbon.

September 4: An Aerobee sounding rocket containing the low energy cosmic ray heavy nuclei experiment was launched from Fort Churchill. Experiment and performance reported good.

— Aerobee 150 sounding rocket launched from Ft. Churchill, Canada, with nuclear emulsion payload to study very-low-energy cosmic ray heavy nuclei. Payload reached 150-mile altitude, was recovered from an inland lake approximately 90 miles from

1963 Continued

launch site. Instrumentation and nuclear emulsions were in excellent condition.

— Construction work on the penthouse on Building 3 to house the AT&T equipment was begun.

— The former Fields and Particles Branch, Space Sciences Division, was reorganized and retitled the Energetic Particles Branch. A new Branch, the Fields and Plasmas Branch, was established in the Space Sciences Division.

September 5: *Syncom II* communications satellite achieved perfect synchronous orbit.

September 10: The German Transportable Ground Station began to receive *Relay I* pointing data and operational traffic.

September 13: *Syncom I* and *Relay I* linked Rio de Janeiro and Lagos, Nigeria, in 20-minute voice conversation, first operation employing both communications satellites in single communications circuit and world's first three-continent telephone conversation. Signal began from USNS *Kingsport* in Lagos Harbor, then to *Syncom I*, which sent it to Lakehurst, N.J., ground station, then to *Relay I* overhead which sent it to Rio de Janeiro ground station. GSFC engineers monitoring the conversation declared quality of transmission to be good.

September 14: U.S.-Scandinavia approval of Memorandum of Understanding for testing of NASA-launched experimental communications satellites was announced by NASA and Scandinavian Committee for Satellite Telecommunication. Vice President Lyndon B. Johnson, on official tour of Scandinavia, received in Copenhagen the Danish Government's note of approval, making the Memorandum effective; Norway had approved Memorandum in note dated September 11 and Sweden, in note dated July 25. Under agreement, Scandinavian Committee would provide ground station to receive multichannel telephone or telegraph signals transmitted from U.S. via orbiting communications satellite.

September 15: Third command and data acquisition station in Tiros meteorological satellite CDA system became operational, the Fairbanks, Alaska, station joining those at Wallops Island, Va., and Pt. Mugu, Calif. CDA stations receive cloud-cover photographs and other data from orbiting Tiros satellites, and relay them to Weather Bureau's National Weather Satellite Center, Suitland, Md., for analysis.

September 17: Opening of U.N. General Assembly transmitted via *Relay I* and *Syncom II* to Europe and Africa.

September 18: First anniversary of orbiting of *Tiros VI* meteorological satellite, its year-long operational lifetime setting new record for weather satellites. On July 31, 1963, *Tiros VI* discovered first hurricane (Arlene) of 1963 season in tropical Atlantic; altogether, *Tiros VI* photographed two hurricanes in Atlantic, two tropical storms in eastern Pacific, eight typhoons in central and western Pacific, as well as sandstorms in Saudi Arabia and ice conditions in southern and northern hemispheres. Along with *Tiros V* it supported Mercury space flights of Astronauts Schirra and Cooper. National Weather Satellite Center issued about 600 weather advisories around the world based on some of the 63,000 cloud-cover pictures from *Tiros VI*.

— Goddard Space Flight Center selected two companies for negotiation of contracts pertaining to Nimbus weather satellite. $252,000 contract to General Electric Company called for development of operating procedures for Nimbus control center as well as training of personnel to operate the center. $165,000 contract to RCA Electron Tube and Semiconductor Division required contractor to furnish solar cells for Nimbus satellites and Nimbus operational system.

September 19: The following offices were established in the GSFC Tracking and Data Systems Directorate: the Systems Analysis Office, the Manned Flight Support Office, The Project Resources Office, and the Directorate Support Office. At the Division level, parts of the Tracking Systems Division and the Space Data Acquisition Division were merged and named Advanced Development Division. The Data Systems Division was expanded with the addition of telemetry and data processing from the Space Data Acquisition Division and the Operations and Support Division and retitled the Network Engineering and Operations Division.

APPENDIX D

The former Manned Space Flight Support Division was renamed Manned Flight Operations Division.

— *Syncom II* 24-hour communications satellite used to relay oceanographic data from research vessel *Geronimo* in Gulf of Guinea off Africa to National Oceanographic Data Center in Washington, which compared the data with its records and sent back to the *Geronimo* the deviations to correct errors. Demonstration via *Syncom II* was performed to determine practicability of providing research ships quickly with information to correct errors. Line of transmission: from *Geronimo* to *Kingsport* in Lagos Harbor, to *Syncom II* some 22,300 miles above Atlantic Ocean, to ground station at Lakehurst, N.J., along ground lines to NODC, and return.

September 20: President John F. Kennedy's speech to the United Nations General Assembly was transmitted to the USNS *Kingsport* via Syncom for further broadcast over the Voice of America network in Nigeria.

September 21: *Tiros VII* meteorological satellite discovered Hurricane Debra, fourth hurricane of season, headed north in Atlantic southeast of Bermuda.

September 23: *Syncom II* communications satellite relayed transmission of speech and teletype between Fort Dix, N.J., and moving USNS *Kingsport* about 40 miles west of Lagos, Nigeria. This was first such transmission via a communications satellite to a moving ship at sea. This was first in series of experiments designed to test shipboard equipment and reception in fringe areas.

September 25: Two similar experiments (one built by U.S., the other by Japan), were launched aboard an Aerobee 150 sounding rocket from Wallops Island. The purpose of the experiments was to make simultaneous measurements in the ionosphere by different methods and then to compare the data obtained. Instruments were supplied by GSFC and the Radio Research Laboratory, Tokyo, Japan. The Japanese scientists' radio-frequency resonance probe was designed to make it possible to measure electron density and temperature simultaneously with one instrument and to process the data faster. The Aerobee reached a peak altitude of 139 statute miles; it impacted in the Atlantic Ocean 80 miles from launch site after 8 minutes of flight. Preliminary data indicated that the experiment succeeded.

September 26: Hurricane Edith was observed between Hispaniola and Puerto Rico by *Tiros VII*. Hurricane Flora was also picked up by this satellite.

— Operations with the *Relay I* satellite continued with successful completion of all scheduled experiments. The spacecraft operations as of this date, for 2,227 orbit revolutions, were: 1,107 wideband experiments; 519 narrowband experiments; 99 demonstrations (TV and narrowband). The transponder had been operated for 225 hours over a period of 560 operations.

— NASA announced first television experiments via *Syncom II* communications satellite had been conducted. Test pattern signals sent Sept. 23 were followed by TV pictures Sept. 24 and 25; because of band-width limitations, no audio was sent. Officials said transmissions were of good quality. Transmissions originated at Fort Dix, N.J., ground station, were sent to *Syncom II* 22,300 miles above the earth, and retransmitted to AT&T ground station at Andover, Me.

September 27: *Explorer XIV* satellite progress report indicated no usable scientific data had been obtained from the scientific satellite since mid-August. In its 10 months of operation since launch into highly elliptical orbit Oct. 2, 1962, *Explorer XIV* sent back more than 6,500 hours of data from the six onboard scientific experiments to chart boundaries of earth's magnetosphere, measure particle population and energies of electrons and protons, and determine how magnetic fields influence these particles. There had been 3,700 hours of data processed through computers and scientific analysis was continuing.

September 28: Aerobee 150A sounding rocket launched from NASA Wallops Station with U.S.-Japanese experiment to measure electron temperatures and densities in the ionosphere by two different methods; Langmuir probe, supplied by NASA Goddard Space Flight Center, and radio-frequency

1963 Continued

resonance probe, developed by Radio Research Laboratory, Tokyo. A 185-pound payload reached 141-mile altitude and transmitted approximately 8 minutes of telemetry before impacting in Atlantic Ocean about 71 miles from launch site. Data obtained from the daytime experiment were compared with data from similar experiment conducted at night, 3 days earlier.

September 29: An Argo D–4 was launched. Based on plotting board information, it achieved an altitude of 1,038 kilometers. Telemetry signals were received for 12 minutes and all experiments functioned normally.

— At the end of the first year of operation of *Alouette I,* all four experiments were performing very well and continued to provide good data. No problems had been encountered in commanding the satellite or in recording of the telemetry transmissions.

October 1: NASA marked its fifth anniversary, with a salute to 23 individuals whose outstanding personal efforts have contributed significantly to the nation's civilian space program. Among cash awards was a $1,500 award to Jesse M. Madley and Xopher W. Mayer at GSFC for the invention of a structural spacer.

October 1 to 3: Youth Science Congress, sponsored by NASA and the National Science Teachers Association, was held at GSFC. Feature event was presentation of 25 award-winning research papers of high school students from Washington, D.C., Maryland, Delaware, Pennsylvania, and New Jersey.

October 8: Explorer XIV energetic particles satellite had ceased useful transmission after almost 10 months of successful operation. Scientists at GSFC said trouble began in August when the satellite's transmitter failed to modulate—translate instrument signals into telemetry code—properly. Intermittent modulation had occurred since then, but little useful data had been received. The satellite signal was still useful for position reference. Some 6,500 hours of data were received from the satellite. While not all the data had been analyzed, Dr. L. Cahill, Univ. of New Hampshire, said a number of new insights had already emerged, among them being: earth's magnetosphere, as shown by mapping charged particles, flared away from the earth in an ogival—pointed arch—shape; confirmation that the vector

The Rosman, N.C., tracking facility.

APPENDIX D

magnetic field changes gently from a dipole configuration to a radial field at increasing distance on the night side of the earth near the equatorial plane; and further evidence probably supporting *Explorer VI*'s finding of a ring current flow on the night side of the earth.

October 9: As of this date, the *Relay I* satellite had continued with successful completion of a majority of the scheduled experiments. Its operations covered 2,334 orbit revolutions with 1,132 wideband experiments; 54 narrowband experiments; 99 demonstrations (TV and narrowband). The transponder had been operated for 234 hours over a period of 582 operations.

October 11: Syncom II operation in orbit remained satisfactory. The N_2 system and H_2O_2 system pressures remained the same. There was no perceptible change in these parameters for the last several weeks. The satellite spin speed continued to decrease.

—*Tiros VI* acquired its last usable pictures, after 338 days of useful life.

October 17: The *Relay I* spacecraft operations as of Oct. 17, for 2,389 orbit revolutions, were as follows: 1,151 wideband experiments; 553 narrowband experiments; 100 demonstrations (TV and narrowband). The transponder had been operated for 239 hours over a period of 593 operations.

October 22: GSFC began negotiations with Republic Aviation Corp. for Phase I contract for Advanced Orbiting Solar Observatory (AOSO). AOSO would be launched into a 300-mile near-polar orbit for observations of x-rays, gamma rays, and ultraviolet emissions of the sun. Phase I calls for one-year development of systems engineering and detailed design of the satellite.

October 26: The Rosman, N.C., tracking and data acquisition facility was dedicated. A key station in NASA's Satellite Tracking and Data Acquisition Network (STADAN), the 85-foot-diameter parabolic antenna at Rosman would be used to track and receive the large flow of telemetered data from the large orbiting observatories and would relay the data to GSFC for processing and analysis.

October 30: Symposium on the Physics of Solar Flares was held at GSFC, sponsored by NASA and the American Astronomical Society.

— *Syncom II* operation in orbit remained satisfactory. The orbital elements were:

Epoch 22 October 1963 _ 0200.00 hours UT
Semi-major axis _____ 26,204.11 miles
Eccentricity _____ 0.00026
Inclination _____ 32.993°
R.A. of ascending node _ 316.603°
Height of perigee _____ 22,233.97 miles
Height of apogee _____ 22,247.58 miles
Anomalistic period _____ 1,436.3957 min.

All telemetry indicated that the 10 instruments aboard the 128-pound satellite were functioning normally.

October 31: A second Aerobee-Hi research rocket in the NASA-French joint program investigating propagation of very-low-frequency waves in the ionosphere was launched from Wallops Station. The 193-pound payload went to an altitude of 115 miles and yielded 7 minutes of telemetry data before impact. The first experiment in this series was conducted on October 17, 1963.

November 1: GSFC awarded contract to Yale Univ. to design and develop a worldwide radio monitoring network for study of planet Jupiter. Four stations would comprise the global network, located at approximately every 90° longitude around the earth—one at GSFC in Greenbelt, Md., and the other three at U.S. satellite tracking stations at Hartesbeesthoek, South Africa; Carnarvon, Australia; and South Point, Hawaii. Primary duty of the stations would be to maintain a 24-hour radio monitor of the mysterious low-frequency radio noises sporadically emitted from the planet. The data should provide information on Jupiter's magnetosphere, the interplanetary medium, and the earth's ionosphere.

November 7: French VLF project: The second Aerobee in the French VLF program launched from Wallops Island Oct. 31, was successful. The experimenter, Dr. Owen Story, indicated in a preliminary appraisal that the data were of excellent quality. The monitoring circuit operated as anticipated. Both firings occurred during periods of ionospheric disturbances due to solar

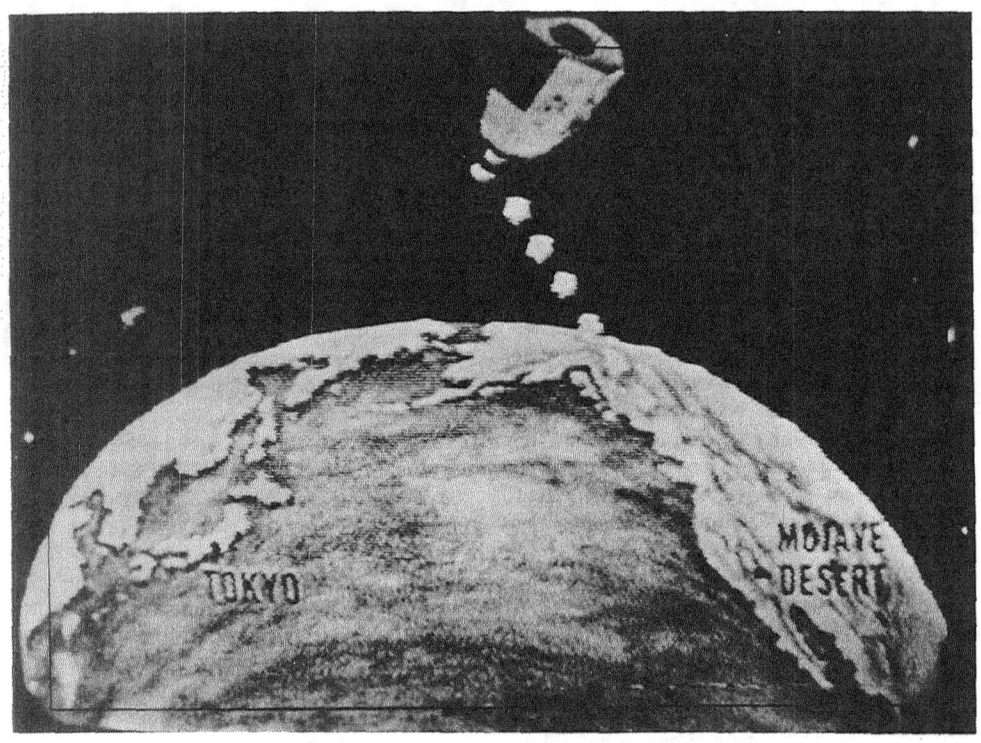

Artist's conception of Relay mission received in Japan.

1963 Continued

flare activity. This was expected to complicate reduction of the data.

November 8: West Germany joined the list of nations participating in satellite communications with the opening of its narrowband station at Raisting, near Munich. A wideband station to permit television transmission was under construction. Raisting became the seventh station in the satellite communications network. Other narrowband stations were at Nutley, N.J., Rio de Janeiro, Brazil; Fucino, Italy. Wideband stations were at Andover, Me.; Goonhilly Downs, U.K.; and Pleumeur-Bodou, France.

November 21: GSFC announced that with a $2 million contract, under final negotiation, Sperry Rand Corp.'s Univac Div. would deliver 11 Model 1218 computer systems to manned space flight tracking stations for operation by July 1964. These computers would automatically summarize telemetry from the spacecraft, provide summaries for display in the Mission Control Center so that the controllers can select and examine certain data on a "real-time" basis, and prepare the telemetry data for final processing in the more elaborate computers at GSFC and MSC. During the Mercury program, controllers at the tracking stations had to select data manually. The computers would be located at Cape Canaveral; Bermuda; Canary Islands; Corpus Christi, Tex.; Guaymas, Mexico; Carnarvon, Australia; Wallops Island, Va.; Greenbelt, Md.; and on two ships used in manned space flight tracking, the *Rose Knot Victory* and the *Coastal Sentry Quebec*.

—First rocket to be launched from India was achieved as the result of the coordinated efforts of France, India, and the U.S. The Nike-Apache was launched from

APPENDIX D

Thumba, the site near the southern tip of India that would become an international rocket launching facility. The Thumba site is located at the earth's magnetic equator, making possible the investigation of important phenomena which could be studied to a greater advantage from this region.

November 22: The first live American television transmission across the Pacific by means of *Relay I* communications satellite was received clearly in Tokyo. Pictures transmitted by the Mojave ground station in California and received at the new Space Communications Laboratory in Ibaraki Prefecture, north of Tokyo, were clear and distinct. The sound transmission was excellent. The transmission was received live from 5:16 a.m. to 5:46 a.m. Viewers in Tokyo saw and heard taped messages from Ryuji Takeuchi, Japanese Ambassador to Washington, and James E. Webb, Administrator of the National Aeronautics and Space Administration. A message of greeting from President John F. Kennedy to the Japanese people, which was to have been the highlight of the program, was deleted when news of the President's death was received shortly before the transmission. In place of the taped 2½-minute appearance of the President, viewers saw brief panoramic views of the Mojave transmitting station and the surrounding desert area. ABC and NBC shared in producing the program.

—A solar array characteristics test was run on the orbiting *Syncom II* synchronous-orbit communications satellite. The test found a power loss of 20 percent from the effects of solar radiation on the solar cells during 4 months in orbit. The test confirmed the desirability of changing the next Syncom satellite.

November 25: President John F. Kennedy was buried in Arlington National Cemetery in a state funeral attended by the largest gathering of foreign dignitaries ever to visit Washington. *Relay I* communications satellite enabled all of Europe, including the U.S.S.R., to view events of the tragic weekend and the funeral ceremonies. The satellite also provided transmission across the Pacific to Japan, where an estimated 95 million persons viewed the ceremonies.

November 27: Explorer XVIII, first of a series of Interplanetary Monitoring Platforms (IMP) to map magnetic fields of space and the effects of solar winds and cosmic rays on the earth's atmosphere, was launched.

—GSFC's Field Projects Branch launched their first Atlas-Centaur. The booster, a new experimental hydrogen fuel 2d-stage rocket for deep space work, was launched from pad 36-A, Cape Kennedy, atop an Atlas 1st stage.

November 29: GSFC announced negotiations with Northrop Electronics for design and construction of a test device to simulate the launch phase of space flight. Final negotiations are expected to lead to a contract estimated at $1,800,000. Called a Launch Phase Simulator (LPS), the device would test unmanned space-flight units and components under the separate or combined conditions of acceleration, vibration, noise, and vacuum. It is designed to duplicate, as nearly as possible, the environmental conditions typical of current launch vehicles.

December 2: House Joint Resolution 787 was submitted to Congress providing for the erection of a memorial statue to the late Dr. Robert H. Goddard, the father of American rocketry.

—Nike-Cajun was launched from Wallops for the "falling sphere" experiment and consisted of ejecting three balloons which were tracked by radar. Data correlation between these two experiments was studied.

December 9: 350 representatives of 55 aerospace firms were briefed on GSFC's requirements for a new "Unified S-Band" method for tracking and communications for Apollo lunar missions.

December 11: On Dec. 11 and Dec. 13 the GSFC Data Operations Branch supported SA-5 network simulations. The nominal SA-5 launch and orbital phase were simulated in real-time using data tapes from the sites. These tapes were generated using the SA-5 nominal insertion conditions. The real-time computing program which reflects the operational Apollo launch and near-earth orbit determination program worked well. This was the first time the three networks, SAO, STADAN, and Manned Flight Networks, were simultaneously controlled from

View of the Space Communications Laboratory, Ibaraki, Japan.

1963 Continued

Goddard. The first simulation served as a training session for the development of standard operating procedures to control all three networks. The second simulation ran very smoothly.

December 18: The final static inflation test with *Echo II* balloon No. 16 was successfully conducted by NASA at Lakehurst, N.J. The balloon burst at a nominal skin stress level of 23,000 psi. Visual inspection of the inflated balloon indicated an improved balloon surface. This balloon was fabricated using the GSFC-developed gore cutting and sealing technique as well as preshrunk material. The balloon was pressurized to 3,400 psi nominal skin stress; relaxed to approximately 500 psi; pressurized to 7,400 psi, relaxed and then inflated to the burst pressure of 23,000 psi nominal skin stress. RF measurements at L-band and C-band were obtained for each of the test pressure levels.

—*Relay I* operations as of this date for 2,800 orbit revolutions were: 1,330 wideband experiments; 720 narrowband experiments; 157 demonstrations (TV and narrowband). The transponder had been operated for 288

Antenna at the Space Communications Laboratory, Ibaraki, Japan.

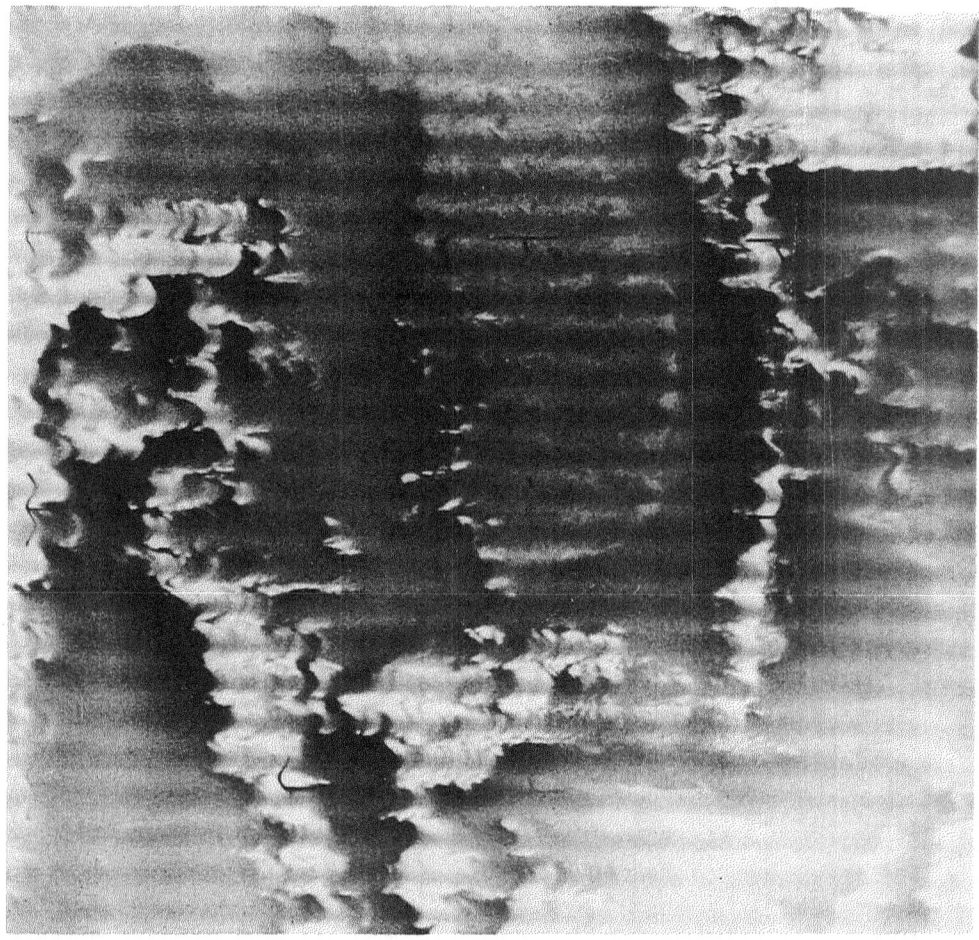

Tiros VIII: The experimental Automatic Picture Transmission camera beamed this photo from more than 400 miles in space to GSFC at 12:30 EST, December 21, 1963. Florida and the Gulf of Mexico are shown.

hours over a period of 720 operations. Since Oct. 1, 1963, 80 hours of radiation data had been taken by *Relay I*.

December 19: All major components of the Univac 490 communications switching system had been installed at the GSFC communications complex. Items remaining to be installed consisted of the Communications Line Terminations (CLT's) and two tape decks for the second system and system transfer switches.

December 21: Tiros VIII was launched at AMR. The satellite contained the first Automatic Picture Transmission camera, permitting rather "inexpensive" readouts at ground receiving stations. The APT system was developed for Nimbus, the advanced meteorological weather satellite.

December 30: Aviation Week and Space Technology magazine gave "Laurels for 1963 to Harry Goett, Jack Townsend, and Bob Gray of NASA's Goddard Space Flight Center for

1963 Continued

their excellent record (100 percent on eight launched in 1963) of successful satellite launchings and operations. This is the second consecutive year that all of Goddard's satellites were successful...."

During December: The Italian Space Commission advised that it planned to ship the San Marco flight spacecraft to Goddard for the following tests: Minitrack compatibility tests at Blossom Point and dynamic balance procedures at the Center's test facilities. The spacecraft was then to be shipped to the Langley Center for mating checks with the vehicle.

—*Ariel I* radiated normal modulation for a 3-month period, Aug. 14 through Nov. 17. During this time, the percentage of sunlight varied between 75 and 63 percent. Good data were obtained for housekeeping and the electron temperature experiment. From Nov. 17, 1963, through Dec. 6, 1963, the percentage of sunlight was above 76 percent and abnormal modulation (an intermittent 312-cps signal) prevailed.

Appendix E

Reports of Procurement Actions 1960-1963

SUMMARY REPORT OF PROCUREMENT ACTIONS

For January 1, 1960, thru June 30, 1960

FROM: Goddard Space Flight Center*

TO: NASA Headquarters

ACTIONS OVER $2500	NUMBER AND DOLLAR VALUE OF CONTRACT TRANSACTIONS							
	Govt. Agency		Small business		Big business		Educational et al.	
	No.	$ Value	No.	$ value	No.	$ value	No.	$ value
1. RESEARCH & DEVELOPMENT								
a. Negotiated			50	1,502,222	69	18,675,289	2	920,787
b. Interdepartmental	102	29,067,246						
Total	102	29,067,246	50	1,502,222	69	18,675,289	2	920,787
2. CONSTRUCTION								
a. Advertised			7	482,618	4	1,558,743		
b. Negotiated			11	199,950	4	94,551		
c. Interdepartmental	9	2,103,340						
Total	9	2,103,340	18	382,568	8	1,653,294	-	-
3. SUPPLY								
a. Advertised			9	202,392	7	54,839		
b. Negotiated			17	295,136	20	473,460		
c. Interdepartmental	67	4,585,142						
d. GSA schedule			4	19,486	11	60,844		
Total	67	4,585,142	30	517,014	38	589,143		
4. TOTAL ACTIONS OVER $2500	178	36,147,504	94	2,401,804	115	20,917,726	2	920,787
5. ALL ACTIONS NOT EXCEEDING $2500	75	40,311	856	343,607	450	259,956		

NASA Form 274 rev. (Jan. 1959)

*This Report includes Space Task Group, Langley

VENTURE INTO SPACE

YEARLY REPORT OF PROCUREMENT ACTIONS

NASA Installation: GODDARD SPACE FLIGHT CENTER   FISCAL YEAR 1961

PART A: PROCUREMENT ACTIONS BY NASA APPROPRIATION

Category (a)	Total No. (b)	Total $ Value (c)	Research & Development No. (d)	Research & Development $ Value (e)	Construction & Equipment No. (f)	Construction & Equipment $ Value (g)	Salaries & Expenses No. (h)	Salaries & Expenses $ Value (i)
1. TOTAL	10723	188,420,999	10426	165,376,920	248	21,921,420	51	1,122,659
2. INTERGOVERNMENTAL	613	42,059,736	579	36,010,924	27	6,021,096	7	27,710
3. NONPROFIT INSTITUTION OR ORGANIZATION	176	2,697,931	173	2,022,181	2	675,000	1	750
4. SMALL BUSINESS - TOTAL	6482	14,466,277	6401	11,112,542	73	3,297,826	8	55,909
a. Advertised	594	1,735,627	554	581,835	40	1,153,792	-	-
b. Negotiated Competitive	2432	5,442,402	2417	3,986,711	15	1,630,696	-	-
c. Negotiated Non-Competitive	3369	7,158,330	3271	5,144,613	18	512,808	8	55,909
d. Government Schedule	159	129,918	159	129,918	-	-	-	-
5. LARGE BUSINESS - TOTAL	3432	129,197,061	3273	116,231,723	146	11,927,498	33	1,038,290
a. Advertised	169	5,591,609	77	399,565	92	5,192,044	-	-
b. Negotiated Competitive	887	64,284,021	875	61,988,687	12	2,295,334	-	-
c. Negotiated Non-Competitive	1889	57,942,090	1843	53,427,951	40	4,448,034	6	76,105
d. Government Schedule	507	1,379,341	559	415,070	2	2,086	27	962,185

PART B: NEGOTIATED PROCUREMENT ACTIONS

Negotiation Authority 10 U.S.C.	Number	$ Value	Negotiation Authority 10 U.S.C.	Number	$ Value
6. TOTAL	8681	137,524,774	2304(a) (9)		
2304(a) (1)	509	9,523,642	(10)	210	14,578,340
(2)			(11)	166	101,939,320
(3)	7738	3,735,543	(12)		
(4)	12	211,012	(13)		
(5)	14	2,436,764	(14)	29	4,725,153
(6)	3	375,000	(15)		
(7)			(16)		
(8)			(17)		

NASA FORM 508 (JUNE 1960)

APPENDIX E

YEARLY REPORT OF PROCUREMENT ACTIONS

NASA Installation __GODDARD SPACE FLIGHT CENTER__   FISCAL YEAR 1962

PART A: PROCUREMENT ACTIONS BY NASA APPROPRIATION

Category (a)	Total No. (b)	Total $ Value (c)	Research & Development No. (d)	Research & Development $ Value (e)	Construction & Equipment No. (f)	Construction & Equipment $ Value (g)	Salaries & Expenses No. (h)	Salaries & Expenses $ Value (i)
1. TOTAL	20717	209,292,154	20432	188,234,691	205	19,824,349	80	1,233,114
2. INTERGOVERNMENTAL	616	5,752,844	612	5,746,868	1	1,736	3	4,240
3. NONPROFIT INSTITUTION OR ORGANIZATION	188	13,209,413	183	12,456,011	2	744,980	3	8,422
4. SMALL BUSINESS - TOTAL	14436	23,299,070	14367	22,274,592	50	835,665	19	118,813
a. Advertised	252	3,194,931	232	2,909,887	17	244,332	3	40,712
b. Negotiated Competitive	3374	5,721,710	3362	5,645,711	11	75,201	1	798
c. Negotiated Non-Competitive	10370	13,676,111	10335	13,143,676	22	516,132	13	16,303
d. Government Schedule	440	636,318	438	575,318			2	61,000
5. LARGE BUSINESS - TOTAL	5477	167,100,827	5270	147,757,220	152	18,241,968	55	1,101,639
a. Advertised	517	7,039,954	428	4,626,221	88	2,410,733	1	3,000
b. Negotiated Competitive	1336	73,554,241	1308	66,829,131	17	6,401,550	11	323,560
c. Negotiated Non-Competitive	2874	82,281,111	2806	75,325,984	41	6,504,601	27	450,526
d. Government Schedule	750	4,225,521	728	975,884	6	2,925,084	16	324,553

PART B: NEGOTIATED PROCUREMENT ACTIONS

Negotiation Authority 10 U.S.C.	Number	$ Value	Negotiation Authority 10 U.S.C.	Number	$ Value
6. TOTAL	18,142	188,442,554	2304(a) (9)		
2304(a) (1)	736	15,934,895	(10)	853	42,056,080
(2)	12	625,380	(11)	235	103,911,638
(3)	16,192	7,683,152	(12)		
(4)	18	291,477	(13)		
(5)	59	8,897,255	(14)	16	4,144,396
(6)	20	4,895,388	(15)		
(7)			(16)		
(8)			(17)	17	2,893

NASA FORM 508 (JUNE 1960)

VENTURE INTO SPACE

CORRECTED COPY

QUARTERLY REPORT OF PROCUREMENT ACTIONS *(Use reverse for remarks)*		FOR THE QUARTER ENDING July thru June FY '63		INCLUDES NASA FORM 507	
				REPORTS NOS. 63-1	THRU. 63-1347
TO: Procurement and Supply Division Headquarters, NASA		FROM: *(NASA Installation)* Goddard Space Flight Center			

CATEGORY (a)	TOTAL		RESEARCH DEVELOPMENT AND OPERATIONS		CONSTRUCTION OF FACILITIES	
	NO. (b)	$ VALUE (c)	NO. (d)	$ VALUE (e)	NO. (f)	$ VALUE (g)
PART A - PROCUREMENT ACTIONS BY NASA APPROPRIATION						
1. TOTAL *(Lines 2 thru 7)*	30,477	303,506,335	30,333	286,272,935	144	17,233,400
2. INTRAGOVERNMENTAL	657	11,863,487	654	8,167,387	3	3,696,100
3. LARGE BUSINESS - TOTAL	8048	232,265,627	7968	220,984,570	80	11,281,057
a. Advertised	354	10,977,133	318	3,676,747	36	7,300,386
b. Negotiated Competitive	2050	116,527,557	2031	116,077,758	19	449,799
c. Negotiated Noncompetitive	5644	104,760,937	5619	101,230,065	25	3,530,872
4. SMALL BUSINESS - TOTAL	21,115	41,261,657	21,059	39,602,072	56	1,659,585
a. Advertised	493	7,399,117	470	6,647,832	23	751,285
b. Negotiated Competitive	4359	6,694,960	4351	6,401,434	8	293,526
c. Negotiated Noncompetitive	16,263	27,167,580	16,238	26,552,806	25	614,774
5. UNIVERSITIES	568	11,567,934	566	11,494,276	2	73,658
6. OTHER NONPROFIT INSTITUTIONS	52	162,170	52	162,170		
7. OUTSIDE U.S. & POSSESSIONS	37	6,385,460	34	5,862,460	3	523,000
8. SMALL BUSINESS SET ASIDES - TOTAL *(included in line 4)*	50	1,298,922	50	985,780	2	313,142
a. Individual Set Asides	50	1,298,922	50	985,780	2	313,142
b. Class Set Asides						

NEGOTIATION AUTHORITY 10 U.S.C.	NO.	$ VALUE	NEGOTIATION AUTHORITY 10 U.S.C.	NO.	$ VALUE
PART B - NEGOTIATED PROCUREMENT ACTIONS					
9. TOTAL	28,973	273,266,598	2304(a) (9)		
2304(a) (1)	1257	12,856,686	(10)	2323	45,414,090
(2)	78	2,858,638	(11)	1396	166,536,433
(3)	21,494	9,737,867	(12)		
(4)	36	448,113	(13)		
(5)	567	11,193,496	(14)	21	4,669,876
(6)	43	6,735,026	(15)		
(7)			(16)		
(8)		—	(17)	1758	12,816,373

APPENDIX E

QUARTERLY REPORT OF PROCUREMENT ACTIONS (Use reverse for remarks)			FOR THE QUARTER ENDING 7/1/63 thru 6/30/64		INCLUDES NASA FORM 507 REPORTS NOS. 64-1 THRU 64-1703	
TO: Procurement and Supply Division Headquarters, NASA				FROM: (NASA Installation) Goddard Space Flight Center SS/SA/T&D/FS/IS		
CATEGORY (a)	TOTAL		RESEARCH DEVELOPMENT AND OPERATIONS		CONSTRUCTION OF FACILITIES	
	NO. (b)	$ VALUE (c)	NO. (d)	$ VALUE (e)	NO. (f)	$ VALUE (g)
1. TOTAL (Lines 2 thru 7)	32,922	370,142,643	28,767	306,687,437	212	59,400,417
2. INTRAGOVERNMENTAL	705	55,404,400	613	13,937,359	11	41,192,329
3. LARGE BUSINESS - TOTAL	9,963	243,168,380	9,055	230,394,362	110	11,875,768
a. Advertised	876	13,099,707	691	8,149,178	73	4,648,384
b. Negotiated Competitive	2,760	113,125,846	2,521	107,548,214	17	5,245,392
c. Negotiated Noncompetitive	6,327	116,942,827	5,843	114,696,970	20	1,981,992
4. SMALL BUSINESS - TOTAL	21,872	49,336,800	18,759	42,132,023	83	4,335,440
a. Advertised	1,116	10,516,818	816	7,185,488	40	2,365,770
b. Negotiated Competitive	5,298	14,320,914	4,316	11,586,770	18	1,238,550
c. Negotiated Noncompetitive	15,458	24,499,068	13,627	23,359,765	25	731,120
5. UNIVERSITIES	143	12,884,535	143	12,884,535	-	--
6. OTHER NONPROFIT INSTITUTIONS	186	497,058	150	473,648	2	10,920
7. OUTSIDE U.S. & POSSESSIONS	53	8,851,470	47	6,865,510	6	1,985,960
8. SMALL BUSINESS SET ASIDES - TOTAL (included in line 4)	477	6,254,181	249	2,779,236	27	2,313,031
a. Individual Set Asides	412	4,496,060	236	2,517,291	6	1,173,930
b. Class Set Asides	65	1,758,121	13	261,945	21	1,139,101
NEGOTIATION AUTHORITY 10 U.S.C.	NO.	$ VALUE	NEGOTIATION AUTHORITY 10 U.S.C.		NO.	$ VALUE
9. TOTAL	30,225	291,121,718	2304(a) (9)			
2304(a) (1)	259	2,649,618	(10)		2,766	48,112,526
(2)	7	100,098	(11)		3,534	180,551,372
(3)	22,059	7,952,478	(12)			
(4)	57	1,100,861	(13)		3	1,364,441
(5)	139	12,887,129	(14)			
(6)	72	8,848,980	(15)			
(7)			(16)			
(8)			(17)		1,329	27,554,215

259

Appendix F
Organization Charts

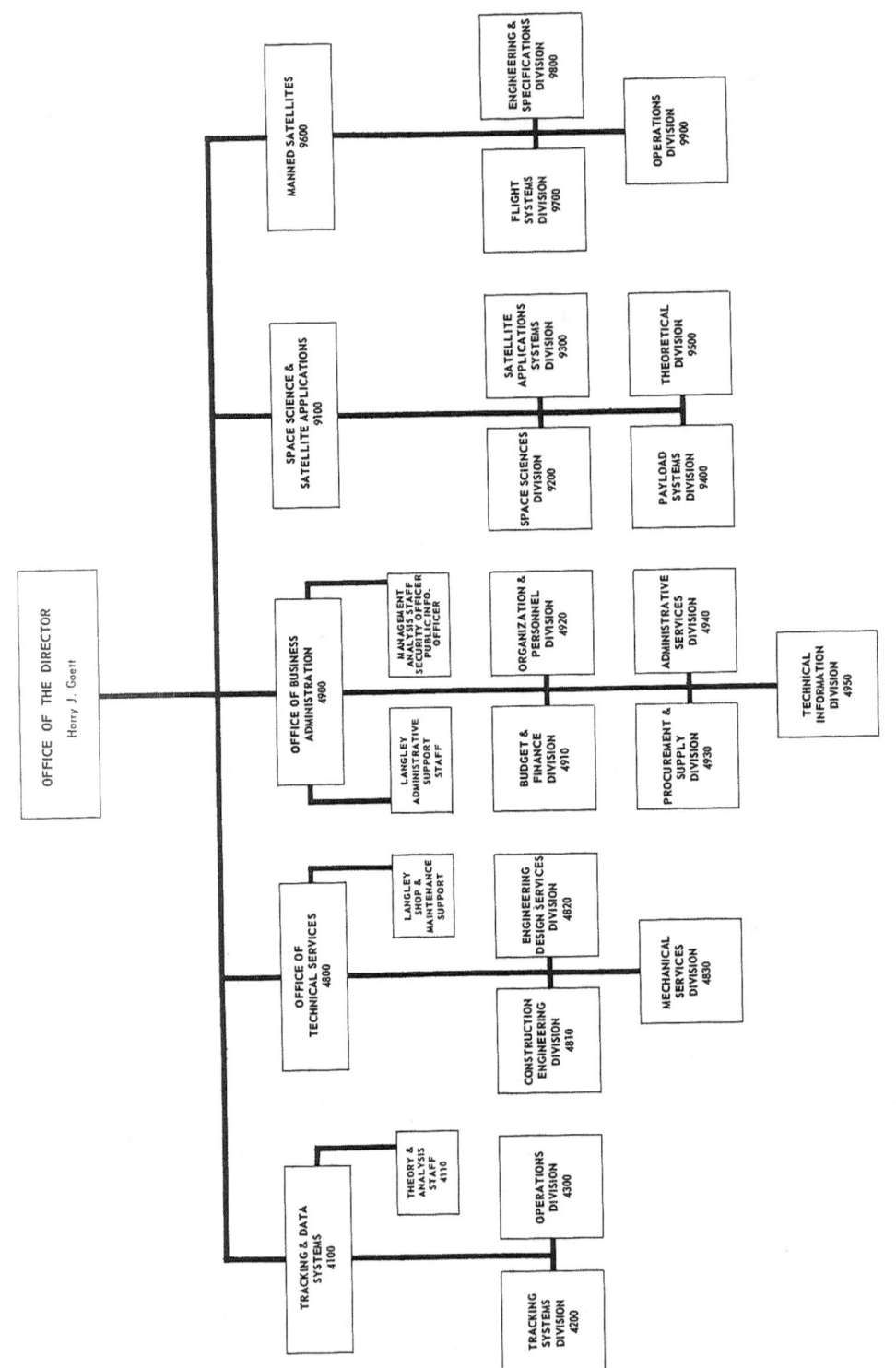

Organization chart, July 1959.

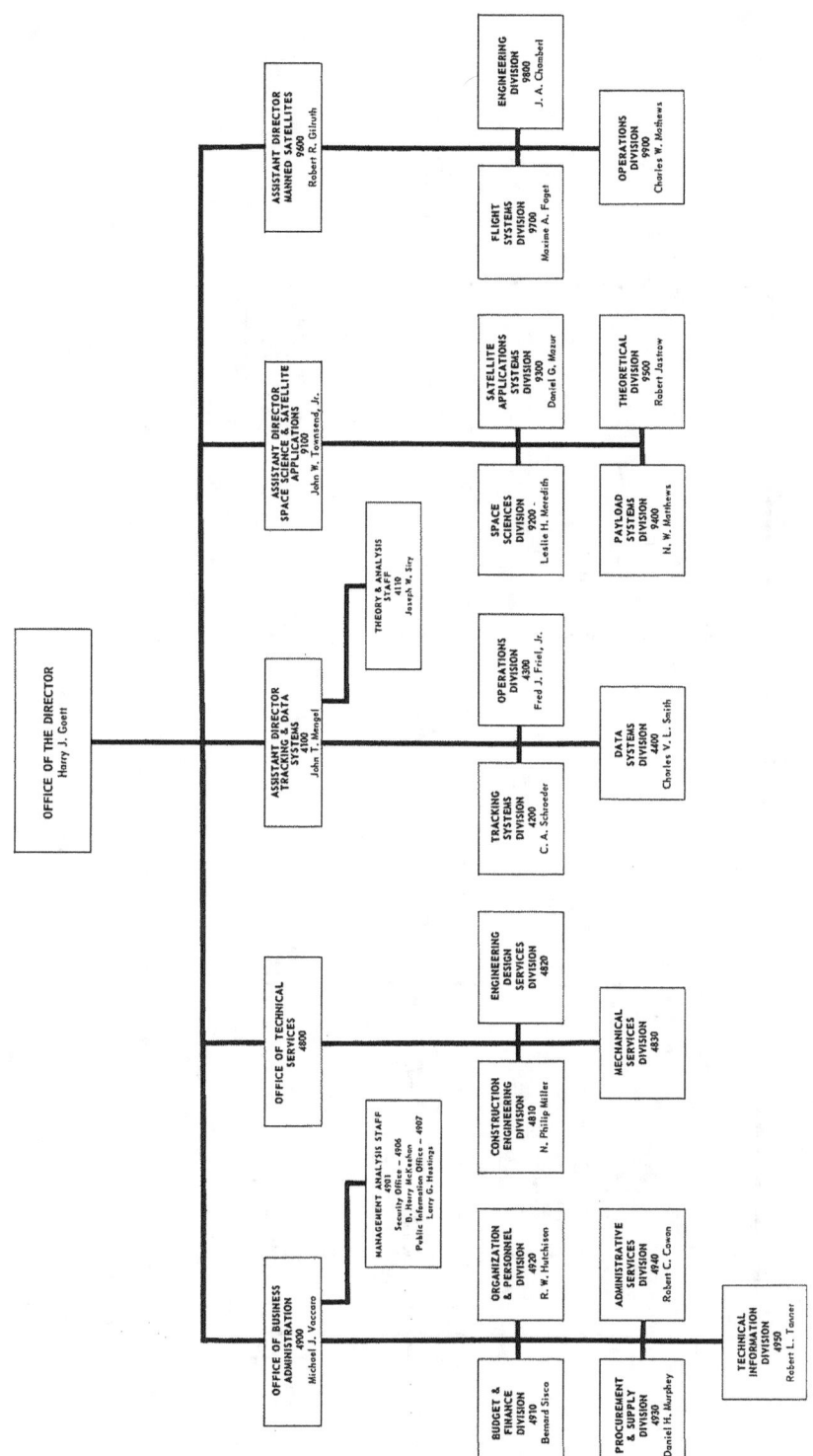

Organization chart, March 1960.

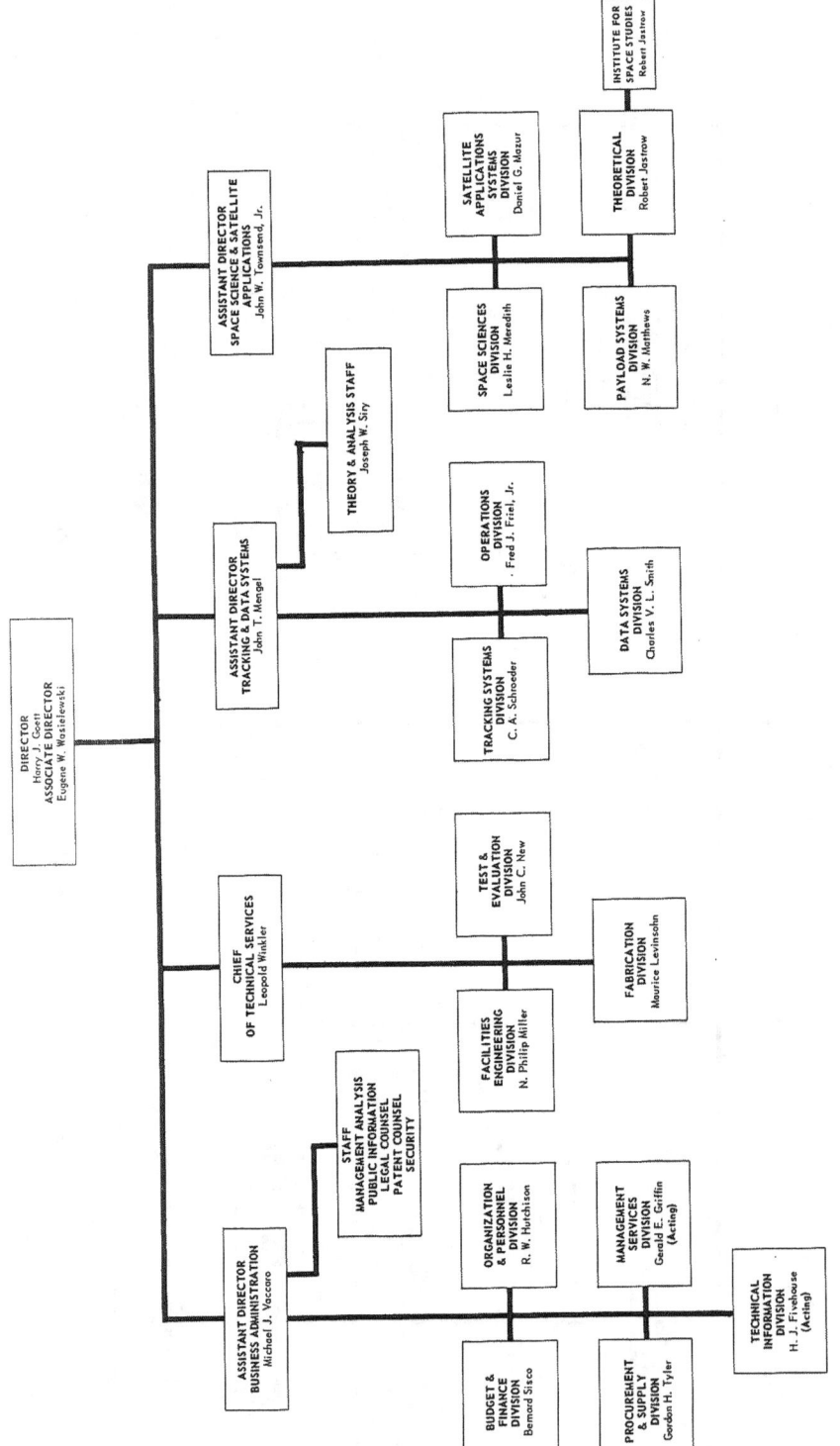

Organization chart, January 1961.

Organization chart, November 1962.

Appendix G

Scientific Exploration of Space and Its Challenge to Education*

Harry J. Goett
First Director, Goddard Space Flight Center

IT IS A PRIVILEGE to be with you in these halls which sparked the mind of the father of the Space Age, Dr. Robert H. Goddard. Those of us engaged in the space program have a very special regard for Dr. Goddard. We see in him the embodiment of the curious and far-seeing scholar who best exemplified the theme of this convocation—the partnership of engineering and science in progress. Two generations ago, well ahead of his time, he gave us the theory and tools with which to reach into the universe in our never ending quest for knowledge. Those who read his reports cannot help but being impressed by the fact that it was due to the unique combination of the scientist and engineer in a single individual that enabled Dr. Goddard to be as far ahead of his time as he was.

We are now on the threshold of the Space Age which will require the same combination of the vision and practical application which characterized Dr. Goddard's work. Just as some 500 years ago man ventured beyond the Mediterranean, leading to the discovery of the New World, so today, man is breaking his earth-bound shackles to venture into space. Aside from the technological advances to which we are witness today, we must expect possibly even greater changes to our political, social, and educational concepts.

Those earlier explorations extended the horizons of the times in a literal sense, but even more important, they opened up new possibilities and concepts. They forced the people out of their established patterns of thought and produced an intellectual ferment and interest in new ideas necessary for the scientific revolution and for the political and social advances of the 18th century. These explorations were the most important events of that time; now some 500 years later, the space program can potentially play that same role.

The challenge posed by the Space Age is therefore addressed not only to the scientist and the engineer who are directly engaged in its projects; more importantly, it is a challenge to our society and, in particular, to its educational processes. The physicist, the astronomer, the geodesist, the meteorologist, the geologist, and the astrophysicist, all have new frontiers open to them. Their job as scientists is to bridge the gap between

---

* Presented at Centennial Convocation Luncheon, Worcester Polytechnic Institute, Worcester, Mass., October 8, 1964.

VENTURE INTO SPACE

the known and the unknown. The question we must ask ourselves is whether they are being educated in such a manner as to prepare them to meet the challenges which the new laboratory of space has opened up to them.

The job of the engineer, in contrast to that of the scientist, is to use the resources of nature for social ends—to bridge the gap between the known and the desired. The laboratory of space has already opened up a new "known" to the engineer in the field of communications and meteorology. The experimental communication satellites—Syncom, Relay, Telstar, and Echo—have answered many of the questions that used to exist relative to the use of satellites for communication. The job of the engineer now is to translate this knowledge into a system that will be better than the under-ocean cables.

The experimental meteorological satellites—Tiros and Nimbus—have demonstrated the utility of cloud pictures taken from satellites as an additional operational tool for weather forecasting. The job of the engineer is to translate this knowledge into a practical and economical operational system.

The second question we must ask is whether the engineer is being educated in such a manner as to enable him to exploit these new developments in space.

We have seen the changes made during the past 30 years to adapt engineering education first to the new field of aeronautical and guided-missile engineering, later to the use of radar, still later to the adaptation of nuclear energy to practical uses. Space explora-

Goddard Center missions.

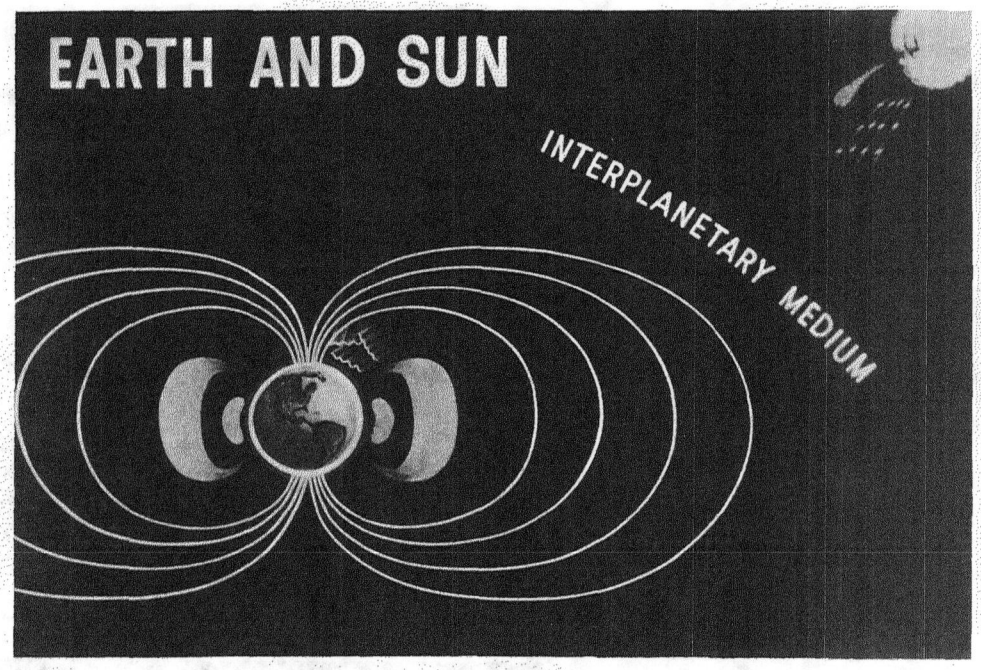

Divisions of space in the earth-sun region.

tion will continue this trend. I think that even closer collaboration than heretofore is going to be required between the scientist and the engineer. Also there is growing a need for closer interdisciplinary collaboration between scientific specialists in various fields. Have our universities who are now training these scientists and engineers reacted to this trend?

To seek a basis for an answer to the questions I have posed, I would like to digress and describe our space efforts from the viewpoint from which I see it. This viewpoint tends to emphasize, as you will see, the involvement of the various scientific disciplines and the close collaboration that is required with the engineer.

The first illustration gives a somewhat kaleidoscopic view of the variety of the projects involved. We have launched some 35 major U.S. satellites for various scientific communication and meteorological purposes. You can see the Syncom, Relay, and Echo communications satellites. Tiros and Nimbus satellites have served as experimental meteorological satellites. The group on the remainder of the illustration are the scientific satellites with which we are literally exploring space. Quite appropriately many of them are named Explorers.

This next picture shows our map of space which is being explored by these satellites. This map might be compared with the maps of the world that were probably available to the early maritime explorers. Space is not an empty void but can be divided into various regions of distinctly different characteristics that are emphasized on this picture. First, there is the near-earth region, the upper atmosphere and the ionosphere. Then, there is the region called the magnetosphere in which the magnetic field lines anchored in the earth extend out to space. They form a gigantic magnetic shield around the earth which makes this region quite different from that on out

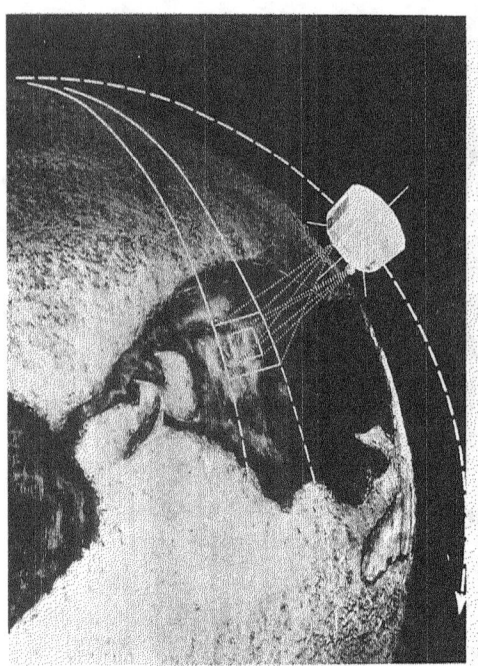

Scope of Tiros photographs.

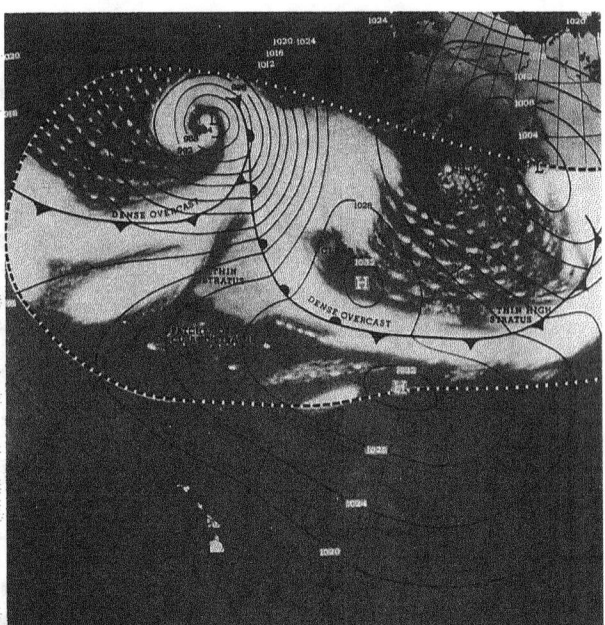

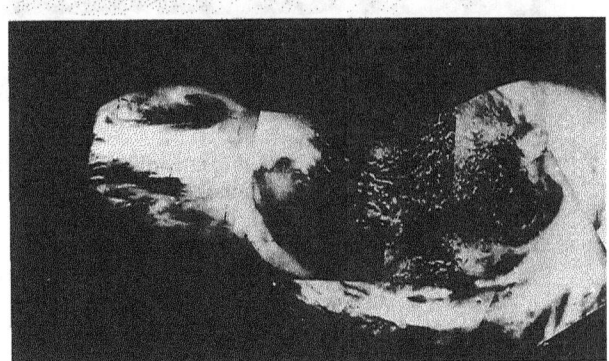

Tiros coverage,
May 20, 1960

further. This region, labeled the "interplanetary medium," is essentially uninfluenced by the earth's magnetic field. Finally, there is the sun which might on our map be given the same prominence as was India on the map of the early explorers, since, as you will see, the sun is the basic cause of many of the variations observed in the other regions of space.

Just as the early explorers initially ventured out only a short distance from their home ports, our first ventures into space were in the near-earth region. Satellites such as Tiros and Nimbus go up into orbits some 300 to 600 miles and look down on the earth as shown in the next illustration. From a satellite such as this, we have obtained data on the upper atmosphere. This next picture is a striking example of the result. On the lower portion, you can see a montage made up of some 64 successive pictures taken by Tiros. You can observe the huge cyclonic disturbance that has been mapped extending

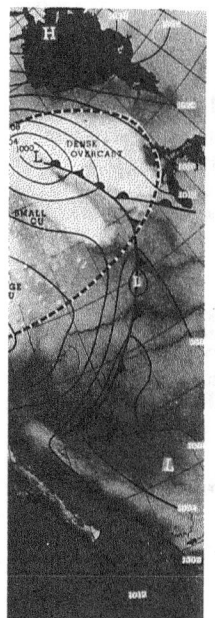
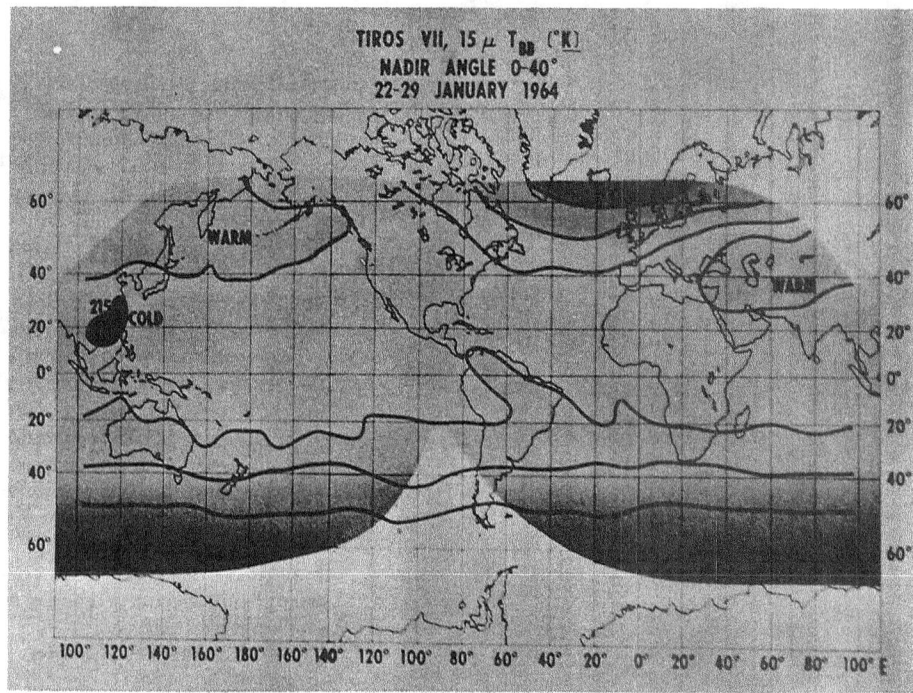

Temperature distribution derived from *Tiros VII*.

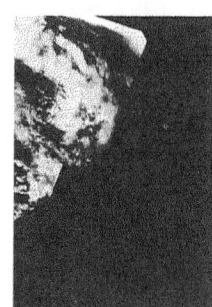

all the way from Wake Island in the Pacific to the Great Lakes. This was the first opportunity for meteorologists to observe weather patterns on such a massive global scale. Cloud pictures such as this are now being used daily by the operational meteorologists in their weather predictions. They are also serving a more basic research purpose in that they give an insight into the dynamics of the weather. We can look forward to much more accurate long-range weather forecasting as our understanding of this phenomenon improves.

Another type of experimental information made available by Tiros to the upper altitude physicists is shown in the next illustration. This is a plot of global temperature distribution obtained from infrared instrumentation. It is especially notable because it depicts the phenomenon of stratospheric warming shown in this region. This phenomenon has been suspected to be the trigger of weather disturbances and to be traceable

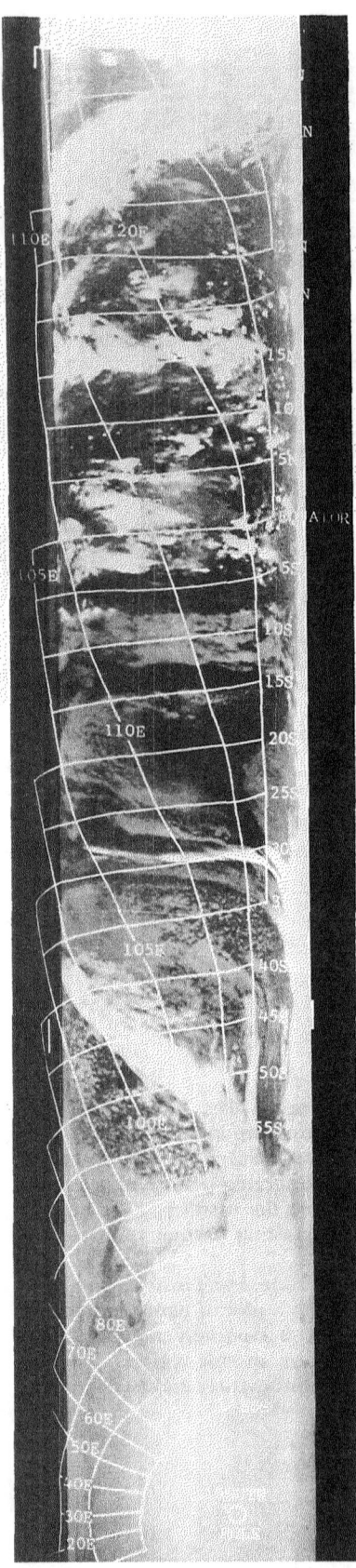

in some manner to solar activity. The data are now being studied by upper altitude physicists in an attempt to obtain a better understanding of this phenomenon, with the eventual hope of using observations such as this for long-range weather prediction.

The picture on the left shows a photograph made by Nimbus infrared techniques. Here you see a strip approximately 1,500 miles extending all the way up from the Antarctic to close to the North Pole. These pictures are less than a month old but already they are under detailed scrutiny by meteorologists who consider them to be a gold mine of data. They give a global picture of the cloud patterns and enable an understanding of cause and effect in the movement of the weather, not heretofore obtainable in observations from the ground. Features observable include:

Antarctic ice shelf
Low pressure system generated at polar fronts
Location of jet streams
Intertropical convergence zone
Volcanoes
Ocean currents

To analyze this global picture some of the following disciplines are involved: meteorologist, upper altitude physicist, geologist, and oceanographer.

The next area of exploration has been the ionosphere—the region of highly ionized particles that exists above what is conventionally considered the upper atmosphere. We have launched several satellites into elliptical orbits; they traverse the altitudes from 200 to 800 miles. These satellites are still pretty close to the earth in terms of our total area of exploration. From such satellites, we measured for the first time the temperature in this region and found that it fluctuated in a 27-day cycle corresponding to the time of rotation of the sun which shows clearly the close link between solar activity and events in our upper atmosphere. These satellites also discovered a helium layer which fluctuates and varies in thickness with solar conditions. Finally, there were measured flows of currents in the ionosphere and observations were made of the patterns of whistlers into outer space.

Our exploration was next pushed out into the

Infrared data obtained from *Nimbus I*. Shown is a "slice" of the globe from the Arctic to the South Pole.

APPENDIX G

magnetosphere. Satellites were sent up to investigate the energetic particle population of the Van Allen radiation belts. This information is not only important in our understanding of sun-earth relationships but also is essential if we are to acquire the engineering information required for the design of communication and meteorological satellites. Their lifetime will be strongly dependent on the radiation environment found in this region.

The next figure is a pictorial description of what this latter group of satellites found. As you can see, there is shown the orbit of the first Interplanetary Monitoring Platform (IMP) which carried it out to 122,000 miles. When it got there, it found that there was a "solar wind" blowing. This wind sometimes is a gentle breeze and on other occasions grows into what might be termed a hurricane. Of course it is not a wind in the conventional sense but is a stream of energetic particles (electrons and protons of varying energy) that are ejected by the sun. During quiet sun conditions, there is gentle breeze and during a solar flare, the number and intensity of these particles increase. It is through this solar wind that the sun has a profound effect on what goes on in our upper atmosphere. The earth is like a rock in a stream with a bow wave in front of it formed by the shock front that marks the boundary of the solar wind and the magnetosphere. Not shown is a wake behind the earth. One of our orbits is shown here—the satellite got in the moon's wake and we were able to observe the effect of the moon on this solar wind.

As you will observe, the sun has been the basic cause of all the phenomena that were observed by these various satellites. Therefore, the results obtained from the Orbiting Solar Observatory were of particular interest since they enabled us to observe what was going on in the sun and causing these variations. Man has been observing the sun for thousands of years. The existence of the early sun worshipers attests to the fact that the importance of the sun to terrestrial conditions has been appreciated for many centuries. However, during all this time, we have been looking at the sun as if through translucent blindfold. The earth's atmosphere and the magnetosphere filter out much of the solar radiation. Thus, earth-based instruments have only been able to observe the sun in a relatively narrow visual band and at radio wavelengths. The Orbiting Solar Observatory was able for the first time to observe the sun in the shorter wave ultraviolet, gamma ray and X-ray regions. New light has been shed

The upper atmosphere.

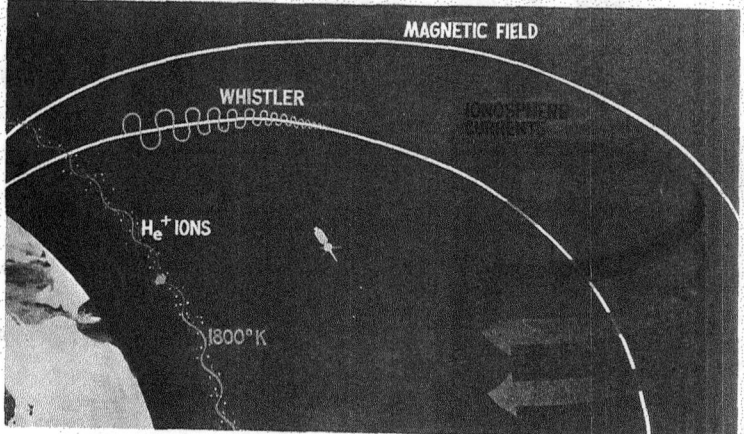

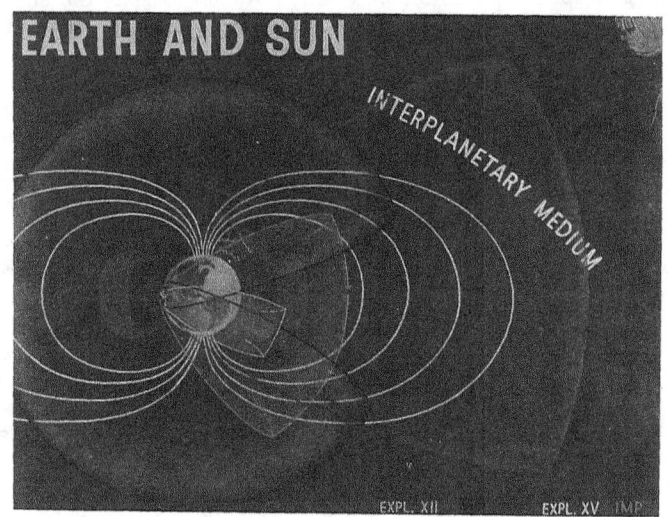

Earth-sun relationships.

The earth and "empty" space.

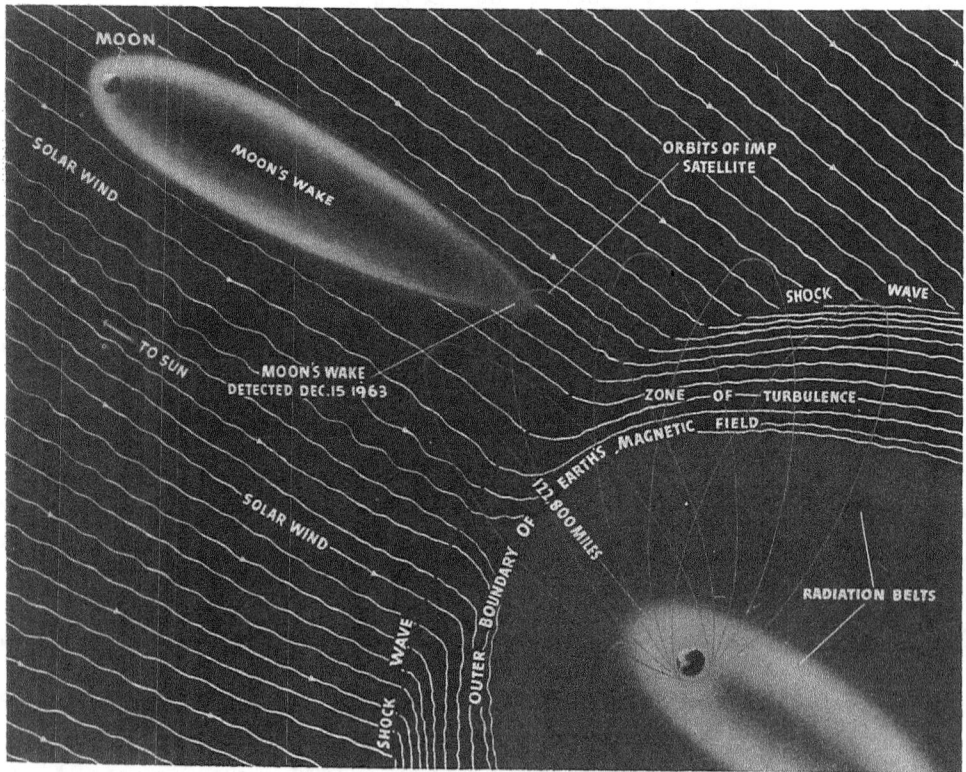

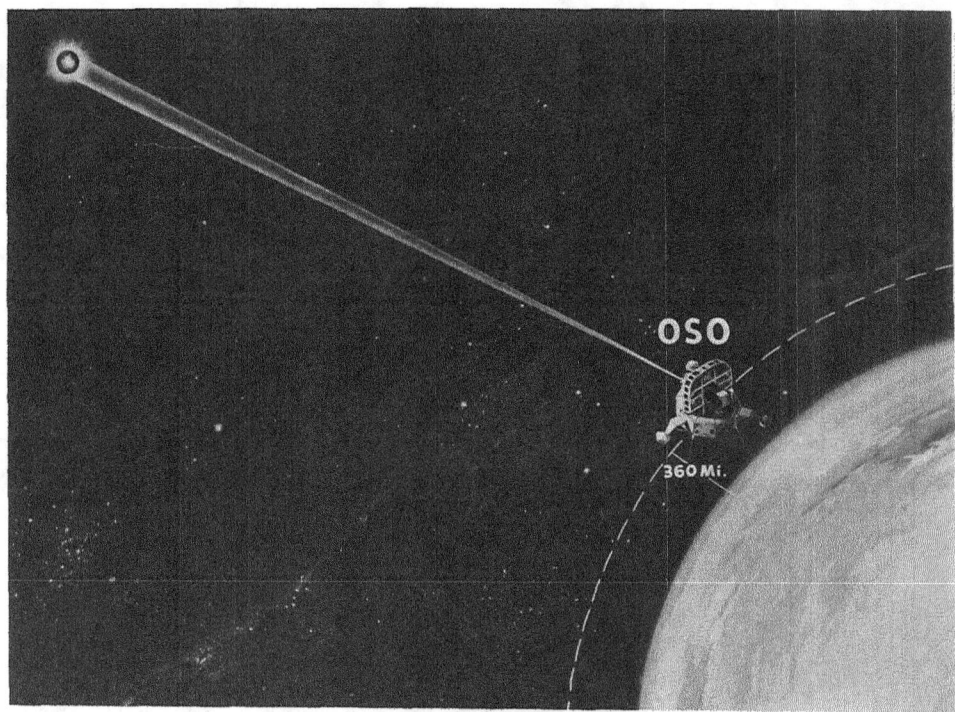

Orbit of *Orbiting Solar Observatory I*.

on the relation between solar flares and the associated variation in the solar wind fluctuations observed from IMP.

When there is a solar flare there apparently is a huge magnetic bottle that is exploded from the sun. Contained within this bottle are energetic particles, electrons, and protons traveling at varying speeds. Anything external to the bottle, such as galactic cosmic rays, cannot penetrate within the bottle and bounces off. This magnetic bottle gradually expands and if headed toward the earth impinges on the earth's magnetosphere and distorts it. Many of the energetic particles bounce off the magnetic shield formed by the magnetosphere, others get injected into the magnetosphere and in due course form the Van Allen belt, as they are commonly called. These particles ride the magnetic lines which exist in the magnetosphere and in due course impinge on the upper atmosphere in the auroral regions and cause changes in temperature that I have previously referred to and in some manner influence our cyclic variations in climate.

My object in giving this broadbrush sketch of the information that has been brought back from space has not been to convey any detailed understanding of the implications of these results. The point that I hope has been conveyed is that it has commanded the efforts of a broad spectrum of scientific disciplines. These include the solar physicist who is involved in the observation of the sun, the nuclear physicist who applies his techniques to the observation of the energetic population in the interplanetary space and the radiation belts; the magnetic field specialist who has been mapping the variation in the magnetic lines of the magnetosphere and studying the collision process with the solar wind; the ionospheric physicist who has been studying the electron distribution in the ionosphere and its variation with the solar cycle; the upper atmosphere physicist

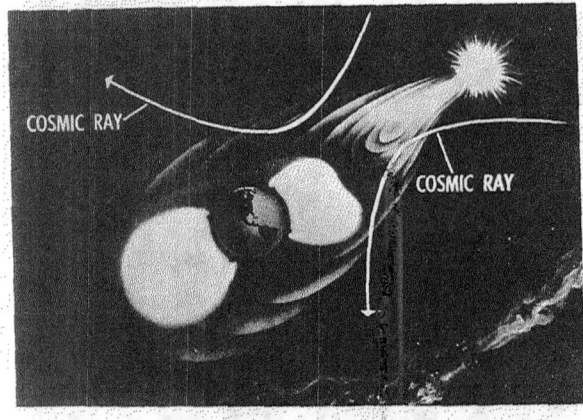

Deflection of cosmic and solar particles by the earth's magnetic field.

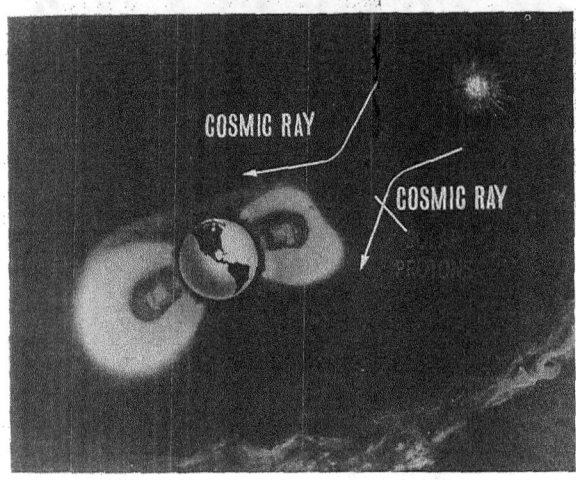

Effect of solar flare on the earth's magnetic field.

who has been trying to correlate observed atmospheric temperature and composition variations with the events in the other regions; and, finally, the meteorologist who has been trying to put together the interrelated effects of all these phenomena on the earth's weather. This interrelation between the various disciplines, I suggest, is a unique feature of space science which should be taken into account in the educational process. It is in distinct contrast to the trend toward specialization that has characterized the past 30 years. The aerodynamicist could concentrate on his wind tunnel with relatively little reference to other disciplines; the nuclear physicist could work with his reactor and the biologist with his microscope in the same relative isolation. Creativity could flourish because it was the era of the individual worker; there were fewer technical committees, budget reviews, administrators and the like; who could kill an idea while it was still in the formative stage.

The new era promised by the Space Age perhaps connotes a return to what was once called natural philosophy. The unifying element of these developments of the space program is a general spirit of inquiry into the nature of the external physical world. It represents a redirection of interest away from the increasingly narrow specialization which has characterized the physical sciences in the last decades.

APPENDIX G

The second distinguishing feature of research in space is the fact that scientists have only been able to make the observations that I have discussed by virtue of the hardware developed by the engineer. This is in distinct contrast to the biologist who could invest in a good microscope and do research comparable with the best.

The control equipment which enables the pointing of instruments with precise accuracy at the sun, the solar power supplies which supply the energy to run the experiments, the communication devices which bring back the information from outer space, all are developments which must come from engineers. In one sense, the normal situation has been reversed. Generally, the engineer exploits and puts to practical use the knowledge acquired by the scientist. But in space research, the scientist seems to be particularly dependent on the engineer to develop new devices and techniques. Engineering, in this context, has become a more creative and trail-blazing profession.

The third and possibly most imposing challenge of the Space Age is the potential feed-back of its developments into the civilian economy. In the long run, the justification of our space budget must stand on this "fallout."

Everyone knows the drive for miniaturization of electronic components to reduce weight and size of space applications. We can foresee the pressure for electronic components to operate at very high temperature. There is a need for new developments for electronic apparatus to operate in the hard vacuum of space. New materials must be developed for the space environment. There are applications for cryogenics and new power sources. There is a special need for insuring long periods of unattended operation of mechanical equipment in space. Development is needed of methods of lubrication in high vacuum and the creation of new sensing and control devices. Means of medical research of man in space must be developed.

These are some of the most immediate returns of space exploration. In due course, they will surely be exploited by our civilian economy. But I submit that the engineering profession is confronted with a new challenge in its job of converting the known into the desired. These developments will not find their way into the civilian economy unless the engineer has the creative initiative necessary for their application.

These three developments of the Space Age present a new challenge to our educational process. They are, in the first place, the interdisciplinary collaboration of various scientific specialists. Secondly, the leadership of the engineer in developing new techniques to enable the scientist to achieve his objectives in space. And thirdly, the creative thinking which is required to apply the new developments of space to our civilian economy.

If I were an educator, I would conclude this talk by suggesting some solutions. However, I claim no competence in the educational field; my job is to produce reliable satellites. There, I will terminate my discussion by suggesting a reexamination of the educational process and its curriculum by those competent in this field to see whether they think it has been properly adjusted to meet these new challenges of the Space Age.

Appendix H
Exhibits

THE EXHIBITS COLLECTED in this appendix are available to the general public. However, it was felt that the value of this work as a historical tool would be much enhanced if it contained some of the pertinent historical documents collected in one place. Therefore, selected key documents considered important for the formation and early years of Goddard Space Flight Center have been assembled in this appendix.

Exhibits

Release by Senator J. Glenn Beall, Maryland	Aug. 1, 1958
NASA General Directive No. 1	Sept. 25, 1958
White House Press Release	Oct. 1, 1958
Executive Order No. 10783	Oct. 1, 1958
NASA Fact Sheet	Oct. 1, 1958
Beltsville Space Center General Notice No. 1	Jan. 15, 1959
NASA General Notice	Jan. 22, 1959
Beltsville Space Center Memorandum for the Record	Feb. 16, 1959
Beltsville Space Center Memordandum for All Concerned	March 6, 1959
NASA Memorandum from the Administrator	May 1, 1959
GSFC Memorandum to Assistant Directors and Division Chiefs	May 1, 1959
NASA News Release No. 59-125	May 1, 1959
GSFC News Release No. 3-10-61-5	March 12, 1959
GSFC News Release No. 3/14/61-1	March 14, 1961
Historical Background and Communications Satellite Act of 1962	(Undated)
Public Law 87-624, Communications Satellite Act of 1962	Aug. 31, 1962

EXHIBIT 1
COPY

Release by Senator J. Glenn Beall, Maryland, at Washington, D.C. office
August 1, 1958

U.S. Senator Beall today announced that the new "outer space agency" will establish its laboratory and plant in Maryland. "The location at Greenbelt, Md., is an ideal location for the new Government agency," Senator Beall said, "because of its accessibility to the nation's capital and its proximity to numerous important highways and others to be built."

Senator Beall has been in consultation on the matter of location with officials of the new agency to be known as the National Aeronautics and Space Administration and expressed himself as "greatly pleased with the decision" to locate in Maryland. The senator pointed out that the space laboratory and plant are to be established on land already owned by the federal government, "thus saving the taxpayers what would have been added expense of land purchase."

"The Greenbelt laboratory will employ 650 technicians, mostly electronic engineers and some chemists," Sen. Beall said, adding that, "all research work in connection with outer space programs will be conducted at the Greenbelt installation."

Sen. Beall said further: "Construction work on the new plant is expected to get underway immediately in view of the fact that legislation authorizing appropriations of $47,800,000 for the construction and installation of the space project center (S4208) was passed by the Senate on Friday and is expected to clear the House on Monday."

APPENDIX H

> **EXHIBIT 2**
> **COPY**

NATIONAL AERONAUTICS AND SPACE ADMINISTRATION
Washington, D.C.

September 25, 1958

NASA GENERAL DIRECTIVE NO. 1

Subject: Proclamation on organization of the National Aeronautics and Space Administration

On this date I have issued the following proclamation, to be published in the Federal Register:

A PROCLAMATION

"1. By virtue of the authority vested in me by the National Aeronautics and Space Act of 1958 (Public Law 85-568, approved July 29, 1958, 72 Stat. 426, 433) I hereby proclaim that as of the close of business September 30, 1958, the National Aeronautics and Space Administration has been organized and is prepared to discharge the duties and exercise the powers conferred upon it by said law.

"2. In accordance with the provisions of the Act, all functions, powers, duties, and obligations, and all real and personal property, personnel (other than members of the Committee), funds, and records of the National Advisory Committee for Aeronautics are hereby transferred to the National Aeronautics and Space Administration.

"3. The existing National Advisory Committee for Aeronautics Committees on Aircraft, Missile and Spacecraft Aerodynamics; Aircraft, Missile and Spacecraft Propulsion; Aircraft, Missile and Spacecraft Construction; Aircraft Operating Problems; the Industry Consulting Committee and the Special Committee on Space Technology and their subcommittees are hereby reconstituted advisory committees to the Administration through December 31, 1958, for the purpose of bringing their current work to an orderly completion.

"4. Existing policies, regulations, authorities, and procedural instructions governing the activities of the National Advisory Committee for Aeronautics, not inconsistent with law, and which are applicable to the activities of the National Aeronautics and Space Administration, shall be continued in effect until superseded or revoked.

"5. The Langley Aeronautical Laboratory, the Ames Aeronautical Laboratory, and the Lewis Flight Propulsion Laboratory are hereby renamed the Langley Research Center, the Ames Research Center and the Lewis Research Center, respectively.

"DONE at the City of Washington, District of Columbia this 25th day of September in the year Nineteen Hundred and Fifty-Eight."

/s/ T. Keith Glennan

> **EXHIBIT 3**
> **COPY**

FOR RELEASE AT 2:00 P.M., EDT   October 1, 1958
James C. Hagerty, Press Secretary to the President

THE WHITE HOUSE

The President today signed an Executive Order transferring certain functions with respect to space activities from the Department of Defense to the new National Aeronautics and Space Administration which comes into existence today under the National Aeronautics and Space Act of 1958 enacted by Congress last July.

Under the Act, the National Aeronautics and Space Administration is to be responsible for aeronautical and space activities sponsored by the United States, except that activities peculiar to or primarily associated with the development of weapons systems, military operations, or the defense of the United States shall be the responsibility of the Department of Defense. The determination as to which agency has responsibility for any such activity is to be made by the President with the advice of the National Aeronautics and Space Council

The present Executive Order transfers from Defense to the National Aeronautics and Space Administration responsibility for non-military space projects such as lunar probes and scientific satellites which have been initiated by the Advanced Research Projects Agency of the Department of Defense pending the establishment of the new civilian agency. It also transfers from Defense certain space-related projects of the Air Force, principally in the field of "super-thrust" propulsion systems which are primarily applicable to future space vehicles. The order also gives to the National Aeronautics and Space Administration responsibility for Project Vanguard, the United States Scientific Satellite program which has heretofore been the responsibility of the Department of the Navy.

The order transfers from Defense to the NASA the amount of $117 million in connection with the Advanced Research Projects Agency and Air Force projects being transferred. This is the same as the amount anticipated to be transferred for these activities in the initial budget estimates for the National Aeronautics and Space Administration submitted to Congress in August. The particular projects to be transferred are to be identified in one or more supplementary Executive Orders. The transfers of funds necessary in connection with Project Vanguard are to be determined by the Director of the Bureau of the Budget.

Details regarding transfers of records, property, facilities, and civilian personnel, in connection with all of the transfers covered in the present Executive Order, are to be carried out as agreed upon between the National Aeronautics and Space Administration and the Department of Defense.

APPENDIX H

> EXHIBIT 4
> COPY

EXECUTIVE ORDER
- - - - - - - - - -
No. 10783
TRANSFERRING CERTAIN FUNCTIONS FROM THE
DEPARTMENT OF DEFENSE TO THE NATIONAL
AERONAUTICS AND SPACE ADMINISTRATION

By virtue of the authority vested in me by the National Aeronautics and Space Act of 1958 (Public Law 85-568; 72 Stat. 426), and as President of the United States, it is ordered as follows:

Section 1. All functions (including powers, duties, activities, and parts of functions) of the Department of Defense, or of any officer or organizational entity of the Department of Defense, with respect to the following are hereby transferred to the National Aeronautics and Space Administration:

(a) The United States Scientific Satellite project (Project Vanguard).

(b) Specific projects of the Advanced Research Projects Agency and of the Department of the Air Force which relate to space activities (including lunar probes, scientific satellites, and super-thrust boosters) within the scope of the functions devolving upon the National Aeronautics and Space Administration under the provisions of the National Aeronautics and Space Act of 1958, and which shall be more particularly described in one or more supplementary Executive orders hereafter issued.

Section 2. (a) The Secretary of the Treasury shall immediately transfer to the appropriation of the National Aeronautics and Space Administration for "Research and Development", from such appropriations of the Department of Defense as the Secretary of Defense shall designate, the following amounts:

   (1) In connection with the transfer of functions provided for in section 1(a) hereof, such amounts as shall be determined by the Director of the Bureau of the Budget pursuant to section 202(b) of the Budget and Accounting Procedures Act of 1950 (31 U.S.C. 581c
(b) and section 1(k) of Executive Order No. 10530 of May 11, 1954.

   (2) In connection with the transfer of functions of the Advanced Research Projects Agency provided for in section 1(b) hereof, $59,200,000.

   (3) In connection with the transfer of functions of the Department of the Air Force provided for in section 1(b) hereof, $57,800,000.

(b) In connection with the transfer of functions provided for in section 1, appropriate transfers of records, property, facilities, and civilian personnel shall be carried out as may be agreed upon from time to time by the National Aeronautics and Space Administration and the Department of Defense.

/s/ DWIGHT D. EISENHOWER

THE WHITE HOUSE
October 1, 1958

> **EXHIBIT 5**
> **COPY**

NATIONAL AERONAUTICS AND SPACE ADMINISTRATION

Washington 25, D.C.

For immediate release
Tuesday, October 1, 1958

FACT SHEET ON THE TRANSFER OF CERTAIN FUNCTIONS FROM DEPARTMENT OF DEFENSE TO THE NATIONAL AERONAUTICS AND SPACE ADMINISTRATION

The National Aeronautics and Space Administration (NASA) was established by Public Law 85-568 on July 29, 1958. In accordance with the Act, T. Keith Glennan, Administrator of NASA and president-on-leave from Case Institute of Technology, issued a proclamation on September 25, 1958, which stated in part. "I hereby proclaim that the National Aeronautics and Space Administration has been organized and is prepared to discharge the duties and exercise the powers conferred upon it . . ."

Prior to the enactment of this legislation, the President wrote to the Secretary of Defense and the Chairman of the National Advisory Committee for Aeronautics concerning his special message to the Congress on April 2, 1958 which recommended establishment of NASA.

In his memorandum, the President said: "The Department of Defense and the National Advisory Committee for Aeronautics should jointly review the pertinent programs currently underway within or planned by the Department, including those authorized by me on March 27, 1958, and should recommend to me as soon as possible which of these programs should be placed under the direction of the new (space) Agency. It should be noted that Public Law 85-325 authorized the Department of Defense for a period of one year; period will come to a close February 12, 1959. Since the new agency will absorb the going organization of the National Advisory Committee for Aeronautics, it should be capable of assuming the appropriate programs prior to the date."

The executive order of the President, which transfers certain functions from the Department of Defense to the NASA, is a result of the above stated review.

In accordance with the executive order, the following projects are being transferred to the NASA by the Department of Defense.

1. The United States Scientific Satellite Project (Project Vanguard). The transfer of this project will include approximately 150 civilian scientific personnel under the direction of Dr. John P. Hagen, Director, Project Vanguard, Naval Research Laboratory.

2. Four Lunar Probes and their instrumentation, and three satellite projects. These projects are being transferred from the Advanced Research Projects Agency. Two of the Lunar Probes are assigned to the Air Force Ballistic Missile Division and two to the Army Ballistic

APPENDIX H

Missile Agency. The three satellite projects are assigned to the Army Ballistic Missile Agency and call for putting into orbit two inflatable spheres—one 12 feet in diameter, and the other 100 feet in diameter—and a cosmic ray satellite.

The total cost of these ARPA programs is approximately $35.5 million, most of which was funded in Fiscal Year 58. The Fiscal Year 59 funds, necessary to complete these programs, is $9.6 million and will be made available to NASA. Additionally, $49.6 million originally designated for scientific projects will be transferred to NASA from ARPA.

3. A number of engine development research programs now being carried on in the Department of the Air Force will be transferred to NASA. These include basic research programs in such areas as nuclear rocket engines, fluorine engines, and the 1 million lb thrust single chamber engine study and development. As specified in the executive order, $57.8 million will be transferred from the Air Force to NASA for these programs.

The Advanced Research Projects Agency will continue work in activities peculiar to, or primarily associated with, the development of weapons systems for the defense of the United States, including the research and development necessary to make effective projects for the defense of the United States. At the present time, while ARPA has underway projects of considerable priority related to anti-missile defense and solid propellants, it is devoting major attention to continuing military projects, which include satellite programs from warning, navigation, communications, meteorology, and other military exploratory space programs. Space components and activities necessary to these projects, such as the development of super thrust boosters and high energy upper stages are also being carried on by ARPA. A program to put man into space is being pursued jointly by ARPA and NASA.

The ARPA military budget for 1959 is approximately $420 million.

> EXHIBIT 6
> COPY

NATIONAL AERONAUTICS AND SPACE ADMINISTRATION
BELTSVILLE SPACE CENTER
Washington 25, D.C.

4102-AN:lds
15 January 1959

GENERAL NOTICE NO. 1

Subj: Designation as Beltsville Space Center

1. By action of the Administrator, four divisions have been designated as comprising the Beltsville Space Center of the National Aeronautics and Space Administration. The official address is as follows:

> National Aeronautics and Space Administration
> Beltsville Space Center
> 4555 Overlook Avenue, S.W.
> Washington 25, D.C.
>
> Telephone: JOhnson 2-6610
> Dial Code: 172
> TWX: WA 134
> Military Routing: RBEPRL
> Messenger Mail: Stop 10

All members of the Beltsville Space Center are requested to use the new address immediately, and to disseminate it as widely as is necessary. The local Post Office and telephone company are being notified of this action.

2. New stationery has been ordered, and will be distributed as soon as it is received. The stationery masthead will give the address as it appears above, and also will list the three scientific divisions and the telephone and TWX numbers. The Vanguard Division may continue to use its Vanguard stationery until the stock on hand is exhausted or until directed otherwise. Vanguard secretaries should type above the printed masthead the words BELTSVILLE SPACE CENTER.

3. The Space Sciences Division and Vanguard Division will be located at NRL. Vanguard will continue to occupy its present spaces, but there may be some adjustments from time to time, with a view toward achieving better use of what space is available. The Space Sciences Division will move into space being prepared for it in Building 5-7 of the NGF annex, adjacent to NRL Building 71. This move will probably take place within the next three months. The Theoretical Division will remain

APPENDIX H

physically with NASA Headquarters until the Beltsville facilities are ready.

4. All divisions of the Beltsville Space Center will receive their administrative support, generally speaking, from the central administrative group being established at NRL under my direction. There will be exceptions to this made necessary by the physical separation of units and individuals. For example, for the immediate future, payrolling, travel, and some supply support for the Theoretical Division will be furnished by Headquarters.

5. The NRL Code Directory will show some listings for the Beltsville Space Center, but unfortunately there is not enough space to show all of the listings we requested. All branch listings could not be accommodated, so no branches will be listed. Service functions will be listed since they are believed to be subject to call frequently enough to be appropriate to the NRL Code Directory. However, we have duplicated, and are distributing with this memorandum, a supplement to the NRL Code Directory. We shall publish a complete directory for your use as soon as we can.

6. The telephone JOhnson 2-6610 is answered at the NRL switchboard by NRL operators. The separate number differs from NRL's only as a matter of identification, but you are requested NOT to use JOhnson 3-6600 unless you are unable to reach the board through JOhnson 2-6610. The change in trunk line number has absolutely no effect on your extension number. The government interdepartmental dial code remains the same as for NRL (172). There will be installed in the near future a two-way tie line between the Headquarters switchboard, and the NRL switchboard, which will permit transfer of incoming calls between the two boards. This means that a party calling JOhnson 2-6610 for a person who is located at Headquarters can have his call transferred to the Headquarters board without having to place a new call, and vice versa.

7. Various notices and instructions will be published from time to time, to advise all personnel of policies and procedures governing the administration of the Beltsville Space Center. Until policies and precedents are established, many actions will be taken on an individual case basis. This is necessary in day-to-day operations; however, the fact that changes in or new procedures are established frequently should come as no surprise under the circumstances.

A formal beneficial suggestion program has not yet been established, but constructive suggestions and the pointing out of deficiencies are always welcome.

/s/ T. E. Jenkins
    Administrative Officer

Distribution:
Construction Division
Space Science
Theoretical Division
Vanguard Division

COPY

NATIONAL AERONAUTICS AND SPACE ADMINISTRATION
BELTSVILLE SPACE CENTER

VANGUARD DIVISION

Code

			Bldg.	Room	Tele
4100	Chief, Vanguard Division	Dr. J.P. Hagen	12	369	447
4103	Asst Chief for Operations	CDR L.I. Baird	12	372	673
4104	Management Coordinator	Mr. H.E. Canney, Jr.	12	371A	533
4105	Science Liaison Coordinator	Mr. R.W. Stroup	12	366	445
4120	Vehicle Systems Branch	Mr. D.G. Mazur	2	254A	502
4130	Space Tracking Systems Branch	Mr. J.T. Mengel	42	319	2427
4140	Theory and Planning Staff	Dr. J.W. Siry	30	322	2347
4150	Vanguard Operations Group Manager	Mr. R.H. Gray	AMR	ULster	3-4515
4162	Administrative Assistant	Mr. R.W. Batchelder	AMR	ULster	3-2618
4170	Payload Systems Branch	Mr. N.W. Matthews	60	406	384
4180	Data Systems Branch	Dr. J.P. Hagen	12	369	477

SPACE SCIENCES DIVISION

4200	Chief, Space Sciences Division	Mr. J.W. Townsend, Jr.	Hq.	T602	128-7815
4203	Consultant	Mr. J.C. Seddon	42	415	345
4210	Fields and Particles Branch	Dr. L.H. Meredith	42	314	594
4220	Planetary Atmosphere Branch	Mr. H.E. LaGow	42	424	358
4230	Astronomy Branch	Dr. J.E. Kupperian, Jr.	30	222	893
4240	Solar Physics Branch	Dr. J.C. Lindsay	Hq.	T605	128-7723
4250	Meteorology Branch	Mr. J.W. Townsend, Jr.	*Hq.	T602	128-7815
4260	Instruments Branch	Mr. K.R. Medrow	42	410	518

THEORETICAL DIVISION

4300	Chief, Theoretical Division	Dr. R. Jastrow	Hq.	T611	128-7721

APPENDIX H

CONSTRUCTION DIVISION

	Chief, Construction Division	Mr. N.P. Miller	Hq.	T204	128-7523

ADMINISTRATIVE OFFICE

4100A	NASA Administrative Officer	Mr. T.E. Jenkins	12	370	668
4100B	Asst Administrative Officer	Mr. A.P. Nagy	12	371A	2504
4100C	Administrative Services Officer	Mr. R.C. Cowan	12	374	2339
4100D	Personnel Clerk	Mrs. M.C. Dytrt	12	364	2593
4100E	Mail, Records & Documents Clerk	Mrs. P.M. Egan	12	367	798
4100F	Travel and Voucher Clerk	Mrs. H.G. Jackson	12	364	2564
4100G	Supply Clerk	Mr. R.J. Fisher	12	374	2445

> **EXHIBIT 7**
> **COPY**

NATIONAL AERONAUTICS AND SPACE ADMINISTRATION
Washington, D.C.

January 22, 1959

NASA GENERAL NOTICE

Subject: Establishment of Beltsville Space Center

    1. The NASA has recently established a new field office to be known as the "Beltsville Space Center." This Center will be operated under the direction of the Director of Space Flight Development—NASA Headquarters (Dr. Abe Silverstein).
    2. Pending completion of permanent facilities at Beltsville, Maryland, program activities of the Space Center will be carried out at Washington, D.C. and at the Langley and Lewis Research Centers.
    3. During this interim period, official communications with staff members attached to the Center should be addressed as follows:

Office	Address
a. Theoretical Division	National Aeronautics and Space Administration
b. Space Sciences Division	1520 H Street, N. W. Washington 25, D. C. (Telephone: Executive 3-3260 - TWX: WA-755)
c. Vanguard Division	National Aeronautics and Space Administration Beltsville Space Center 4555 Overlook Avenue, S. W. Washington 25, D. C. (Telephone: JOhnson 2-6610 - TWX: WA-134)
d. Langley Space Task Group	National Aeronautics and Space Administration Langley Research Center Langley Field, Virginia (Telephone: PArk 3-3325 - TWX: HA-198)

APPENDIX H

c. Lewis Space Task Group
) National Aeronautics and Space
) Administration
) Lewis Research Center
) 21000 Brookpark Road
) Cleveland 35, Ohio
) (Telephone: WInton 1-6620 -
) TWX: CV-520)

/s/ Albert P. Siepert
Director of Business Administration

EXHIBIT 8
COPY

NATIONAL AERONAUTICS AND SPACE ADMINISTRATION
BELTSVILLE SPACE CENTER
4555 OVERLOOK AVENUE, S.W.
WASHINGTON 25, D.C.

SPACE SCIENCES DIVISION
THEORETICAL DIVISION
VANGUARD DIVISION

TELEPHONE: JOHNSON 2-6610
TWX: WA 134
4100A-19:TEJ: vmb
February 16, 1959

MEMORANDUM FOR THE RECORD

Subject: Organization and functions of the Beltsville Space Center; report of observations and stating of implications drawn during conference concerning, held in Mr. Wyatt's office, February 12, 1959

1. Conference Participants:
Robert Gilruth, Assistant Director and Director, Project Mercury
John P. Hagen, Chief, Vanguard Division
Robert Jastrow, Chief, Theoretical Division
Thomas E. Jenkins, Administrative Officer
John W. Townsend, Jr., Chief, Space Sciences Division
Abe Silverstein, Director, Space Flight Development
DeMarquis Wyatt, Assistant to Director of Space Flight Development
Herbert S. Fuhrman, Administrative Assistant to Dr. Silverstein
Al Hodgson, Director of Management

2. Purpose of Meeting: To survey the organization and functions of the Beltsville Space Center as it currently exists; the functions to be performed by the Center in the National Space Program; the organization these functions will require for successful program execution; and to take a very rough cut at the overall staffing problem.

3. Based on this discussion, I am attempting in this memorandum to synthesize the guidelines which can be drawn from the discussion. Attached are organization charts which evolve from the current Space Center groups at scattered locations into an integrated organization at one center physical location.

4. Major functions of the Space Center. It was generally agreed that the Space Center will probably perform five major interrelated functions on behalf of NASA as follows:
    (1) Project Management. It is generally agreed that Space Center

APPENDIX H

technical staff will perform management functions in connection with NASA space projects, both of the contract and "in-house" variety (or more likely, combination of the two). A project in this sense is a complete application of science and space technology to a given objective such as Man-in-Space, Meteorological Satellite, Lunar Probes, et cetera. In many cases, the feasibility of projects will be studied by the Headquarters technical staff with the input from the Space Center until the gross outline of the job becomes evident. Detailing of the plan of attack and execution of the project will rest primarily with the Space Center provided, of course, the systems responsibility for the project is in-house. The Space Center will also germinate ideas for projects and forward proposals to Headquarters for evaluation and authorization. Management of a project by the Space Center will be augmented by the Headquarters technical staff which will assign a senior project coordinator to expedite the work of the project and to help resolve management problems requiring Headquarters attention and support.

(2) <u>Research.</u> The Space Center will perform research (a), as required to meet the gaps in current knowledge to form a base for future ideas and applications and (b), research-in-space in all the scientific disciplines as dictated by the expanding knowledge of space phenomena.

(3) <u>Development and Fabrication.</u> The Space Center will design and develop, including prototype fabrication, components and systems which advance the state of the art of space technology and for specific project application. This in-house development effort is required to translate the ideas of scientists and engineers into hardware where the contracting process would prove infeasible. For instance, the development of payload systems of a highly experimental nature requiring many stages of development, each dependent upon advanced techniques which in themselves are developed as the hardware takes shape, can best be done with in-house development and fabrication facilities available on the spot and under the detailed direction of the staff scientists and engineers. In both industry and government experience shows that this type of effort is very difficult to "farm out" on contract.

(4) <u>Advance Planning.</u> The Space Center is and will continue to be staffed by leaders in their respective fields who will germinate many of the ideas necessary to the future program of space science and exploration. As the practitioners of the art, they will, like many good contractors, develop as a fruitful source of ideas. In fact, the current National Space Program, as expressed in various NASA documents, is in large part a synthesis of ideas emanating from the NASA staff. All of the ideas expressed in the space program are not, of course, exclusive with NASA personnel, but the program does express the broad competence of the staff. It is expected, as is currently the case, that Space Center staff will continue to participate in the formulation of the program and in this respect act as a technical staff to Headquarters.

(5) <u>Operations.</u> The Space Center will be in charge of technical operations in the field in all programs assigned to the Space Center. These programs will include in-house space center projects plus projects carried mostly on contract from Headquarters but technically monitored by the Space Center. Space Center technical teams at the launching

sites will, of course, be under the direction and control of project and functional heads within the Space Center. Space communications and tracking will be a prime operational responsibility of the Center. This function includes data handling, both orbital and telemetered scientific.

5. It was concluded that to satisfy the questions in the mind of the Administrator we should immediately come up with (a) an interim plan of organization and operations to assure full effectiveness and integration into the program of the groups which now exist for the Space Center and (b) a tentative ultimate plan of organization and functions to show what the Space Center may look like three years from now.

6. Functional versus Project Organization within the Space Center. These two fundamental bases for organization were discussed at some length, in terms of what would be the best form of organization for the Space Center.

It was concluded that each of these bases of organization has merit and each has some drawbacks. Both in industry and government, outstanding examples of each can be cited.

(1) Functional Organization. By this is meant organizing the Space Center into sub-elements identified with across-the-board scientific or space technology disciplines. For instance, guidance and control, astronomy, data handling, scientific instrumentation, and propulsion. In an R&D institution, this form of organization is necessary to advancing the basic state of various disciplines and arts, and functional applied areas. Specialists in a given area are grouped and are interested in evolving new knowledge and techniques without reference to a preconceived specific application, but with the full knowledge that whatever gaps are covered will benefit the program across the board. This type of organization is generally stable and develops specialists who can be tapped as consultants by the applied projects. Effort organized along this line, when staffed by competent people who are really research people, creates the store of knowledge which is absolutely required if program stagnation is to be prevented.

The disadvantage of this type of organization is that unless it is modified or subverted for this purpose, it results in diffuse project management when a major project is planned in the organization.

(2) Project Organization. A project manager should have under his control or as his responsibility all of the necessary resources prepared to fulfill the project's mission. In a functional organization dependent, of course, on personalities and other factors, the project manager is handicapped from "managing" and "directing" and is apt to become something of a coordinator-expediter. If the institution is charged with carrying out several major projects, the competition among projects for specialists' services, in-house central technical service support, and even use of certain contractors must be delineated by management by formal assignment and delegation. The easiest way to accomplish this is to assign resources including personnel to the project under the direct control of the project director. The "project" should use central services where they exist and it is feasible.

APPENDIX H

It is concluded from the above that the Beltsville organization should be based on functional lines with organizational flexibility allowing the Center to organize for major or "macro" projects. Small or "micro" projects can and should be carried in the functional divisions.

7. **Program Requirements.** The program currently assigned and contemplated for the Space Center has not yet been detailed in full. However, the major efforts which are listed below indicate a very large and active program which will require a great deal of contracting plus a heterogeneous, competent, and sizable Space Center staff. Some of these assignments, active or to become active in the next eighteen months, are:
   (1) <u>Science Program</u>
       155 sounding rockets of various types
       30 satellites
       15 probes
   (2) <u>Application Satellites</u>
       — Communications and Meteorological, 6 or 7 vehicles,
       — Geodetic—number of satellites unknown at this time,
       — Other Vanguard Division follow-on programs—vehicles and numbers as yet undetermined.
   (3) <u>Project Mercury</u>
       — Phase I
       — Phase II
   (4) <u>Vehicles and Engines</u>
       Vega, Centaur, Thor-Vanguard (Thor-Delta), Juno V (1.5 meg engine), Nova (1.6 meg engine known currently as 1.0 meg engine).
   (5) <u>Global Range and Operations</u>
       — Space Communications and Tracking
       — Operational Telemetering
       — Ranges—Wallops, AMR, PMR

8. **Numbers of Personnel Required.** Currently, all segments of the organization, in terms of personnel numbers, are merely nuclei of the ultimate staffs needed to do the jobs to be done. There is the feeling that budget planning for the next fiscal year does not include adequate numbers for the Space Center to carry out its mission. In general the program managers, that is Division Heads, feel that if adequate numbers of competent personnel can be brought on the staff, this is more important than studying at this point how the Space Center is to be organized. This feeling is not to be interpreted as a cry for anarchy or lack of organization, but as an expression that thinking about the Space Center sometimes seems to ignore the fact that the Space Center is already very much a going concern. Geographical dispersion and logistic support currently supplied in part by NRL, Langley, and Lewis tend to cloud thinking on immediate future requirements. The fact that the physical plant does not exist should not deny the existence of the going major components of the organization. The general feeling was expressed that we are headed for trouble in 1960 unless the complement figures, in particular, are raised above levels being cited currently.

9. <u>Space Center functions in terms of existing nucleus capabilities.</u>
The Space Center is not starting from scratch in organizing a technical

competence. The five groups at the Division level now in operation have
a nucleus capability to perform work organized functionally or by project
in the following areas:
- (1) Space Sciences,
- (3) Vehicle Systems,
- (3) Theoretical Research and Support,
- (4) Instrumentation Support,
- (5) Payload Systems,
- (6) Data Handling and Techniques,
- (7) Communications and Tracking,
- (8) Operations.

10. My suggestions for organization of the Space Center:
- (1) To fully integrate it into the National Space Program,
- (2) To fully capture and utilize the nuclei of existing capabilities, and
- (3) To fulfill the potential which the senior personnel of the Center feel can be brought into being realistically,

are attached hereto with explanatory notes.

/s/ T. E. JENKINS
Administrative Officer

APPENDIX H

> **EXHIBIT 9**
> *COPY*

NATIONAL AERONAUTICS AND SPACE ADMINISTRATION
BELTSVILLE SPACE CENTER

6 March 1959

MEMORANDUM FOR ALL CONCERNED

FROM: T. E. Jenkins, Administrative Officer

SUBJ: BSC Divisions' Authorized Personnel Complements, FY 1959;
    summary of recent revisions

1. To sum up the current situation with respect to Beltsville Space Center division approved personnel complements (ceiling points):
    a. We requested a total of 342 ceiling points to cover the Vanguard Division, including 42 for my Administrative staff;
    b. Dr. Silverstein approved an allocation against the Vanguard money of 400 but to include 90 for Space Sciences Division. Dr. Silverstein also promised Mr. Townsend he could get an additional 10 from the 400 if he needs it by 30 June. Space Sciences Division then can go to a total of 100;
    c. Dr. Jastrow requested relief for the Theoretical Division of 10 points. Abe approved 6 points for Jastrow (plus 30 points carried against S&E for a total of 36);
    d. In summary, the total of 400 ceiling points authorized against Vanguard funds reserved on Headquarters' books is tentatively allocated as follows:

        Total authorized 400
            Space Sciences Division (Townsend) 100
            Theoretical Division (Jastrow) 6
            Vanguard Division (Hagen) 252
            Beltsville Administrative Office (Jenkins) 42
                Total 400

2. Vanguard Division, in all likelihood, will require more than 252 points prior to 30 June 1959. In the financial summary we presented to Dr. Silverstein there is money reserved to cover some additional points. In fonecon with Dr. Silverstein I agreed it was too early to forecast how successful Vanguard recruiting efforts will be, but all indications are that they will be highly successful. Currently, physical work space is the limiting factor. Vanguard will not exceed this ceiling without specifically getting an approval by the Director, Space Flight Development.

3. <u>Controls.</u> I am taking steps through the Beltsville personnel office to control accessions within the above figures. <u>By this memorandum Division Heads (Hagen, Jastrow, Townsend) are advised that all PAR Form 52s must be submitted to the Beltsville Personnel Office, Attention Code</u>

<u>4100B (Al Nagy) for approval within authorized complement and money before commitments to hire can be considered firm with the candidate.</u> If the total ceiling of 400 proves inadequate I will request additional authorizations from Dr. Silverstein based on division justifications.

/s/ T. E. JENKINS
Administrative Officer

Distribution:

Headquarters:
  A. Silverstein
  A. Siepert
  D. Wyatt
  A. Hodgson
  R. Ulmer (3)
  R. Lacklen (2)
  H. Fuhrman

Beltsville:
  J. Hagen
  J. Townsend
  R. Jastrow
  A. Nagy
  B. Sisco
  L. Best
  B. Cowan

APPENDIX H

> EXHIBIT 10
> COPY

NATIONAL AERONAUTICS AND SPACE ADMINISTRATION
1520 H STREET NORTHWEST
WASHINGTON 25, D.C.

OFFICE OF THE ADMINISTRATOR   May 1, 1959

MEMORANDUM from the Administrator

Subject: Functions and Authority — Goddard Space Flight Center (GSFC)

1. Purpose of this Memorandum.
   a. To establish the Goddard Space Flight Center, Greenbelt, Maryland
   b. To provide a statement of functions and authority for the Goddard Space Flight Center.
2. Functions. The Goddard Space Flight Center is assigned the following functions:
   a. Conducting advanced planning and theoretical studies leading to the development of payloads for scientific and manned space flights.
   b. Conducting necessary supporting research in scientific payloads, applications systems, instrumentation, communications, guidance, and vehicles.
   c. Developing payloads for approved scientific programs, applications programs, and manned space flights.
   d. Developing, subject to specific approval in each case, vehicles to launch payloads.
   e. Supervising Goddard Space Flight Center flight operations, integrating the activities of all participants as necessary to accomplish missions successfully.
   f. Supervising tracking, data acquisition, communications and computing operations for the provision of orbital and reduced flight data from satellites and space vehicles for NASA space flight programs assigned to or monitored by the Center and for other space programs as requested by the Director of Space Flight Development.
   g. Interpreting results of flight programs for which the Goddard Space Flight Center is responsible.
   h. Furnishing technical management of projects, including monitoring of contractors, to insure timely and economical accomplishment of objectives.
   i. Exercising such procurement and contract administration authority as may be delegated by the Director, Office of Business Administration.
   j. Upon request by the Director of Space Flight Development, providing support for space program activities of other

organizations, e.g., military establishments of Department of Defense.

k. Reporting on the status of approved programs and recommending changes or modifications as necessary to meet program goals and schedules.

l. Providing necessary administrative and management support as required for carrying out assigned programs.

3. <u>Responsibility of Director, GSFC.</u> The Director, GSFC, reports directly to the Director of Space Flight Development and is responsible for the exercise of the functions assigned to the Goddard Space Flight Center.

4. <u>Scope of Authority.</u> The Director, GSFC, is authorized and directed to take such action as is necessary to carry out the responsibilities assigned to him within the limitations of this and other official NASA communications and issuances.

5. <u>Relationships with Other Officials.</u> In performing the functions assigned to him, the Director, GSFC, is responsible for recognizing the responsibility and authority of heads of divisions and offices, Headquarters and for assuring that actions he may take are properly coordinated with Headquarters groups having joint interests and are in accordance with NASA policies.

6. <u>Approval of Organization.</u> The basic organization of the GSFC is outlined on the attached organization chart. Modifications or changes in basic organization structure are subject to the approval of the Director, Space Flight Development and the Administrator, NASA.

/s/ T. Keith Glennan
Administrator

Encl
 Chart

APPENDIX H

> **EXHIBIT 11**
> **COPY**

NATIONAL AERONAUTICS AND SPACE ADMINISTRATION
GODDARD SPACE FLIGHT CENTER

4100A-89:TEJ:vmb
May 1, 1959

MEMORANDUM TO ASSISTANT DIRECTORS AND DIVISION CHIEFS

Subject: Organization and Functions of the Goddard Space Flight Center

1. Effective May 1, 1959 the Beltsville Space Center became the Goddard Space Flight Center. The change in name accompanies the first formal announcement of the Space Flight Center organization, mission, functions, and appointments of many of the key personnel.

2. Attached is the official statement on organization and functions accompanied by the official organization chart for the Space Flight Center. As with most documents of this type, the wording in some cases may carry different meaning to different people. Therefore, it should be stated that the spelling out of the mission was not done to be restrictive so far as activities conducted by the Space Center are concerned. The mission is broad and the functions certainly spell out more responsibilities than we can fully carry out or will be able to carry out until our capabilities have been built up much beyond the point where they now exist.

3. There are many administrative details to be sorted out and published to gear internal operations to the new organization structure. Such workaday things as time cards, directories, job order numbers, delegations of authority, et cetera are in the process of being published. We did as much preplanning work as possible, but since all organization details were not nailed down until a very late hour it will take us a couple of days to publish the necessary instructions. In the meantime, you are advised that all previous instructions remain in effect until superseded. For instance, job order and account numbers to be used on time cards, stubs, et cetera will remain the same until changed by specific notice. This paragraph is not to be interpreted in any way as abrogating the authority of any appointed official.

(signed)
T. E. JENKINS
Administrative Officer

```
       COPY
```

NATIONAL AERONAUTICS AND SPACE ADMINISTRATION
WASHINGTON 25, D.C.

ORGANIZATION OF ACTIVITIES
OF GODDARD SPACE FLIGHT CENTER

Mission and Functions

1. Operating as an integral part of the NASA space flight program, under the overall guidance and direction of the Director of Space Flight Development, NASA Headquarters, the Goddard Space Flight Center will be a major field arm of NASA. It will carry out assigned missions in the planning, research, developmental, and operational phases of the nation's space flight program. Specifically, personnel of the Center shall:

a) Conduct advanced planning and theoretical studies leading to the development of payloads for scientific and manned space flights.

b) Conduct necessary supporting research in scientific payloads, applications systems, instrumentation, communications, guidance, and vehicles.

c) Develop payloads for approved scientific programs, applications programs, and manned space flights.

d) Develop, subject to specific approval in each case, vehicles to launch payloads.

e) Supervise Goddard Space Flight Center flight operations, integrating the activities of all participants as necessary to accomplish missions successfully.

f) Supervise tracking, data acquisition, communications and computing operations for the provision of orbital and reduced flight data from satellites and space vehicles for NASA space flight programs assigned to or monitored by the Center and for other space programs as requested by the Director of Space Flight Development.

g) Interpret results of flight programs for which the Goddard Space Flight Center is responsible.

h) Furnish technical management of projects, including monitoring of contractors, to insure timely and economical accomplishment of objectives. In this connection, the Goddard Space Flight Center will execute contracts, in accordance with delegated authority, for performance of work necessary to accomplish program objectives.

i) Upon request by the Director of Space Flight Development, provide support for space program activities of other organizations, e.g., military establishments of Department of Defense.

2. These missions will be carried out through the execution of major functions as described below:

a) <u>Advance Planning.</u> The Goddard Space Flight Center staff will include leaders in the several space science fields who will be responsible for the formulation of ideas and concepts which are essential to the effectiveness of the space flight program. This advance planning will be of value to the Space Flight Center in carrying out its mission. Also, the Space Flight Center will assist and advise the Office of Space Flight Development, NASA Headquarters, in the formulation of the national space flight program.

b) <u>Research.</u> The Space Flight Center will perform two broad types of research: scientific and engineering research necessary to serve as a base for future ideas and applications; and research in the disciplines of the space sciences as a basis for acquiring ever-increasing knowledge of space phenomena.

c) <u>Development and Fabrication.</u> The Goddard Space Flight Center will design and develop, including prototype fabrication, components and systems which advance the state-of-the-art of space technology or which have direct project application. A part of this activity will be in-house, while the remainder will be on contract. The development and fabrication activities to be handled in-house will normally be those which cannot be handled as economically, expeditiously, or effectively by a contractor, either because they are so intimately related to other activities being carried out at the Center or because they involve matters so fundamental to major in-house project operations of the Center that they cannot readily be separated therefrom.

d) <u>Operations.</u> The Space Flight Center personnel will be in charge of technical operations in the field for all programs assigned to the Center. These programs will include in-house projects as well as those on contract for which the Center has the monitoring responsibility. Space Flight Center technical teams at launches will be under the technical direction and control of project and functional heads within the Center. Space communications and tracking, as well as data reduction and interpretation, will be operational responsibilities of the Space Flight Center. For other NASA space flight programs, the Center, upon request of the Director of Space Flight Development, will assume operational responsibility for tracking, data acquisition, communications and computing operations and for the provision of orbital and reduced flight data from satellites and space vehicles.

e) <u>Project Management.</u> The Space Flight Center will be responsible for the management of projects and parts of projects as necessary for the accomplishment of its assigned missions. A project in this sense is the application of scientific and engineering technology to a specified objective, such as man-in-space, meteorological satellites, lunar probes, and so forth. Ideas for projects may originate in Headquarters, in the NASA field organizations, or from other sources. Projects will be undertaken following authorization by Headquarters. Projects will be carried out on either an in-house or a contract basis. Monitoring responsibility may from time to time be assigned to the Space Flight Center on contracts let by NASA Headquarters. Project officers will be designated from the NASA Headquarters staff to advise and assist the Goddard Space Flight Center in projects execution and in resolving management problems which require Headquarters attention and support.

Relationships

1. **With Office of Space Flight Development, NASA Headquarters**

   a) The Goddard Space Flight Center will participate with the Office of Space Flight Development in the definition and establishment of national program objectives.

   b) Within the defined national program objectives, the Goddard Space Flight Center will carry the full responsibility for accomplishing the missions assigned to it, subject to such technical advice and assistance as the Office of Space Flight Development, NASA Headquarters, may provide.

   c) The Goddard Space Flight Center will keep the Office of Space Flight Development fully informed at all times concerning problems arising and progress made in connection with the Center's assigned responsibilities.

   d) The Goddard Space Flight Center will be responsible for its own internal administration, in accordance with the provisions of applicable statutes and policies, procedures, and regulations issued by NASA Headquarters.

2. **With JPL**

   The missions and functions of the Goddard Space Flight Center and the Jet Propulsion Laboratory complement each other. There will be no contractual relationships between the two, but there will be frequent interchange of data and technical information to the extent warranted by the common interests of both or as necessary for the operations of either. The relationships between these two organizations, both active in the space flight development field, will be essentially the same as those which exist between the NASA Research Centers.

3. **With NASA Research Centers**

   The Goddard Space Flight Center will have no formal relationships with the NASA Research Centers except through NASA Headquarters. The channel of communications in matters requiring formal action will normally be: Goddard Space Flight Center to Office of Space Flight Development, NASA Headquarters, to Office of Aeronautical and Space Research, NASA Headquarters, to the Research Centers. Such matters originating in the Research Centers will be referred to the Goddard Space Flight Center through the same offices, but in opposite sequence. Interchange of information and data, however, may be carried on freely between the Space Flight Center and the Research Centers without regard to these channels, except that information copies shall be furnished to the Office of Space Flight Development and the Office of Aeronautical and Space Research, NASA Headquarters.

4. **With Contractors**

   The Goddard Space Flight Center will be authorized to enter into contractual arrangements with private contractors in accordance with applicable NASA policies, regulations, and procedures. Subject to the limitations imposed by such policies, regulations, and procedures, the Goddard Space Flight Center will have the authority to develop contract specifications, including engineering design where appropriate, invite bids, evaluate proposals, make awards and monitor performance.

5. **With Governmental Agencies**

   There will be contractual relationships between the Goddard Space

APPENDIX H

Flight Center and other governmental agencies as authorized by NASA Headquarters.

Organization of the Goddard Space Flight Center

1. Attached is an organization chart reflecting the components which are proposed as major segments of the Goddard Space Flight Center. This organization provides:

    a) Integration of the missions and functions of the Goddard Space Flight Center with the national space program objectives.

    b) Coordination of the direction of the scientific and technical functions of the Center with the supporting engineering and business management activities.

    c) A staffing pattern that anticipates and provides for the orderly integration of the personnel and functions now separately located at Langley Research Center, Lewis Research Center, and Naval Research Laboratory.

    d) A management structure capable of full execution of both technical and support activities with the same degree of delegated responsibility that characterizes the operation of other major NASA field establishments.

2. The scientific and technical programs of the Center are grouped under three Assistant Directors.

    a) The Assistant Director, Space Science and Satellite Applications, will be responsible for:

        1) Space Sciences Division — responsible for conduct of a broad program of basic research in the space sciences through the use of experiments carried in rockets, earth satellites, and space probes.

        2) Theoretical Division — responsible for conduct of a broad program of study and research in phases of theoretical physics, mathematics, and mechanics associated with the exploration of space.

        3) Payload Systems Division — responsible for integration of scientific experiments and equipments into complete earth satellites and space probes; conducts research and development in the fields of space environment and special electronic devices such as telemetering systems, command receivers, control circuitry, and power supplies; designs, fabricates, constructs, and tests satellite and probe space systems.

        4) Satellite Applications Systems Division — responsible for research and development on applications systems, including: applications satellites, such as the meteorological, communications, and geodetic satellites; space techniques such as advanced control and stabilization systems; and space and booster vehicle systems.

    b) The Assistant Director, Tracking and Data Systems, will be responsible for:

        1) Tracking Systems Division — responsible for research and development on new systems for tracking, data acquisition, and communications between satellites and/or space vehicles and ground receiving stations; provides evaluation capability for all system proposals plus calibration capability for systems which are in operation; provides technical and engineering

support for all modifications and enlargements of technical equipment at the field stations, by furnishing prototype equipment, contract monitoring, and field installation and technical inspection services.

2) Data Systems Division — responsible for the application of data reduction techniques to: launching of vehicles, orbit calculation, and satellite and space probe scientific data; conducts research and development on new data handling techniques as required for Space Flight Center programs, and furnishes theoretical support for all phases of data techniques and analyses; responsible for data reduction and computation activities connected with NASA space flight programs assigned to, or monitored by, personnel of the Center; provides computation service for all areas of the Center as required, as requested by Director of Space Flight Development, assumes operational responsibility for data reduction and computation activities for other NASA or Department of Defense programs.

3) Operational Support Division — responsible for establishment and continuous operation of global tracking, data acquisition, and communications network, furnishing personnel and technical and logistic support therefore; operates control center of the Goddard Space Flight Center, furnishes continuing technical support of space flight operations, including range support and operations necessary to initiate communication, tracking and data sequences after vehicle launch; coordinates operations to assure utilization of the net and adequate service to user teams, both inside and outside NASA.

c) The Assistant Director, Manned Satellites, will be responsible for:

1) Flight Systems Division — responsible for conceptual design, integration and performance of complete manned space flight systems, including on-board equipment and subsystems; carries out preliminary design and performance analyses and specifies requirements for advanced systems; monitors or performs the required basic and applied research in support of space flight systems.

2) Operations Division — responsible for establishing operational procedures associated with the launch, in-orbit, reentry and recovery phases of manned space flight missions; accomplishes detailed trajectory analyses associated with each mission in the Mercury Program to ensure that test plans and test schedules are met; conducts necessary pre-flight checking and in-flight monitoring in order to maintain flight safety; selects and trains flight crews appropriate for each flight test.

3) Engineering and Specifications Division — responsible for engineering studies of proposed manned space flight vehicles and components to determine their feasibility; prepares specifications and cost estimates; supplies and monitors technical information required for the procurement and administration of contracts.

3. There is also included in the organization a Special Projects Group, currently identified as the Lewis Space Task Group, which has program

APPENDIX H

responsibility for engineering development and design of a special high energy rocket stage.

4. The Office of Business Administration will provide the Center with the full range of central administrative support functions, including: personnel administration; budget and financial administration; plant and personnel security programs; management analysis activities; administrative services, including plant and fire protection; procurement and supply, including contract administration; and technical information, including library services.

5. The Office of Technical Services initially will include only the construction group which is already in existence. As construction progresses, however, and the Space Flight Center becomes a physical reality, it will be necessary to staff this Office to perform the full range of its functions. These will include: fabrication and shop operations; construction and repair; buildings and grounds maintenance; utilities operations, automotive operations, and industrial safety.

/s/ Abe Silverstein
Director of Space Flight Development

Attachment
1 Chart
Date: 1 May 1959

EXHIBIT 12
COPY

NATIONAL AERONAUTICS AND SPACE ADMINISTRATION
Washington 25, D. C.

NASA RELEASE NO. 59-125
EX 3-3260
Ext. 7827

FOR RELEASE:
Friday P.M.
May 1, 1959

NASA'S NEW SPACE PROJECTS FACILITY NAMED GODDARD SPACE FLIGHT CENTER

T. Keith Glennan, NASA Administrator, announced today the Government's space projects center at Greenbelt, Maryland, will be named the Goddard Space Flight Center in commemoration of Robert H. Goddard, American pioneer in rocket research.

The Goddard Space Flight Center, under the overall guidance of the Director of Space Flight Development at NASA Headquarters, will perform basic space research and will be responsible for the development of satellites, space probes and vehicles, tracking, communications, and data reduction systems. In addition, the facility will eventually be a command control center for NASA space flight operations.

The organization of NASA's new Space Center includes a director, not yet appointed; three major research and development groups, each headed by an assistant director; and business administration and technical services departments.

John W. Townsend, Jr., formerly Chief of NASA's Space Sciences Division, has been appointed Assistant Director for Space Science and Satellite Applications. Divisions reporting to him are: Space Sciences, Theoretical, Satellite Applications Systems, and Payload Systems. The Vanguard Operations Coordinating Group also reports to Townsend. Beginning today, the staff of the Vanguard Division will be integrated into other major NASA space flight projects.

John T. Mengel, former head of the Space Tracking Systems Branch in the Vanguard Division, has been named Assistant Director for Tracking and Data Systems. Reporting to him are the Tracking Systems, Data Systems, and Operational Support Divisions.

Robert R. Gilruth is the Center's Assistant Director for Manned Satellites. He currently heads the Mercury manned space flight project. Divisions under his direction are: Flight Systems, Engineering and Specifications, and Operations.

Michael J. Vaccaro, formerly assistant head of the Administrative Management Office and Personnel Director at the Lewis Research Center, Cleveland, Ohio, has been appointed Business Manager of the Space Center. The head of Technical Services has not been announced.

The Goddard Space Flight Center will be built on an approximately

APPENDIX H

550-acre tract acquired from the Government's Beltsville Agricultural Center, north of Washington, D. C. Located east of the Baltimore-Washington Parkway, the site is bounded on the south by Glendale Road.

The contract for the first two major buildings at the Center — Space Projects Building and Research Projects Laboratory—was let April 10, 1959, to Norair Engineering Corporation of Washington, D.C., at a total cost of $2,882,577. These two-story buildings, scheduled for completion in mid-1960, will total about 100,000 square feet of laboratory and office floor space. They will house a staff of about 450. The remainder of the staff of the Goddard Space Flight Center will be housed at the U.S. Naval Research Laboratory in Washington, and at the Langley Research Center, Langley Field, Virginia, until the completion of the facility.

EXHIBIT 13
COPY

NATIONAL AERONAUTICS AND SPACE ADMINISTRATION
NEWS RELEASE

GODDARD SPACE FLIGHT CENTER
GREENBELT, MD.

OFFICE OF PUBLIC INFORMATION
PHONE: GRanite 4-9000, Ex. 555

RELEASE NO. 3-10-61-5

FOR RELEASE
SUNDAY A.M.'s
March 12, 1961

    The National Aeronautics and Space Administration will dedicate its $27-million Goddard Space Flight Center at Greenbelt, Md., on March 16. The Dedication date marks the 35th anniversary of the world's first flight of a liquid-propelled rocket engine by Dr. Robert H. Goddard, father of American rocketry, for whom the Center has been named.
    The Center, which will be completed by the end of next year, is the first completely new complex to be constructed and staffed for the peaceful exploration of space since NASA was established in October, 1958.
    Center personnel conceive, develop and fabricate satellite and sounding rocket instrumentation that probe space in the immediate vicinity of the earth. The Center also has the world-wide responsibility for tracking, communications and data analysis for both manned and unmanned spacecraft.
    Several hundred invited guests are expected to attend the Dedication ceremony which will begin at 2 P.M. The ceremony will be followed by a guided one and one-half hour tour of the Center's facilities. The Dedication will not be open to the public. Open house for employees and their families will be held Saturday, March 18 from 10:30 to 3:30 P.M. It is planned to hold a public open house later as the Center nears completion.
    The Center will be dedicated by Dr. Detlev Wulf Bronk, President of the National Academy of Sciences. He will be introduced by James E. Webb, NASA Administrator. An honored guest will be Mrs. Esther Goddard, widow of the Clark University rocket pioneer. She will be presented a Congressional Gold Medal, authorized by the 86th Congress. The presentation will be made jointly by Senator Robert S. Kerr (D-Okla.), Chairman of the Senate Committee on Aeronautical and Space Sciences, and

Overton Brooks (D-La.), Chairman of the House Committee on Science and Astronautics.

Mrs. Goddard and Dr. Bronk will unveil a sculpture of Dr. Goddard which later will be placed in the Center's Administration Building. It was created by Washington sculptor, Joseph Anthony Atchison, noted for his creative work in the Shrine of the Immaculate Conception in Washington, the World Flight Memorial for the Smithsonian Institution, and the Second Inaugural Medal of Franklin D. Roosevelt.

Invocation for the ceremonies will be delivered by The Rev. Kenneth B. Wyatt, pastor of the Greenbelt Community Church. The benediction will be said by Father Victor J. Dowgiallo, pastor of St. Hugh's Catholic Church, Greenbelt.

The tours, both for invited guests and the "open house" for employees and their families, will embrace the full spectrum of the Center's activities. Included will be a Control Room demonstration with simulation of a pre-rocket launch and countdown procedures, followed by a simulated satellite injection into orbit.

Lectures will be given on the Center's operation of global satellite tracking networks, including Minitrack and Project Mercury. The Center's cooperative role for the international exploration of space will be explained. Guests will see an animated miniature tracking station, and a scale model of the forthcoming S-51 spacecraft being instrumented by the United Kingdom which will be flown by Goddard.

There will be an extensive display of spacecraft instrumentation along with many of the Goddard Center's family of sounding rockets used for scientific experiments. Included will be an Iris Sounding Rocket and an Aerobee 150A with a new attitude (or pointing) control system. A similar rocket will be fired March 15 for the first test of controlling a rocket's three axes of pitch, yaw and roll. The experiment also will carry equipment aimed at measuring gamma ray intensities and the solar illumination.

Other models will include the Tiros weather satellite; the P-14 magnetometer (or radiation-counting) spacecraft; the S-3 energetic particles measurement satellite; the Explorer VIII and Vanguard I. Guests will also see a demonstration of a micrometeorite detector.

Tours will be conducted through the Center's extensive laboratory facilities where guests will see vacuum, vibration and spin-balancing equipment used to simulate space environmental conditions.

> EXHIBIT 14
> *COPY*

NATIONAL AERONAUTICS AND SPACE ADMINISTRATION
NEWS RELEASE

GODDARD SPACE FLIGHT CENTER
GREENBELT, MD.

OFFICE OF PUBLIC INFORMATION
PHONE: GRanite 4-9000, Ex. 555

RELEASE NO. 3/14/61-1

GODDARD SPACE FLIGHT CENTER

FACT SHEET

The National Aeronautics and Space Administration Goddard Space Flight Center is the first completely new scientific center created since the NASA was established October 1, 1958. It is the nation's newest facility devoted exclusively to the peaceful exploration of space.

Formally organized on May 1, 1959, it was named for the late Dr. Robert H. Goddard, recognized as the Father of American Rocketry. He designed, developed and flew the world's first liquid-fuel rocket.

The Center was dedicated March 16, 1961, the Thirty-fifth Anniversary of that launching by Dr. Goddard.

The Goddard Space Flight Center is one of ten field laboratory facilities of the N.A.S.A., and is one of several integrated units under the direction of the *Office of Space Flight Programs*. The Center carries out assigned missions in the theoretical, planning, research, development and operational phases of space flight, utilizing laboratory studies and experiments, sounding rockets, earth satellites and space probes.

Organization

<u>Dr. Harry J. Goett is Director</u> of the Goddard Space Flight Center. Principal operating executive is <u>Eugene W. Wasielewski, Associate Director.</u> Other major officials and their duties are:

<u>Mr. John W. Townsend, Jr., Assistant Director, Space Sciences and Satellite Applications,</u> supervises four divisions:

a. The Space Sciences Division, headed by Dr. Leslie H. Meredith, conducts basic research in the space sciences through the use of experiments carried in rocket sondes, earth satellites, and space probes. It supports the NASA National Sounding Rocket Program and provides management and contract monitoring.

b. The Theoretical Division, headed by Dr. Robert Jastrow, studies and conducts research in theoretical physics, mathematics, and mechanics associated with space exploration. Included are special analytical problems involving the use of large computers. The Division is now organizing an Institute for Space Studies in New York City, where it will draw on talent from universities and research groups.

c. The Payload Systems Division, headed by N. Whitney Matthews, is responsible for the integration of experiments and equipments into complete earth satellites and space probes; and for the basic satellite structure, thermal balance, and integrity of the entire system through all anticipated environmental conditions.

d. The Satellite Applications Division, headed by D. G. Mazur, is concerned with research, preliminary design and project management of meteorological, communications, and geodetic satellites.

Mr. John T. Mengel, Assistant Director, Tracking and Data Systems. This office supervises three divisions:

a. Tracking Systems Division, headed by Clarence A. Schroeder, which concerns itself with research and development of new tracking, data acquisition and communications systems between space vehicles and ground receiving stations.

b. The Data Systems Division, headed by Dr. Charles V. L. Smith, applies data reduction techniques to launchings of vehicles, orbit calculations, and satellite and space probe findings.

c. The Operational Support Division, headed by Fred S. Friel, establishes and operates NASA's global tracking, data acquisition, and communications network.

d. The Theory and Analysis Staff, headed by Dr. Joseph W. Siry, provides orbital and system analysis support for tracking and data systems.

Dr. Michael J. Vaccaro, Assistant Director, Business Administration. This office provides business management support functions, including personnel, budget and finance, security, procurement and supply, administrative, and public information.

Mr. Leopold Winkler, Chief, Office of Technical Services. This office is charged with fabrication and shop operation, construction and repair, buildings and grounds maintenance, utilities and automotive operation, and industrial safety.

The Goddard Mission

Specific areas of Goddard's responsibilities are:
- Advanced planning and theoretical studies leading to development of spacecraft for manned and unmanned scientific space investigations. This work includes formulation of concepts and ideas essential to the effectiveness of the NASA program.
- Supporting research in spacecraft, applications systems,

instrumentation, communications, guidance, and rocket vehicles as assigned.

* Development and fabrication of spacecraft for scientific and applications programs and manned space flight. The Center designs, develops, and fabricates prototypes of components and systems to advance space technology or to foster practical applications. Although the Center directs all such work, most of it is contracted out to industry and universities. For reasons of economy, urgency, efficiency or effectiveness, about fifteen percent of such activities are performed internally by the Center.

* Development and supervision of worldwide tracking, data acquisition, communications and computing operations for all NASA space programs except deep space probes.

* Interpretation of results of experiments under Goddard management.

* Management of projects, including technical direction and the execution and monitoring of contracts.

Staff and Facilities

Located on a 550-acre tract of land near Greenbelt, Maryland, about fifteen miles from Washington, D.C., the physical plant when completed in 1962 will consist of eight facilities costing approximately $27 million. Facilities now in operation are: Space Projects Building, Research Projects Laboratory and Central Flight Control and Operations.

First elements of the Goddard Staff were drawn from the Vanguard Project Team of the U.S. Naval Research Laboratory. The complement now numbers more than 1,300 and will total 1,800 by the end of 1961. It is expected to rise to about 2,000 when the Center construction is completed.

Facilities at the Center will enable scientists and engineers to subject payloads to the complete range of flight environments without having to risk an actual launch with untried hardware.

As one example, the Center's environmental test facilities include centrifuges, dynamic balancing machines, vibration machines, and thermal vacuum chambers. Of the last, there will be four when the present construction program is completed—including some of the largest chambers in the country. The largest will measure thirty feet by forty feet and will be capable of producing the complete range of near-space vacuum and temperature conditions.

Scientific Exploration in Space

Goddard Center programs embrace unmanned scientific research and exploration of space; study of the earth's upper atmosphere; study of the earth itself from the space viewpoint; unmanned technological utilization of space for practical purposes, such as weather forecasting and global telephone, radio and television communications; and near-space tracking and data handling.

During the present decade, in the execution of this program, the Center plans to launch at least ninety-six scientific satellites and twenty-eight applications satellites.

Launching of sounding rockets and space probes will total

hundreds. Purchases of sounding rockets in the current fiscal year include 22 Aerobees, 18 Nike-Asps, 20 Nike-Cajuns, 4 Argo D-4's (Javelins), 2 Argo D-8's (Journeyman), 5 Iris. Frequency of rocket launchings is expected to increase steadily over the next decade, particularly in support of the Tiros-Nimbus weather satellite programs.

The Center plays a major role in NASA's international cooperation for the peaceful exploration of space. Scientists of twenty-one nations are participating in the Tiros meteorological satellite experiments; Canada is designing the payload for a swept-frequency topside ionosphere sounder satellite to be launched and tracked by Goddard; United Kingdom will supply the experiments for an international ionosphere satellite to be built and launched by Goddard; and a number of scientists and technicians are in cooperative study and training at the Center.

Accomplishments

Since its organization, the Goddard staff has made significant contributions to knowledge of space and the earth.

The VANGUARD III satellite, designed and launched by the Goddard Staff, provided information on the distribution and intensity of the earth's magnetic fields; detailed location of the lower edge of the Van Allen Great Radiation Belt, and made an accurate count of micrometeorite impacts.

The EXPLORER VII satellite, in which Goddard played a key role in cooperation with the NASA Marshall Space Flight Center and the Jet Propulsion Laboratory, provided valuable information on radiation balance, Lyman-Alpha x-rays, heavy primary rays, micrometeorite impacts, solar exposure, temperature, magnetic storms and detected large-scale weather patterns.

The PIONEER V space probe experiment, jointly conducted by Goddard and the Space Technology Laboratories, transmitted valuable data on solar flares, particle energies and their distribution, and magnetic fields. The probe, second U.S. spacecraft to orbit the sun, transmitted data to earth from a record distance of 22,500,000 miles.

The TIROS I and II meteorological satellite experiments, launched by the Delta rocket developed under Goddard supervision, provided many thousands of photographs of the earth's cloud cover and mapped radiation and heat balance on a global scale.

ECHO I, the first passive communications satellite, was launched in cooperation with the NASA Langley Research Center. This 100-foot inflatable sphere proved the feasibility of communications over long distances by bouncing radio signals off its reflective surface. Messages were transmitted across the continent and across the Atlantic and photographs were sent by the same means.

The EXPLORER VIII satellite, another joint project of Goddard, the Jet Propulsion Laboratory and Marshall Space Flight Center, carried out the first intensive direct measurements of the earth's ionosphere, by measuring concentrations of charged particles and their temperatures.

In its sounding rocket programs, Goddard made the first measurements of auroral absorption events and solar proton beams; first flew an alkali vapor magnetometer to measure the earth's magnetic field at altitudes above 600 miles, and obtained the first ultraviolet stellar spectra.

The NERV, or Nuclear Emulsion Recovery Vehicle, launched by sounding rocket from the Pacific Missile Range to a height of 1,260 miles, was recovered from the ocean. The experiment produced exact measurements of the lower Van Allen Great Radiation Belt.

EXHIBIT 15
COPY

HISTORICAL BACKGROUND AND COMMUNICATIONS SATELLITE ACT OF 1962

On August 31, 1962, the President signed H.R. 11040, and the Communications Satellite Act of 1962 became law. At the time of signing, the President congratulated the Congress for "a step of historical importance." He stated further: "It promises significant benefits to our own people and to the whole world. Its purpose is to establish a commercial communications system, utilizing space satellites which will serve our needs and those of other countries and contribute to world peace and understanding."

Major steps in the development of this legislation were as follows:
  a. June 15, 1961, the President asked the Chairman of the National Aeronautics and Space Council to have recommendations prepared for communications satellite policy. Under direction of the Council staff, interagency meetings were held; policy recommendations were drafted; and those recommendations were acted upon unanimously by the Council.
  b. July 14, 1961, the President approved and released the policy statement, which stressed the public interest objectives in obtaining a global system as soon as technically feasible. This policy stated that private ownership and operation of the U.S. portion of the system is favored, provided that the public interest is adequately protected through opportunities for foreign participation, non-discriminatory use of and equitable access to the system, and effective competition in the acquisition of equipment and in the structure of ownership and control.
  c. In the fall of 1961, the President requested the staff of the Council to draft recommendations in order that the communications satellite policy could be effectively implemented. Under the direction of the Council staff, interagency drafting sessions were held, and the proposed bill was prepared and transmitted to the President.
  d. February 7, 1962, the President sent the proposed legislation to the Congress and, in his accompanying message, urged that it be given prompt and favorable consideration.
  e. Extensive hearings were held in the Congress. Six different committees called witnesses and participated in a thorough examination of the communications satellite policy and proposed legislation. After such committee actions, explanation and debate took place prior to votes in both the House and the Senate. The House passed a bill by a 354 to 9 vote on May 3; the Senate passed its corresponding version of a bill by a 66 to 11 vote on August 11; and the House acted to accept the Senate bill by a 377 to 10 vote on August 27.
  f. August 31, 1962, the bill was signed by the President and became law.

g. October 4, 1962, the President nominated 13 distinguished citizens to be Incorporators, with the statutory responsibility for taking the necessary actions to establish a Communications Satellite Corporation.

The Incorporators, under interim appointments, have held a number of meetings to consider and initiate the steps required to organize the corporation and to apply for a charter under the District of Columbia Business Incorporation Act, as provided under the terms of the Communications Satellite Act.

The Communications Satellite Act of 1962 incorporates the major objectives of the President's policy statement of July 24, 1961. It provides authority for the creation of a private corporation to serve as the United States portion of any global system. It will be privately financed and the essential business management will be in the hands of 12 directors elected by the stockholders and 3 directors appointed by the President and confirmed by the Senate. At the same time that the benefits of profit-making incentives and private management are obtained, the Act is most careful to identify national policy objectives in relation to the use of commercial communications satellites and to provide the machinery within Government for the regulation of and assistance to the corporation. In such a framework, it is expected that the services the corporation provides and the way it conducts business will be wholly responsive to the several objectives of the Act.

EXHIBIT 16
COPY

Public Law 87-624
87th Congress, H. R. 11040
August 31, 1962

An Act

76 STAT. 419.

To provide for the establishment, ownership, operation, and regulation of a commercial communications satellite system, and for other purposes.

Be it enacted by the Senate and House of Representatives of the United States of America in Congress assembled,

TITLE I—SHORT TITLE, DECLARATION OF POLICY AND DEFINITIONS

SHORT TITLE

SEC. 101. This Act may be cited as the "Communications Satellite Act of 1962".

DECLARATION OF POLICY AND PURPOSE

SEC. 102. (a) The Congress hereby declares that it is the policy of the United States to establish, in conjunction and in cooperation with other countries, as expeditiously as practicable a commercial communications satellite system, as part of an improved global communications network, which will be responsive to public needs and national objectives, which will serve the communication needs of the United States and other countries, and which will contribute to world peace and understanding.

(b) The new and expanded telecommunication services are to be made available as promptly as possible and are to be extended to provide global coverage at the earliest practicable date. In effectuating this program, care and attention will be directed toward providing such services to economically less developed countries and areas as well as those more highly developed, toward efficient and economical use of the electromagnetic frequency spectrum, and toward the reflection of the benefits of this new technology in both quality of services and charges for such services.

(c) In order to facilitate this development and to provide for the widest possible participation by private enterprise, United States participation in the global system shall be in the form of a private corporation, subject to appropriate governmental regulation. It is the intent of Congress that all authorized users shall have nondiscriminatory access to the system; that maximum competition be maintained in the provision of equipment and services utilized by the system; that the corporation created under this Act be so organized and operated as to maintain and strengthen competition in the provision of communications services to the public; and that the activities of the corporation created under this Act and of the persons or companies participating in the ownership of the corporation shall be consistent with the Federal antitrust laws.

(d) It is not the intent of Congress by this Act to preclude the use of the communications satellite system for domestic communication services where consistent with the provisions of this Act nor to preclude the creation of additional communications satellite systems, if required to meet unique governmental needs or if otherwise required in the national interest.

319

DEFINITIONS

SEC. 103. As used in this Act, and unless the context otherwise requires—

(1) the term "communications satellite system" refers to a system of communications satellites in space whose purpose is to relay telecommunication information between satellite terminal stations, together with such associated equipment and facilities for tracking, guidance, control, and command functions as are not part of the generalized launching, tracking, control, and command facilities for all space purposes;

(2) the term "satellite terminal station" refers to a complex of communication equipment located on the earth's surface, operationally connected with one or more terrestrial communication systems, and capable of transmitting telecommunications to or receiving telecommunications from a communications satellite system.

(3) the term "communications satellite" means an earth satellite which is intentionally used to relay telecommunication information;

(4) the term "associated equipment and facilities" refers to facilities other than satellite terminal stations and communications satellites, to be constructed and operated for the primary purpose of a communications satellite system, whether for administration and management, for research and development, or for direct support of space operations;

(5) the term "research and development" refers to the conception, design, and first creation of experimental or prototype operational devices for the operation of a communications satellite system, including the assembly of separate components into a working whole, as distinguished from the term "production," which relates to the construction of such devices to fixed specifications compatible with repetitive duplication for operational applications; and

(6) the term "telecommunication" means any transmission, emission or reception of signs, signals, writings, images, and sounds or intelligence of any nature by wire, radio, optical, or other electromagnetic systems.

(7) the term "communications common carrier" has the same meaning as the term "common carrier" has when used in the Communications Act of 1934, as amended, and in addition includes, but only for purposes of sections 303 and 304, any individual, partnership, association, joint-stock company, trust, corporation, or other entity which owns or controls, directly or indirectly, or is under direct or indirect common control with, any such carrier; and the term "authorized carrier", except as otherwise provided for purposes of section 304 by section 304(b)(1), means a communications common carrier which has been authorized by the Federal Communications Commission under the Communications Act of 1934, as amended, to provide services by means of communications satellites;

(8) the term "corporation" means the corporation authorized by title III of this Act.

(9) the term "Administration" means the National Aeronautics and Space Administration; and

(10) the term "Commission" means the Federal Communications Commission.

TITLE II—FEDERAL COORDINATION, PLANNING, AND REGULATION

IMPLEMENTATION OF POLICY

SEC. 201. In order to achieve the objectives and to carry out the purposes of this Act—

(a) the President shall—

(1) aid in the planning and development and foster the execution of a national program for the establishment and operation, as expeditiously as possible, of a commercial communications satellite system;

(2) provide for continuous review of all phases of the development and operation of such a system, including the activities of a communications satellite corporation authorized under title ᵀII of this Act;

(3) coordinate the activities of governmental agencies with responsibilities in the field of telecommunication, so as to insure that there is full and effective compliance at all times with the policies set forth in this Act;

(4) exercise such supervision over relationships of the corporation with foreign governments or entities or with international bodies as may be appropriate to assure that such relationships shall be consistent with the national interest and foreign policy of the United States;

(5) insure that timely arrangements are made under which there can be foreign participation in the establishment and use of a communications satellite system;

(6) take all necessary steps to insure the availability and appropriate utilization of the communications satellite system for general governmental purposes except where a separate communications satellite system is required to meet unique governmental needs, or is otherwise required in the national interest; and

(7) so exercise his authority as to help attain coordinated and efficient use of the electromagnetic spectrum and the technical compatibility of the system with existing communications facilities both in the United States and abroad.

(b) the National Aeronautics and Space Administration shall—

(1) advise the Commission on technical characteristics of the communications satellite system;

(2) cooperate with the corporation in research and development to the extent deemed appropriate by the Administration in the public interest;

(3) assist the corporation in the conduct of its research and development program by furnishing to the corporation, when requested, on a reimbursable basis, such satellite launching and associated services as the Administration deems necessary for the most expeditious and economical development of the communications satellite system;

(4) consult with the corporation with respect to the technical characteristics of the communications satellite system;

(5) furnish to the corporation, on request and on a reimbursable basis, satellite launching and associated services required for the establishment, operation, and maintenance of the communications satellite system approved by the Commission; and

(6) to the extent feasible, furnish other services, on a reimbursable basis, to the corporation in connection with the establishment and operation of the system.

(c) the Federal Communications Commission, in its administration of the provisions of the Communications Act of 1934, as amended, and as supplemented by this Act, shall—

(1) insure effective competition, including the use of competitive bidding where appropriate, in the procurement by the corporation and communications common carriers of apparatus, equipment, and services required for the establishment and operation of the communications satellite system and satellite terminal stations; and the Commission shall consult with the Small Business Administration and solicit its recommendations on measures and procedures which will insure that small business concerns are given an equitable opportunity to share in the procurement program of the corporation for property and services, including but not limited to research, development, construction, maintenance, and repair.

(2) insure that all present and future authorized carriers shall have nondiscriminatory use of, and equitable access to, the communications satellite system and satellite terminal stations under just and reasonable charges, classifications, practices, regulations, and other terms and conditions and regulate the manner in which available facilities of the system and stations are allocated among such users thereof;

(3) in any case where the Secretary of State, after obtaining the advice of the Administration as to technical feasibility, has advised that commercial communication to a particular foreign point by means of the communications satellite system and satellite terminal stations should be established in the national interest, institute forthwith appropriate proceedings under section 214(d) of the Communications Act of 1934, as amended, to require the establishment of such communication by the corporation and the appropriate common carrier or carriers;

(4) insure that facilities of the communications satellite system and satellite terminal stations are technically compatible and interconnected operationally with each other and with existing communications facilities;

(5) prescribe such accounting regulations and systems and engage in such ratemaking procedures as will insure that any economies made possible by a communications satellite system are appropriately reflected in rates for public communication services;

(6) approve technical characteristics of the operational communications satellite system to be employed by the corporation and of the satellite terminal stations; and

(7) grant appropriate authorizations for the construction and operation of each satellite terminal station, either to the corporation or to one or more authorized carriers or to the corporation and one or more such carriers jointly, as will best serve the public interest, convenience, and necessity. In determining the public interest, convenience, and necessity the Commission shall authorize the construction and operation of such stations by communications common carriers or the corporation, without preference to either;

(8) authorize the corporation to issue any shares of capital stock, except the initial issue of capital stock referred to in section 304(a), or to borrow any moneys, or to assume any

Pub. Law 87-624
76 STAT. 423.

obligation in respect of the securities of any other person, upon a finding that such issuance, borrowing, or assumption is compatible with the public interest, convenience, and necessity and is necessary or appropriate for or consistent with carrying out the purposes and objectives of this Act by the corporation;

(9) insure that no substantial additions are made by the corporation or carriers with respect to facilities of the system or satellite terminal stations unless such additions are required by the public interest, convenience, and necessity;

(10) require, in accordance with the procedural requirements of section 214 of the Communications Act of 1934, as amended, that additions be made by the corporation or carriers with respect to facilities of the system or satellite terminal stations where such additions would serve the public interest, convenience, and necessity; and

(11) make rules and regulations to carry out the provisions of this Act.

TITLE III—CREATION OF A COMMUNICATIONS SATELLITE CORPORATION

CREATION OF CORPORATION

SEC. 301. There is hereby authorized to be created a communications satellite corporation for profit which will not be an agency or establishment of the United States Government. The corporation shall be subject to the provisions of this Act and, to the extent consistent with this Act, to the District of Columbia Business Corporation Act. The right to repeal, alter, or amend this Act at any time is expressly reserved.

PROCESS OF ORGANIZATION

SEC. 302. The President of the United States shall appoint incorporators, by and with the advice and consent of the Senate, who shall serve as the initial board of directors until the first annual meeting of stockholders or until their successors are elected and qualified. Such incorporators shall arrange for an initial stock offering and take whatever other actions are necessary to establish the corporation, including the filing of articles of incorporation, as approved by the President.

DIRECTORS AND OFFICERS

SEC. 303. (a) The corporation shall have a board of directors consisting of individuals who are citizens of the United States, of whom one shall be elected annually by the board to serve as chairman. Three members of the board shall be appointed by the President of the United States, by and with the advice and consent of the Senate, effective the date on which the other members are elected, and for terms of three years or until their successors have been appointed and qualified, except that the first three members of the board so appointed shall continue in office for terms of one, two, and three years, respectively, and any member so appointed to fill a vacancy shall be appointed only for the unexpired term of the director whom he succeeds. Six members of the board shall be elected annually by those stockholders who are communications common carriers and six shall be elected annually by the other stockholders of the corporation. No stockholder who is a communications common carrier and no trustee for such a stockholder shall vote, either directly or indirectly, through the votes of subsidiaries or affiliated companies, nominees, or any persons subject to his direction or control, for more than three candidates for membership on the board. Subject to such limitation, the articles of incorporation to be filed by the incorporators designated under section 302 shall provide for cumulative voting under section 27(d) of the District of Columbia Business Corporation Act (D.C. Code, sec. 29–911(d)).

(b) The corporation shall have a president, and such other officers as may be named and appointed by the board, at rates of compensation fixed by the board, and serving at the pleasure of the board. No individual other than a citizen of the United States may be an officer of the corporation. No officer of the corporation shall receive any salary from any source other than the corporation during the period of his employment by the corporation.

FINANCING OF THE CORPORATION

SEC. 304. (a) The corporation is authorized to issue and have outstanding, in such amounts as it shall determine, shares of capital stock, without par value, which shall carry voting rights and be eligible for dividends. The shares of such stock initially offered shall be sold at a price not in excess of $100 for each share and in a manner to encourage the widest distribution to the American public. Subject to the provisions of subsections (b) and (d) of this section, shares of stock offered under this subsection may be issued to and held by any person.

(b)(1) For the purposes of this section the term "authorized carrier" shall mean a communications common carrier which is specifically authorized or which is a member of a class of carriers authorized by the Commission to own shares of stock in the corporation upon a finding that such ownership will be consistent with the public interest, convenience, and necessity.

(2) Only those communications common carriers which are authorized carriers shall own shares of stock in the corporation at any time, and no other communications common carrier shall own shares either directly or indirectly through subsidiaries or affiliated companies, nominees, or any persons subject to its direction or control. Fifty per centum of the shares of stock authorized for issuance at any time by the corporation shall be reserved for purchase by authorized carriers and such carriers shall in the aggregate be entitled to make purchases of the reserved shares in a total number not exceeding the total number of the nonreserved shares of any issue purchased by other persons. At no time after the initial issue is completed shall the aggregate of the shares of voting stock of the corporation owned by authorized carriers directly or indirectly through subsidiaries or affiliated companies, nominees, or any persons subject to their direction or control exceed 50 per centum of such shares issued and outstanding.

(3) At no time shall any stockholder who is not an authorized carrier, or any syndicate or affiliated group of such stockholders, own more than 10 per centum of the shares of voting stock of the corporation issued and outstanding.

(c) The corporation is authorized to issue, in addition to the stock authorized by subsection (a) of this section, nonvoting securities, bonds, debentures, and other certificates of indebtedness as it may determine. Such nonvoting securities, bonds, debentures, or other certificates of indebtedness of the corporation as a communications common carrier may own shall be eligible for inclusion in the rate base of the carrier to the extent allowed by the Commission. The voting stock of the corporation shall not be eligible for inclusion in the rate base of the carrier.

(d) Not more than an aggregate of 20 per centum of the shares of stock of the corporation authorized by subsection (a) of this section which are held by holders other than authorized carriers may be held by persons of the classes described in paragraphs (1), (2), (3), (4), and (5) of section 310(a) of the Communications Act of 1934, as amended (47 U.S.C. 310).

(e) The requirement of section 45(b) of the District of Columbia Business Corporation Act (D.C. Code, sec. 29-920(b)) as to the percentage of stock which a stockholder must hold in order to have the rights of inspection and copying set forth in that subsection shall not be applicable in the case of holders of the stock of the corporation, and they may exercise such rights without regard to the percentage of stock they hold.

(f) Upon application to the Commission by any authorized carrier and after notice and hearing, the Commission may compel any other authorized carrier which owns shares of stock in the corporation to transfer to the applicant, for a fair and reasonable consideration, a number of such shares as the Commission determines will advance the public interest and the purposes of this Act. In its determination with respect to ownership of shares of stock in the corporation, the Commission, whenever consistent with the public interest, shall promote the widest possible distribution of stock among the authorized carriers.

PURPOSES AND POWERS OF THE CORPORATION

SEC. 305. (a) In order to achieve the objectives and to carry out the purposes of this Act, the corporation is authorized to—

(1) plan, initiate, construct, own, manage, and operate itself or in conjunction with foreign governments or business entities a commercial communications satellite system;

(2) furnish, for hire, channels of communication to United States communications common carriers and to other authorized entities, foreign and domestic; and

(3) own and operate satellite terminal stations when licensed by the Commission under section 201(c)(7).

(b) Included in the activities authorized to the corporation for accomplishment of the purposes indicated in subsection (a) of this section, are, among others not specifically named—

(1) to conduct or contract for research and development related to its mission;

(2) to acquire the physical facilities, equipment and devices necessary to its operations, including communications satellites and associated equipment and facilities, whether by construction, purchase, or gift;

(3) to purchase satellite launching and related services from the United States Government;

(4) to contract with authorized users, including the United States Government, for the services of the communications satellite system; and

(5) to develop plans for the technical specifications of all elements of the communications satellite system.

(c) To carry out the foregoing purposes, the corporation shall have the usual powers conferred upon a stock corporation by the District of Columbia Business Corporation Act.

TTLE IV—MISCELLANEOUS

APPLICABILITY OF COMMUNICATIONS ACT OF 1934

SEC. 401. The corporation shall be deemed to be a common carrier within the meaning of section 3(h) of the Communications Act of 1934, as amended, and as such shall be fully subject to the provisions of title II and title III of that Act. The provision of satellite terminal station facilities by one communication common carrier to one or more other communications common carriers shall be deemed to be a common carrier activity fully subject to the Communications Act. Whenever the application of the provisions of this Act shall be inconsistent with the application of the provisions of the Communications Act, the provisions of this Act shall govern.

NOTICE OF FOREIGN BUSINESS NEGOTIATIONS

SEC. 402. Whenever the corporation shall enter into business negotiations with respect to facilities, operations, or services authorized by this Act with any international or foreign entity, it shall notify the Department of State of the negotiations, and the Department of State shall advise the corporation of relevant foreign policy considerations. Throughout such negotiations the corporation shall keep the Department of State informed with respect to such considerations. The corporation may request the Department of State to assist in the negotiations, and that Department shall render such assistance as may be appropriate.

SANCTIONS

SEC. 403. (a) If the corporation created pursuant to this Act shall engage in or adhere to any action, practices, or policies inconsistent with the policy and purposes declared in section 102 of this Act, or if the corporation or any other person shall violate any provision of this Act, or shall obstruct or interfere with any activities authorized by this Act, or shall refuse, fail, or neglect to discharge his duties and responsibilities under this Act, or shall threaten any such violation, obstruction, interference, refusal, failure, or neglect, the district court of the United States for any district in which such corporation or other person resides or may be found shall have jurisdiction, except as otherwise prohibited by law, upon petition of the Attorney General of the United States, to grant such equitable relief as may be necessary or appropriate to prevent or terminate such conduct or threat.

(b) Nothing contained in this section shall be construed as relieving any person of any punishment, liability, or sanction which may be imposed otherwise than under this Act.

(c) It shall be the duty of the corporation and all communications common carriers to comply, insofar as applicable, with all provisions of this Act and all rules and regulations promulgated thereunder.

REPORTS TO THE CONGRESS

SEC. 404. (a) The President shall transmit to the Congress in January of each year a report which shall include a comprehensive description of the activities and accomplishments during the preceding calendar year under the national program referred to in section 201(a)(1), together with an evaluation of such activities and accomplishments in terms of the attainment of the objectives of this Act and any recommendations for additional legislative or other action which the President may consider necessary or desirable for the attainment of such objectives.

(b) The corporation shall transmit to the President and the Congress, annually and at such other times as it deems desirable, a comprehensive and detailed report of its operations, activities, and accomplishments under this Act.

(c) The Commission shall transmit to the Congress, annually and at such other times as it deems desirable, (i) a report of its activities and actions on anticompetitive practices as they apply to the communications satellite programs; (ii) an evaluation of such activities and actions taken by it within the scope of its authority with a view to recommending such additional legislation which the Commission may consider necessary in the public interest; and (iii) an evaluation of the capital structure of the corporation so as to assure the Congress that such structure is consistent with the most efficient and economical operation of the corporation.

Approved August 31, 1962, 9:51 a.m.

Appendix I
Robert H. Goddard Contributions and Memorabilia

Robert H. Goddard's Basic Contribution to Rocketry and Space Flight

First American to explore mathematically the practicality of using rocket propulsion to reach high altitudes and to traject to the moon (1912)

First to receive a U.S. patent on the idea of a multistage rocket (1914)

First to prove, by actual static test, that rocket propulsion operates in a vacuum, that it needs no air to push against (1915–1916)

First to develop suitable pumps for liquid-fuel rockets (1923)

First to develop and successfully fly a liquid-fuel rocket (March 16, 1926)

First to launch a scientific payload (a barometer, a thermometer, and a camera) in a rocket flight (1929)

First to use vanes in the rocket thrust for guidance (1932)

First to develop gyro control apparatus for rocket flight (1932)

First to fire a liquid-fuel rocket faster than the speed of sound (1935)

First to launch successfully a rocket with a motor pivoted in gimbals controlled by a gyro mechanism (1937)

Robert H. Goddard Memorabilia

National Air Museum, Smithsonian Institution, Washington, D.C.—Exhibit of the four complete extant Goddard rockets, made in 1926, 1935, 1938, and 1941; also some rocket parts, an oil portrait of Dr. Goddard, and a few personal memorabilia.

Institute of the Aerospace Sciences, New York City—Goddard collection of early rocket literature, one of the best in the U.S., which was transferred to the Library of Congress. The Institute had, from 1950 to 1959, an exhibit of numerous devices and parts developed and used by Dr. Goddard from 1918–1945, on long-term loan to the Roswell Museum, Roswell, New Mexico.

Roswell Museum, Roswell, New Mexico—Largest exhibit of parts of Goddard rockets, housed in the Goddard Wing of museum (dedicated April 25, 1959), including piping, drawings, murals of life-size photographs of four Smithsonian rockets, etc. On the grounds of the museum is the launching tower used by Dr. Goddard, with a copy of the 1940 rocket in it. His observation tower is also on display.

Clark University, Worcester, Mass.—Physics Department owned rocket parts, with additional items given by Mrs. Goddard. Also life-size photos, used as murals, of the four Smithsonian-held rockets, and bronze tablet at entrance to Physics Building, a gift of the class of 1959.

Worcester Polytechnic Institute, Worcester, Mass.—Collection of solid-propellant rockets, some of which were developed in WPI building and grounds. Also set of four murals of life-size Smithsonian rockets.

Robert H. Goddard Professorships, at Guggenheim Jet Propulsion Centers, at Princeton University, and at the California Institute of Technology, established by The Daniel and Florence Guggenheim Foundation.

Goddard Award, given by the American Institute of Aeronautics and Astronautics annually to one selected by its Directors as having made the greatest contribution to rocket development during the year; the oldest award in rocketry.

Goddard Power Plant, multimillion dollar plant at the Naval Proving Ground, Indian Head, Maryland, in memory of Dr. Goddard's work there in 1920–1922.

WPI 1908–Goddard Memorial Fund—Established by the class of 1908 of Worcester Polytechnic Institute in June 1958; income to be used for prize or scholarship.

Hill Transportation Award, of the Institute of the Aerospace Sciences—Carrying $5,000 and citation, accepted for Dr. Goddard by Mrs. Goddard at the annual dinner of IAS, January 1959.

Golden Replica of 1926 Goddard Rocket—Accepted for Dr. Goddard by Mrs. Goddard at the Missile Industry Conference at Washington, D.C., June 1958. Now on view at the Goddard Space Flight Center, Greenbelt, Md.

Goddard Memorial Dinner—Sponsored annually since 1958 on the anniversary of the first liquid-fuel flight, in Washington, D.C., by the National Space Club.

Goddard Trophy and Goddard Scholarship—Given annually at the Goddard Memorial Dinner at Washington, D.C., sponsored by the National Space Club.

Air Force Academy Goddard Award, Colorado—Established by the American Ordnance Association, to the cadet with the highest standing in mathematics in each graduating class.

Langley Medal—Dr. Goddard was the ninth recipient of this coveted medal from the Regents of the Smithsonian Institution, Washington, D.C. Presented to Mrs. Goddard, in Washington, June 28, 1960.

American Rocketry Society Goddard Memorial—Granite marker at site of the first flight of a liquid-propellant rocket, March 16, 1926, at Auburn, Mass., with a granite tablet beside road nearby, explaining significance of the marker. Dedicated July 13, 1960.

Congressional Medal—Presented posthumously to Dr. Goddard on March 16, 1961, at the dedication of the Goddard Space Flight Center, Greenbelt, Md.

Goddard Institute for Space Studies, extension in New York City of the Goddard Space Flight Center; established January 1961.

Goddard Alumni Award—Established by the alumni of Worcester Polytechnic Institute, June 1961, awarded annually to an outstanding alumnus.

Robert H. Goddard Squadron—Air Force Association, Vandenberg Air Force Base, California, established 1961.

Robert H. Goddard Industrial Center, Worcester, Mass.—Dedicated June 19, 1961.

Goddard Science Symposium of the American Astronautical Society—Annually on March 16, in Washington, D.C.

Robert H. Goddard Historical Essay Competition of the National Space Club—Prize of $200 and trophy awarded annually for the best essay on the historical development of rocketry and astronautics; established March 1962.

Robert H. Goddard Achievement Award—In Civil Air Patrol Cadet Aerospace Education and Training Program, National Headquarters, Ellington Air Force Base, Tex.

APPENDIX I

Robert H. Goddard Memorial Library, Clark University, Worcester, Mass., depository of Dr. Goddard's papers, established 1964.

Robert H. Goddard Memorial—Tower and rocket at Fort Devens, Mass., at the site of the Goddard testing tower, 1929–1930. Dedicated May 1963.

Appendix J
Selected Bibliography

Books

AARONS, J. (ed.), *Radio Astronomical and Satellite Studies of the Atmosphere* (New York, 1963).

AKENS, DAVID S., *Historical Origins of George C. Marshall Space Flight Center* (Huntsville, Ala., 1960).

American Institute of Aeronautics and Astronautics, *AIAA Unmanned Spacecraft Meeting*, AIAA Publication CP-12 (New York, 1965).

BERGAUST, ERIC, *Reaching for the Stars* (Garden City, N.Y., 1960).

BERKNER, LLOYD V. (ed.), *Manual on Rockets and Satellites* (New York, 1958).

—— AND ODISHAW, HUGH, *Science in Space* (New York, 1961).

BESTEV, ALFRED, *The Life and Death of a Satellite* (Boston, 1966).

BLANCO, V. M., AND McCUSKEY, S. W., *Basic Physics of the Solar System* (Reading, Mass., 1961).

BOEHM, J., FICHTNER, H. J., AND HOBERG, O. A., "Explorer Satellites Launched by Juno 1 and Juno 2 Space Carrier Vehicles," in *Astronautical Engineering and Science*, Ernst Stuhlinger et al., eds. (New York, 1963).

BOYD, R. L. F., *Space Research by Rocket and Satellite* (New York, 1960).

BRANDT, J. C., AND HODGE, P. W., *Solar System Astrophysics* (New York, 1964).

CARTER, L. J. (ed.), *The Artificial Satellite: Proceedings of the Second International Congress on Astronautics*, British Interplanetary Society (London, 1951).

CHAMBERLAIN, JOSEPH W., *Physics of the Aurora and Airglow* (New York, 1961).

CHAPMAN, JOHN L., *Atlas: The Story of a Missile* (New York, 1960).

CHAPMAN, SIDNEY, *Solar Plasma, Geomagnetism and Aurora* (New York, 1964).

DEUTSCH, ARMIN J., AND KLEMPERER, WOLFGANG B. (eds.), *Space Age Astronomy* (New York, 1962).

EMME, EUGENE M., *Aeronautics and Astronautics, 1915–60* (Washington, D.C., 1961).

—— (ed.), *The History of Rocket Technology* (Detroit, 1964).

Franklin Institute: *Earth Satellites as Research Vehicles*, Monograph 2 (Philadelphia, 1956).

FRUTKIN, ARNOLD, *International Cooperation in Space* (Englewood Cliffs, N.J., 1965).

GARTMAN, HEINZ, *The Men Behind the Space Rockets* (New York, 1956).

GATLAND, KENNETH W., *Project Satellite* (London, 1958).

GLASSTONE, S., *Sourcebook on the Space Sciences* (Princeton, 1965).

GODDARD, ROBERT H., *Rocket Development: Liquid Fuel Rocket Research, 1929–1941* (Englewood Cliffs, N.J., 1961).

GRIMWOOD, JAMES M., *Project Mercury: A Chronology*, NASA SP-4001 (Washington, D.C., 1963).

HALE, EDWARD E., *The Brick Moon and Other Stories* (Boston, 1899).

HAVILAND, ROBERT P., AND HOUSE, C. M., *Handbook of Satellites and Space Vehicles* (Princeton, 1965).

HESS, WILMOT N. (ed.), *Introduction to Space Science* (New York, 1965).

JASTROW, ROBERT (ed.), *The Exploration of Space* (New York, 1960).

—— AND CAMERON, A. G. W. (eds.), *Origin of the Solar System* (New York, 1963).

JETER, IRVING E. (ed.), *Scientific Satellites, Mission and Design* (North Hollywood, Calif., 1963).

JONES, BESSIE Z., *Lighthouse of the Skies: A History of the Smithsonian Astrophysical Observatory* (Washington, D.C., 1965).

JOHNSON, FRANCIS S. (ed.), *Satellite Environment Handbook* (Palo Alto, 1965). Second edition.

KALLMAN BIJL, HILDE (ed.), *Space Research* (New York, 1960).

KAULA, WILLIAM M., *Theory of Satellite Geodesy* (New York, 1965).

KING-HELE, D. G., *Satellites and Scientific Research* (New York, 1965).

——, MULLER, P., AND RIGHINI, G., *Space Research V* (New York, 1965).

KURNOSOVA, L. V. (ed.), *Artificial Satellites* (New York, vols. 1–6, 1960–1961).

LASSER, DAVID, *The Conquest of Space* (New York, 1931).

LEGALLEY, DONALD P. (ed.), *Space Science* (New York, 1963).

—— AND ROSEN, ALAN, *Space Physics* (New York, 1964).

—— AND MCKEE, JOHN W., *Space Exploration* (New York, 1964).

LEHMAN, MILTON, *This High Man: The Life of Robert H. Goddard* (New York, 1963).

LEY, WILLY, *Rockets, Missiles, and Space Travel* (3 rev. ed., New York, 1961).

MCMAHON, A. J., *Astrophysics and Space Science: An Integration of Sciences* (Englewood Cliffs, N.J., 1965).

MOORE, PATRICK, *Earth Satellites* (New York, 1956).

MORGENTHALER, GEORGE (ed.), *Unmanned Exploration of the Solar System* (New York, 1965).

MUELLER, I. I., AND ROCKIE, J. D., *Gravimetric and Celestial Geodesy* (New York, 1966).

MULLER, P. (ed.), *Space Research IV* (New York, 1964).

NAUGLE, JOHN E., *Unmanned Space Flight* (New York, 1965).

NEWELL, HOMER E. (ed.), *Sounding Rockets* (New York, 1959).

——, *Express to the Stars* (New York, 1960).

ODISHAW, HUGH (ed.), *Research in Geophysics* series; Vol. 1, *Sun, Upper Atmosphere and Space* (Cambridge, 1964).

—— AND RUTTENBERG, S. (eds.), *Geophysics and the IGY*, American Geophysical Union, Monograph 2 (Washington, 1958).

PENDRAY, G. EDWARD, *The Coming Age of Rocket Power* (New York, 1945).

PRIESTER, WOLFGANG (ed.), *Space Research III* (New York, 1963).

ROSEN, MILTON, *The Viking Rocket Story* (New York, 1955).

ROSHOLT, ROBERT L., *An Administrative History of NASA, 1958–1963*, NASA SP–4101 (Washington, D.C., 1966).

SCHWIEBERT, ERNEST G., *A History of the U.S. Air Force Ballistic Missiles* (New York, 1965).

SHTERNFELD, ARI, *Artificial Satellites* (Washington, 1958).

SMITH-ROSE, R. L. (ed.), *Space Research VI* (Washington, 1966).

STEHLING, KURT R., *Project Vanguard* (Garden City, N.Y., 1961).

SWENSON, LOYD S., GRIMWOOD, JAMES M., AND ALEXANDER, CHARLES C., *This New Ocean: A History of Project Mercury*, NASA SP–4201 (Washington, D.C., 1966).

SULLIVAN, WALTER, *Assault on the Unknown: The International Geophysical Year* (New York, 1961).

THOMAS, SHIRLEY, *Satellite Tracking Facilities: Their History and Operation* (New York, 1963).

VALLEY, SHEA L. (ed.), *Handbook of Geophysics and Space Environments*, U.S.A.F. Office of Aerospace Research (Hanscom Field, Mass., 1965).

VAN ALLEN, JAMES A. (ed.), *Scientific Uses of Earth Satellites* (Ann Arbor, 1958).

VAN DE HULST, H. C., DE JAGER, C., AND MOORE, A. F. (eds.), *Space Research II* (New York, 1961).

VEIS, G. (ed.), *The Use of Artificial Satellites for Geodesy* (New York, 1963).

WILLIAMS, BERYL, AND EPSTEIN, SAMUEL, *The Rocket Pioneers on the Road to Space* (New York, 1955).

APPENDIX J

Official Reports

ADAMS, JAMES L., *Space Technology*, Vol. II, "Spacecraft Mechanical Engineering," NASA SP-66 (Washington, 1965).

ALEXANDER, J. K., AND STONE, R. H., *A Satellite System for Radio-astronomical Measurements at Low Frequencies*, NASA TM X-55089 (Washington, 1964).

ASHBY, JOHN, "A Preliminary History of the Evolution of the Tiros Weather Satellite Program," NASA HBN-45.

AUCREMANNE, MARCEL, JR. (ed.), *The Ionosphere Beacon Satellite, S-45*, NASA TN D-695 (Washington, 1961).

Bell Laboratories: *Final Report on Bell Telephone Laboratories Experiments on Explorer XV*, NASA CR-67106 (New York, 1964).

BLUMLE, L. J., FITZENREITER, R. J., AND JACKSON, J. E., *The National Aeronautics and Space Administration Topside Sounder Program*, NASA TN D-1913 (Washington, 1963).

BOECKEL, JOHN H., *The Purpose of Environmental Testing for Scientific Satellites*, NASA TN D-1900 (Washington, D.C., 1963).

BOURDEAU, ROBERT E., ET AL., *The Ionosphere Direct Measurements Satellite Instrumentation (Explorer VIII)*, NASA TN D-414 (Washington, 1962).

CASPER, JONATHAN D., *History of Alouette: NASA Case-Study of An International Program*, NASA HHN-42, 1964, revised 1965.

COFFEE, CLAUDE W., BRESSETTE, WALTER E., AND KEATING, GERALD M., *Design of the NASA Lightweight Inflatable Satellite for the Determination of Atmospheric Density at Extreme Altitudes*, NASA TN D-1243 (Washington, 1962).

CORLISS, WILLIAM R., *The Evolution of the Manned Space Flight Network*, NASA GHN-4 (Greenbelt, Md., 1967).

———, *The Evolution of the Satellite Tracking and Data Acquisition Network (STADAN)*, NASA GHN-3, X-202-67-26 (Greenbelt, Md., 1967).

———, *Scientific Satellites*, NASA SP-133 (Washington, D.C., 1967).

CORTRIGHT, EDGAR M., *Unmanned Spacecraft of the United States* (Washington, 1964).

D'AIUTOLO, CHARLES T. (ed.), *The Micrometeoroid Satellite Explorer XIII (1961 Chi)*, NASA TN D-2468 (Washington, 1964).

FRANTA, ALLEN L., *Integrating Spacecraft Systems*, NASA TN D-3049 (Washington, 1966).

GIACCONI, R., *An X-Ray Telescope*, NASA CR-41 (Washington, 1965).

GODDARD, MRS. ROBERT H., "Account of Dr. Goddard's World 1917-18," in *Congressional Record*, September 9, 1959.

——— AND PENDRAY, G. EDWARD, "Biographical Data: Dr. Robert H. Goddard," reprinted in *Congressional Record*, May 6, 1960.

HABIB, E. J., KEIPERT, F. A., AND LEE, R. C., *Telemetry Processing for NASA Scientific Satellites*, NASA TN D-3411 (Washington, D.C., 1966).

HAYES, E. NELSON, *The Smithsonian's Satellite-Tracking Program: Its History and Organization*, in *Annual Report*, Smithsonian Institution (Washington, D.C., 1962).

HEPPNER, JAMES P., NESS, NORMAN F., SKILLMAN, THOMAS L., AND SCEARCE, CLELL S., *Goddard Space Flight Center Contributions to 1961 Kyoto Conference on Cosmic Rays and the Earth Storm* (Washington, D.C., 1961).

HESS, WILMOT N., MEAD, G. D., AND NAKADA, M. P., *Bibliography of Particles and Fields Research*, NASA X-640-65-37 (Greenbelt, Md., 1965).

LUDWIG, GEORGE H., *Particles and Fields Research in Space*, NASA SP-11 (Washington, D.C., 1962).

———, *The Orbiting Geophysical Observatories*, NASA TN D-2646 (Washington, D.C., 1963).

NASA, *Astronautics and Aeronautics, 1963: A Chronology on Science, Technology, and Policy*, SP-4004, prepared by the NASA Historical Staff (Washington, D.C., 1964).

———, *Goddard Projects Summary: Satellites and Sounding Rockets*, Goddard Space Flight Center (Greenbelt, Md.; published periodically).

———, *Ariel I, The First International Satellite, Experimental Results*, Goddard Space Flight Center (Washington, D.C., 1966).

———, *Ariel I, The First International Satellite*, NASA SP-43 (Washington, D.C., 1963).

———, *Goddard Space Flight Center Contributions to COSPAR Meeting, May 1962*, NASA TN D-1669 (Greenbelt, Md., 1962).

———, *Goddard Space Flight Center Contributions to COSPAR Meeting, June 1963*, G-545 (Washington, D.C., 1963).

———, *Orbiting Solar Observatory*, Goddard Space Flight Center, NASA SP-57 (Washington, D.C., 1965).

———, *Historical Sketch of NASA*, NASA EP-29, prepared by the NASA Historical Staff (Washington, D.C., 1965).

———, *Juno II Summary Project Report, Vol. I, Explorer VII Satellite. Vol. II, The S-46 Satellite*, NASA TN D-608 (Washington, 1961).

———, *Launch Vehicles of the National Launch Vehicle Program*, NASA SP-10 (Washington, D.C., 1962).

———, *Proceedings of the NASA-University Conference on the Science and Technology of Space Exploration*, NASA SP-11 (Washington, 1962).

———, *Semiannual Reports to Congress, October 1, 1958* and *October 1, 1959* (Washington, D.C., 1959 and 1960).

———, *Significant Achievements in Ionospheres and Radio Physics, 1958-1964*, NASA SP-95 (Washington, 1966).

———, *Significant Achievements in Particles and Fields, 1958-1964*, NASA SP-97 (Washington, 1966).

———, *Significant Achievements in Satellite Geodesy, 1958-1964*, NASA SP-94 (Washington, 1966).

———, *Significant Achievements in Satellite Meteorology, 1958-1964*, NASA SP-96 (Washington, 1966).

———, *Significant Achievements in Solar Physics, 1958-1964*, NASA SP-100 (Washington, 1966).

———, *Significant Achievements in Space Astronomy, 1958-1964*, NASA SP-91 (Washington, 1966).

———, *Space Measurements Survey, Instruments and Spacecraft, October 1957-March 1965*, ed. by Dr. Henry L. Richter, NASA SP-3028 (Washington, D.C., 1966).

———, *The Observatory Generation of Satellites*, NASA SP-30 (Washington, 1963).

———, *United States Space Science Program: Report to COSPAR, Sixth Meeting, Warsaw, Poland, June 1963* (Washington, D.C., 1963).

United States Space Science Program: Report to COSPAR, Seventh Meeting, Florence, Italy, May 1964 (Washington, D.C., 1964).

National Academy of Sciences-National Research Council, *United States Space Science Program: Report to COSPAR, Fifth Meeting, Washington, D.C., May 1962* (Washington, D.C., 1962).

National Research Council, *A Review of Space Research*, Publication 1079 (Washington, 1962).

———, *Proposed United States Program for the International Geophysical Year* (Washington, 1956).

———, *Space Research, Directions for the Future* (Washington, D.C., 1966).

NEW, JOHN C., *Achieving Satellite Reliability through Environmental Tests*, NASA TN D-1853 (Washington, 1963).

———, *Scientific Satellites and the Space Environment*, NASA TN D-1340 (Washington, D.C., 1962).

SOUTHWICK, A. B., "The Memorial Which Dr. Goddard Would Have Liked Best of All," Worcester *Evening Gazette*, May 9, 1958, reprinted in *Congressional Record*, September 9, 1959, p. A7904.

STAFFORD, WALTER H., AND CROFT, ROBERT M., *Artificial Earth Satellites and Successful Solar Probes, 1957-1960*, NASA TN D-601 (Washington, 1961).

STERHARDT, J. A., *NASA Sounding Rocket Program: Summary of Sounding Rocket Flights*, NASA X-721-66-515 (Greenbelt, Md., 1966).

TIMMINS, ALBERT R., AND ROSETTE, KENNETH L., *Experience in Thermal-Vacuum Testing Earth Satellites at Goddard Space Flight Center*, NASA TN D-1748 (Washington, D.C., 1963).

U.S. Congress, *NASA Authorization for Fiscal Year 1961—Part I*, 86th Congress, 2nd Session, Testimony by Homer E. Newell, Jr., Office of Space Flight Programs (Washington, D.C., 1961).

House, Committee on Science and Astronautics, *Aeronautical and Astronautical Events of 1961*, prepared by the NASA Historical Staff (Washington, D.C., 1962).

APPENDIX J

House, Committee on Science and Astronautics, *Astronautical and Aeronautical Events of 1962*, prepared by the NASA Historical Staff (Washington, D.C., 1963).

Senate, Committee on Aeronautical and Space Sciences, *Documents on International Aspects of the Exploration and Use of Outer Space, 1954–1963*, Staff Report (Washington, D.C., 1963).

Space Handbook: Astronautics and Its Applications (Washington, 1959).

Virginia Polytechnic Institute, *Conference on Artificial Satellites*, NASA CR–60131 (Blacksburg, Va., 1963).

Speeches

BOUSHEY, BRIG. GEN. HOMER A., "Vignettes of Dr. Robert H. Goddard," Address at Third Annual Goddard Memorial Dinner, February 17, 1960.

GLENNAN, DR. T. KEITH, "The Nation's Program in Space Exploration," Speech at the Economic Club, Worcester, Mass., February 15, 1960.

——, Speech, Science, Engineering and New Technology Committee, Oregon State Department of Planning and Development, Portland, Ore., October 12, 1960.

GOETT, DR. HARRY J., "Scientific Exploration of Space," Address to Franklin Institute, Philadelphia, Pa., May 9, 1962.

——, "Scientific Exploration of Space and Its Challenge to Education," Address at Centennial Convocation, Worcester Polytechnic Institute, Worcester, Mass., October 8, 1964.

JASTROW, ROBERT, "Results of Experiments in Space," 25th Wright Brothers Lecture (IAS), Washington, D.C., December 18, 1961.

WEBB, JAMES E., Address before the American Institute of Aeronautics and Astronautics, New York, October 21, 1963.

——, Address to Webb School Alumni Association Testimonial Dinner, Los Angeles, Calif., April 26, 1962.

Articles

ARNOLDY, R. L., HOFFMAN, R. A., AND WINCKLER, J. R., "Observations of the Van Allen Radiation Regions During August and September 1959," Part I, *Journal of Geophysical Research*, LXV (May 1960), 1361–1376.

AUGENSTEIN, B. W., "Scientific Satellite—Payload Considerations," RAND Corp., RM–1459, 1955.

BELLER, WILLIAM S., "New Delta May Prove Most Economical," *Missiles and Rockets*, XVII (Aug. 16, 1965), 24–29.

BOURDEAU, ROBERT E., "Ionospheric Research from Space Vehicles," *Space Science Reviews*, I (1962), 683–718.

——, "Research Within the Ionosphere," *Science*, CXLVIII (April 30, 1965), 585–594.

BOYD, R. L. F., "In Space: Instruments or Man?" *International Science and Technology*, No. 41 (May 1965), 65–70.

——, "Techniques for the Measurement of Extraterrestrial Soft X-Radiation," *Space Science Reviews*, IV (Feb. 1965), 35–90.

BURGESS, ERIC, "The Establishment and Use of Artificial Satellites," *Aeronautics*, XXI (Sept. 1949), 70–82.

CAHILL, LAURENCE J., "Magnetic Fields in Interplanetary Space," *Science*, CXLVII (Feb. 26, 1965), 991–1000.

——, "The Magnetosphere," *Scientific American*, CCXII (March 1965), 58–68.

CANNEY, H. E., AND ORDWAY, F. I., "The Uses of Artificial Satellite Vehicles," *Astronautica Acta*, II (1956), 147–179; (1957), 1–15.

CHAZY, J., "Sur les Satellites Artificiels de la Terre," *Comptes Rendus*, CCXXV (Sept. 22, 1947), 469.

CLARKE, ARTHUR C., "Extraterrestrial Relays," *Wireless World* (Oct. 1945).

COLEMAN, P. J., JR., SONETT, C. P., JUDGE, D. L., AND SMITH, E. J., "Some Preliminary Results of the Pioneer V Magnetometer Experiment," *Journal of Geophysical Research*, LXV (June 1960), 1856–1857.

CROSS, C. A., "The Fundamental Basis of Power Generation in a Satellite Vehicle,"

Journal of the British Interplanetary Society, XI (1952), 117–125.

DICKE, R. H., AND PEEBLES, P. J., "Gravitation and Space Science," *Space Science Reviews*, IV (June 1965), 419–460.

EDSON, J. B., AND SNODGRASS, R. J., "Prelude to Missilry," *Ordnance*, XLIII (July–Aug. 1958), 67–70.

EHRICKE, KRAFFT A., "The Satelloid," *Astronautica Acta*, II (1956), 63–100.

EHRLICH, EUGENE, "NASA Particles and Fields Spacecraft," AIAA Paper 64-337 (1964).

EMME, EUGENE M., "Yesterday's Dream . . . Today's Reality; A Biographical Sketch of the American Rocket Pioneer, Dr. Robert H. Goddard," *The Airpower Historian*, V (Oct. 1960), 216–221.

FAN, C. Y., MEYER, PETER, AND SIMPSON, J. A., "Experiments in the Eleven-Year Change of Cosmic-Ray Intensity Using a Space Probe," *Physical Review Letters*, V (Sept. 1960), 272–274.

FINDLAY, JOHN W., "Radio Astronomy from Space Vehicles," *Astronautics and Aeronautics*, IV (Oct. 1966), 10–14.

FRIEDMAN, HERBERT, "The Next 20 Years of Space Science," *Astronautics and Aeronautics*, III (Nov. 1965), 40–47.

——, "X-Ray Astronomy," *Scientific American*, CCX (June 1964), 36–45.

GATLAND, K. W., KUNESCH, A. M., AND DIXON, A. E., "Minimum Satellite Vehicles," *Journal of the British Interplanetary Society*, X (1951), 287.

GODDARD, ROBERT H., "An Autobiography," *Astronautics*, IV (April 1959), 24 ff.

——, "A Method of Reaching Extreme Altitudes," Smithsonian Miscellaneous Publication No. 2540 (1919), reprinted by the American Rocket Society, 1946.

——, "Liquid-Propellant Rocket Development," Smithsonian Miscellaneous Publication No. 3381 (March 1936), reprinted by the American Rocket Society, 1946, and in *The Air Power Historian*, V (July 1958), 152–160.

GOLDMAN, D. T., AND SINGER, S. F., "Studies of a Minimum Orbital Unmanned Satellite of the Earth (MOUSE), Part III," *Astronautica Acta*, III (1957), 110–129.

HABER, HEINZ, "Space Satellites, Tools of Earth Research," *National Geographic Magazine*, CIX (April 1956), 487–509.

HAGEN, JOHN P., "The Viking and the Vanguard," in Emme, Eugene M. (ed.), *The History of Rocket Technology* (1964), 122–141.

HAGERMANN, E. R., "Goddard and His Early Rockets," *Journal of the Astronautical Sciences*, VIII (Summer 1961), 51–59.

HALL, R. CARGILL, "Early U.S. Satellite Proposals," in Emme, Eugene M. (ed.), *The History of Rocket Technology* (1964), 67–93.

——, "Origins and Development of the Vanguard and Explorer Satellite Programs," *The Air Power Historian*, XI (Oct. 1964), 101–112.

HEPPNER, J. P., "The World Magnetic Survey," *Space Science Reviews*, II (1963), 315–354.

HINES, COLIN O., "Sounding Rocket Resurgence," *Astronautics and Aeronautics*, IV (1966), 8–13.

——, "The Magnetopause: A New Frontier in Space," *Science*, CXLI (July 12, 1963), 130–136.

HINTEREGGER, H. E., "Absolute Intensity Measurements in the Extreme Ultraviolet Spectrum of Solar Radiation," *Space Science Reviews*, IV (June 1965), 461–497.

HOOVER, GEORGE W., "Instrumentation for Space Vehicles," American Rocket Society Paper 157-54 (1954).

JASTROW, ROBERT, AND CAMERON, A. G. W., "Space: Highlights of Recent Research," *Science*, CXLV (Sept. 11, 1964), 1129–1139.

KALLMANN, H. K., AND KELLOGG, W. W., "Scientific Uses of an Earth Satellite," RAND Corp., RM-1500 (1955).

KRULL, A. R., "A History of the Artificial Satellite," *Jet Propulsion*, XXVI (May 1956), 369–383.

KUPPERIAN, JAMES E., AND ZEIMER, ROBERT R., "Satellite Astronomy," *International Science and Technology* (March 1962), 48–56.

LAGOW, H. E., "Instrumenting Unmanned Satellites," American Rocket Society Paper 281-55 (1955).

LEHMAN, M., "The Strange Story of Doctor Goddard," *Reader's Digest*, LXVII (Nov. 1955), 147–152.

APPENDIX J

LEY, WILLY, "The Satellite Rocket," *The Technology Review*, LII (Dec. 1949), 93.

LUDWIG, GEORGE H., "Cosmic-Ray Instrumentation in the First U.S. Earth Satellite," *Reviews of Scientific Instruments*, XXX (April 1959), 223.

MALINA, FRANK J., "Origins and First Decade of the Jet Propulsion Laboratory," in Emme, Eugene M. (ed.), *The History of Rocket Technology* (1964), 63-65.

MAXWELL, W. R., "Some Aspects of the Origins and Early Development of Astronautics," *Journal of the British Interplanetary Society*, XVIII (1962), 415-425.

MAYO-WELLS, WILFRID J., "The Origins of Space Telemetry," *Technology and Culture*, IV (Fall 1963), 499-514.

MCGUIRE, JAMES B., SPANGLER, EUGENE R., AND WONG, LEM, "The Size of the Solar System," *Scientific American*, CCIV (April 1961), 64-72.

NESS, NORMAN F., "Earth's Magnetic Field: A New Look," *Science*, CLI (March 4, 1966), 1041-1052.

———, "The Earth's Magnetic Tail," *Journal of Geophysical Research*, LXX (July 1, 1965), 2989-3005.

NEWELL, HOMER E., "The Satellite Project," *Scientific American*, CXCIII (Dec. 1955), 29-33.

NEWTON, ROBERT R., "Geodesy by Satellite," *Science*, CXLIV (May 15, 1964), 803-808.

OBERTH, HERMANN, "From My Life," *Astronautics*, IV (June 1959), 38-39, 100 f.

O'BRIEN, BRIAN J., "Review of Studies of Trapped Radiation with Satellite-Borne Apparatus," *Space Science Reviews*, I (1962), 415-484.

ORDWAY, FREDERICK I., "Project Vanguard—Earth Satellite Vehicle Program," *Astronautica Acta*, III (1957), 67-86.

PENDRAY, G. EDWARD, "Pioneer Rocket Development in the United States," *Technology and Culture*, IV (Fall 1963), 384-392.

PIERCE, JOHN R., "Satellite Science and Technology," *Science*, CXLI (July 19, 1963), 237-244.

RAND Corp., "Preliminary Design of An Experimental World-Circling Spaceship," (May 1946).

ROGERSON, JOHN B., "The Orbiting Astronomical Observatories," *Space Science Reviews*, II (1963), 621-652.

SCHUESSLER, RAYMOND, "How America Muffed Space Supremacy," *American Mercury*, XC (May 1960), 25-30.

SINGER, S. F., "A Minimum Orbital Instrumented Satellite—Now," *Journal of the British Interplanetary Society*, XIII (1954), 74-79.

———, "Research in the Upper Atmosphere with Sounding Rockets and Earth Satellite Vehicles," *Journal of the British Interplanetary Society*, XI (1952), 61-73.

———, "Studies of a Minimum Orbital Unmanned Satellite of the Earth (MOUSE)," *Astronautica Acta*, I (1955), 171-184; and II (1956), 125-144.

SPITZER, LYMAN, JR., "The Beginnings and Future of Space Astronomy," *American Scientist*, L (Sept. 1962), 473-484.

STAMBLER, IRWIN, "The Explorers," *Space/Aeronautics*, XLI (Feb. 1964), 38-46.

———, "The Orbiting Observatories," *Space/Aeronautics*, XLII (Sept. 1964), 34-42.

———, "The OGO," *Space/Aeronautics*, XXXIX (Feb. 1963), 70-76.

STONE, ROBERT G., "RAE—1500-ft. Antenna Satellite," *Astronautics and Aeronautics*, III (March 1965), 46-49.

STUHLINGER, ERNST, "Army Activities in Space—A History," *Transactions of the IRE*, MIL-4 (April-July 1960), 64-69.

THOMAS, J. O., "Canadian Satellites: The Topside Sounder Alouette," *Science*, CXXXIX (Jan. 18, 1963), 229-232.

THOMAS, SHIRLEY, "Robert H. Goddard," *Men of Space*, Vol. I (1960).

———, "Harry J. Goett," *Men of Space*, Vol. VII (1965).

TOUSSEY, RICHARD, "The Extreme Ultraviolet Spectrum of the Sun," *Space Science Reviews*, II (1963), 3-69.

TRENT, N., "Early Days in Rocketry," *Christian Science Monitor* (July 24, 1963).

VON BRAUN, WERNHER, "The Explorers," *Astronautica Acta*, V (1959), 126-143.

———, "The Redstone, Jupiter, and Juno," in Emme, Eugene M. (ed.), *The History of Rocket Technology* (Detroit, 1954), 144-145.

WHIPPLE, FRED L., "Scientific Value of Artificial Satellites," *Journal of the Franklin Institute* (Aug. 1956).

INDEX

Aberdeen Proving Ground, Md., 203
ABMA. *See* Army Ballistic Missile Agency
Ad Hoc Carrier Committee, 218, 219
"Administration of Scientific Research by Federal Agencies" (Executive Order 10521), 204
Advanced Orbiting Solar Observatory. *See* AOSO
Advanced Research Projects Agency (ARPA), 30, 80, 81, 138 fn. 64, 207, 208, 282, 283, 284–285
Advent, Project, 234
Aerobee, 13, 16, 17, 125, 127, 204, 205, 227, 229 illus., 245, 247, 249, 315
    Applied Physics Laboratory, 15, 104
Aerobee 150, 127, 238, 242, 245, 247
Aerobee 150A, 35, 127, 217, 224, 226, 227, 238, 241, 247, 311
Aerobee 300, 127
Aerobee 300A, 127, 239
Aerobee-Hi, 125, 211, 213, 249
    University of Rochester's Institute for Optics, 213
Aerobee Jr., 238
Aerojet-General Corp., 14, 19, 204, 218
AFCRL. *See* USAF Cambridge Research Laboratories
Agena, 228, 231
    Lockheed Corp., 231
Air Force Ballistic Missile Division, 284
Air Force Long-Range Proving Ground, 204
Air Rescue Service, 226
Air Research and Development Command, 209
Alaska Data Acquisition Facility, 245
Albus, James S., 161, 221
Alexander, M., 159, 161, 163, 165
Allegany Ballistics Laboratory (Vanguard, Delta), 19, 218
Alouette Topside Sounder program (S–27), 215, 231, 236, 248
Alouette I, 110–112, 112 illus., 168–171, 227, 228, 236, 241
American Astronomical Society, 249
American Broadcasting Co. (ABC), 251
American Geophysical Union, 239
American Physical Society, 220

American Rocket Society, 15, 206
    Truax, Cdr. Robert, 15
American Telephone & Telegraph (AT&T), 109, 117, 132, 166, 174, 216, 224, 246, 247
American University, 240
Ames Research Center (ARC), 27, 29, 155, 163, 165, 171, 179, 211, 217, 281
"A Method of Reaching Extreme Altitudes," 7, 203
AMR (Atlantic Missile Range), 60, 107, 208, 213, 217, 219, 243, 250, 253, 295
Anacostia, D.C., 31
Anderson, K. A., 177
Andover, Maine, 224, 230, 247
Antigua, West Indies Federation, 65
Antofagasta, Chile, 65, 66, 218
AOMC (Army Ordnance Missile Command), 208
AOSO (Advanced Orbiting Solar Observatory), 226, 234, 249
    Cervenka, A. J., 234
    Republic Aviation Corp., 249
APL. *See* Applied Physics Laboratory
Apollo, 132, 251
Applied Physics Laboratory (APL), 204
Applied Sciences Laboratory, 57
Argentine Comisión Nacional de Investigaciónes Espaciales (CNIE), 215
Argo D–4 rocket (Javelin), 218, 219, 241, 248, 315
Argo D–8 (Journeyman), 315
Ariel I (UK–1), 35, 106, 106 illus., 124, 166–167, 222, 223, 224, 227, 236, 254
Arking, Albert, 60 illus.
Army Ballistic Missile Agency (ABMA), 19, 30, 206, 208, 209, 284–285
Army-Navy Research Development Board, 204
Army Ordnance, 203, 204
    Rocket Development Branch, 204
Army Ordnance Missile Command (AOMC), 208
Army Signal Corps, 30
    Research and Development Laboratory, 30
ARPA. *See* Advanced Research Projects Agency

ARS. *See* Astronautical Research and Development Agency
Arthur Venneri Co., 55, 56
AST (Aerospace Technologist), 51
Astrobee "1500," 238
Astronautical Research and Development Agency (ARS), 206
Astronautics Engineer Achievement Award, 221
  Stroud, William G., 221
AT&T. *See* American Telephone & Telegraph
Atchison, Joseph Anthony, 33, 34, 34 illus., 311
Atlantic Missile Range. *See* AMR
Atlas, 14, 16, 81, 231, 251
Atlas-Agena B, 231
Atomic Energy Commission (AEC), 206, 232
Auburn, Mass., 8, 203, 330
Automatic picture transmission (APT), 253
Azikiwe, Governor General Dr. Nnamdi, 244, 244 illus.

Bacon, Francis, 145
Bader, M., 163, 171
Baird, L. I., 288
Balewa, Sir Abubaker Tafawa, 118, 245
Ball Brothers, 47
Banana River, Fla., 204
Barnes, Richard, 233
Bartol Research, 157
Batchelder, R. W., 288
Bauer, Dr. S. J., 164–165, 241
Baumann, Robert C., 31, 166
Beall, Sen. J. Glenn, 280
Behring, W., 165
Bell Telephone Laboratories, Inc., 47, 69, 89, 155, 159, 167, 171, 205, 229
Belrose, J. S., 169
Beltsville Agricultural Research Center, 28, 31, 309
Beltsville, Md., 54, 57, 72, 290
Beltsville Space Center. *See* Goddard Space Flight Center
Bendix Corp., 20, 47, 69, 224
Berg, O., 161
Berg, Cdr. W. E., 18 illus.
Bermuda, 245, 247, 250
Best, L., 298
Beta II. See Vanguard I
"Big Shot" (*see also* Echo), 226
Blagonravov, Anatoly A., 236
Bloemfontein, South Africa, 234
Bloom, S., 165
Blossom Point, Md., 66, 254
Blumle, L., 169
Bourdeau, Robert E., 134 illus., 158–159, 165, 166, 177, 228
Boyd, R. L. F., 167
Boyden Observatory, 234

Brace, L., 175
Bridge, H. S., 161, 177
Bridger, J. M., 18 illus.
Bristol, England, 240
British National Committee on Space Research, 106
Bronk, Dr., Detlev W., 32, 214, 310–311
Brooklyn Polytechnic Institute, 59
Brooks, Rep. Overton, 31, 32, 311
Brown, W., 167, 171
Buckley, Edmond C., 221
Budget and Accounting Procedures Act of 1950, 283
Bureau of the Budget, 31, 206, 282, 283
Burns & Roe, Inc., 69
Butler, H. I., 158
Butler, Paul, 162, 176

Cahill, Dr. L., 163, 171
Cahill, William, 49
Cain, J., 159
California Institute of Technology (*see also* JPL), 14, 208
Calvert, W., 169
Camp Devens, Mass., 8
Canadian Black Brant sounding rocket, 224, 242
Canadian Defence Research Board, 110, 227, 231
Canadian Defence Research Telecommunications Establishment (DRTE), 112, 155, 169, 209
Canadian Department of Transport, 245
Canary Islands, 111 illus., 250
Canberra, Australia, 235, 236
Canney, H. E., Jr., 228
Cape Canaveral, 19, 116, 122, 127, 204, 216, 220, 221, 222, 223, 224, 226, 227, 231, 234, 238, 240, 241, 250
Cape Kennedy, 251
Carnarvon, Australia, 236, 249, 250
Case Institute of Technology, 284
CDA. *See* Command Data Acquisition
Centaur, 29, 251, 295
Central Flight Control and Range Operations Laboratory, 55
Central Radio Propagation Laboratory, 155, 169
Cerenkov detector, 98, 104
Cervenka, A. J., 234
Chubb, T., 157
Churchill Research Range, 242
Churchill, Sir Winston, 238, 239 illus.
City College of New York, 59
Clark, G., 163
Clark, Dr. John F., ix, 134, 134 illus.
Clark University, 6, 7, 11, 310

INDEX

Cleveland, Ohio, 346
CNES. *See* French National Center for Space Studies
Coleman, P. J., 157
Columbia University, 59
Command, 68, 69, 70
Command Data Acquisition (CDA), 246
"Commercial Applications of Space Communications Systems" report, 218
Committee on Space Research. *See* COSPAR
Committee on Special Capabilities, 16
    Stewart, Homer J., 16
Communications, 70, 72, 73, 313, 314–315
Communications Line Terminations (CLT), 253
Communications Satellite Act of 1962, 110, 223, 317–318
    Kerr, Sen. Robert S., 220
    Miller, Rep. George, 220
Communications Satellite Corp., 31, 132, 234
Congress of Quantitative Electronics, 332
Congressional Medal (*see also* R. H. Goddard honors), 214, 310
Construction Engineering, 56
Cooper, Gordon, 109, 110, 240, 246
COSPAR (Committee on Space Research), 106, 208, 223
Covington, Ozro M., 226
Cowan, R. C., 289, 298

Daniel and Florence Guggenheim Foundation, 8
Darcey, R. J., 172, 174
Darwin, Australia, 236
Data processing, 77, 78
D'Aiutola, C. T., 162
Davis, L., 163, 171
D.C. Council of Engineering and Architectural Society, 234
Defence Research Telecommunications Establishment. *See* Canadian Defence Research Telecommunications Establishment
De Gaulle, Gen. Charles, 241
Delta, 25, 29, 60, 63, 121–122, 123, 124 illus., 209, 218, 219, 227, 231, 234, 315
    Aerojet (*see also* Allegany Ballistics Laboratory, Douglas Aircraft), 218
    Schindler, William, 223, 235 illus.
Department of the Air Force, 282, 283, 285
Department of Defense, 15, 16, 17, 18, 65, 80, 205, 206, 207, 208, 209, 214, 215, 218, 222, 232, 234, 282, 283, 284–285, 317, 335, 338
    Committee on Special Capabilities, 16
Department of Scientific and Industrial Research, England, 155, 169
Department of the Navy, 282
Desai, U., 171
Digital Solar Aspect Sensor, 221

Discoverer I, 81, 208
Discoverer II, 82, 209
Discoverer-Thor, 81
Distinguished Service Award of Prince Georges County, 222
    Sisco, Bernard, 222
District of Columbia Business Corporation, 318
District of Columbia Incorporation Act, 318
DOD. *See* Department of Defense
Donegan, James, 74 illus.
Donley, J., 159, 165
Doppler, 152
Dossin, Dr. Francois V., 243, illus., 245
Douglas Aircraft Co., 121, 209, 218
Dowgiallo, Father Victor J., 311
DRTE. *See* Canadian Defence Research Telecommunications Establishment
Druyvesteyn, 107
Dryden, Dr. Hugh L., 28, 32, 208, 214, 231, 233, 235, 236
Dyke, W., 157
Dynamic test chambers, 56
Dytrt, M. C., 289

"Early Bird," 118, 132
Eastern Pyrenees Department, 231
East Grand Forks, Minn., 66
Echo, 90 illus., 123, 211, 226, 227, 286, 293, 294
    Schjeldahl Co., 215
Echo I, 89–91, 93, 95, 110, 123, 158–159, 211, 215, 217, 218, 221, 222, 226, 227, 315
    Results, 90–91
Echo II, 215, 221, 252
Egan, P. M., 289
EGO (Eccentric Geophysical Observatory), 223, 232
Einstein, Albert, 152
Eisenhower, Dwight D., 24, 81, 204, 206, 207, 208, 213, 283
Electric-field meter (*Explorer VIII*), 91, 92, 93
Electron Density Profile Probe (EDPP). *See* P–21 and P–21a
Elliot, H., 167
Energetic particle counter, 111
Energetic Particles Satellite (*Explorer XIV*, which see), 112, 113 illus.
Engineers, Scientists, and Architects Day, 231
EOGO (Eccentric Orbiting Geophysical Observatory). *See* EGO
Esselen Park, Union of South Africa, 65, 66
Executive Order 10521, 204
Explorer I, 22, 80, 206, 207, 207 illus., 220
Explorer III, 80, 206, 207
Explorer IV, 207
Explorer VI (S–2), 79, 82, 83, 83 illus., 113, 156–157, 209, 249
    University of Chicago, 82, 83, 157
    University of Minnesota, 82, 83, 157

343

Explorer VII, 85, 156–159, 211, 219, 315
Explorer VIII (S–30), 35, 91, 92, 93, 158–161, 213, 215, 311, 315
Explorer IX (S–56a), 95, 160–161, 214, 220, 221, 223
Explorer X (P–14), 35, 95, 96, 96 illus., 97, 160–161, 215, 218, 219, 220, 311
    Rossi, Dr. Bruno, 161, 220
Explorer XI (S–15; Gamma-ray Astronomy Satellite), 97, 98, 162–163, 215
Explorer XII (S–3), 100, 100 illus., 102, 113, 114, 123, 162–163, 217, 218, 219, 221, 236, 311
Explorer XIII (S–55a), 102, 162–165, 218
Explorer XIV (S–3a), 112, 113, 113 illus., 123, 170–171, 227, 228, 230, 231, 236, 245, 247, 248
    Marcotte, Paul G., 170, 231
Explorer XV (S–3b), 114, 123, 170–171, 227, 228, 236
Explorer XVI (S–55b), 124 illus., 229, 233
Explorer XVII (S–6), 115, 116, 117 illus., 123, 174–175, 238, 239
Explorer XVIII (IMP), 118, 119, 120, 123, 124, 176–179, 232, 251, 273
Explorer Satellites, 123, 131

Fairbanks, Alaska, 66, 67, 67 illus., 219, 223, 246
    University of Alaska, 67, 223
Fan, C. Y., 157
Fanfani, Premier Amintore, 237 illus.
Farley, T. A., 157
Fazio, G., 165
FCC. *See* Federal Communications Commission
Federal Civil Servant of the Year-State of Maryland Award, 222
    Robert W. Hutchison, 222
Federal Communications Commission (FCC), 214, 215, 218, 222
Federal Women's Award, 238, 240
    Pressly, Eleanor G., 238, 240
    Roman, Nancy, 240
Field Projects Branch, 60
Financial Management Division, 58, 245
Fisher, R. J., 289
Fitzenreiter, R., 169
Flight Research Center, 27
Forbush decrease, 86, 87, 88
Fort Churchill, Canada, 125, 211, 228, 229, 233, 238, 245
Fort Dix, N.J., 247
Fort Monmouth, N.J., 30, 89
Fort Myers, Fla., 66
French National Assembly, 231
French National Center for Space Studies, 236
French VLF program, 249
    Storey, Dr. Owen, 249
Friedman, H., 157

Friel, Fred S., 313
Fritz, Dr. Sigmund, 219
Frost, K., 165
Fuhrman, Herbert S., 288, 292
Fucino, Italy, 230
Future for Science in Space, 267–277

Gagarin, Y., 43
Galileo, 151
Gamma-ray telescope *(see also Explorer XI)*, 98
Gaspé Peninsula, 95
Gegenschein effect, 120
Geiger counter *(Explorer VI)*, 83
Geiger-Mueller tube *(Explorer VII)*, 85, 101
Geiger tubes *(Explorer VII, Pioneer V)*, 85, 87
Gemini, 132, 231
General Dynamics Astronautics (Atlas), 231
General Electric, 47, 220, 226, 246
Geneva, Switzerland, 209, 236
Geophysics Corp., of America, 219, 223, 241
Geronimo, 247
Gilmore Creek, 67
Gilpatric, Roswell *(see also* Department of Defense*)*, 214
Gilruth, Dr. Robert R., 29, 209, 292, 308
Glenn, John, 221
Glenn L. Martin Co. *See* Martin Co.
Glennan, Dr. T. Keith, 27, 29, 39, 65, 208, 209, 211, 281, 284, 300, 308
Goddard Institute for Space Studies, 59, 215, 221, 313
Goddard Memorial Scholarship Award, 240
Goddard, Mrs. Esther, v illus., vii, 8, 31, 32, 33, 34 illus., 310–311
Goddard, Dr. Robert H., vii, ix, 3 illus., 1–11, 10 illus., 11 illus., 29, 203, 204, 209, 212, 251, 267, 308, 310–311, 312
    "A Method of Reaching Extreme Altitudes," 7, 203
    Auburn, Mass., 8, 203
    Camp Devens, Mass., 8
    Clark University, 6, 7, 11, 310
    Liquid-fuel experimentation, 6–7, 203, 310, 312
    "Liquid Propellant Rocket Development," 204
    Roswell, N. Mex., 8, 203, 204
    Worcester Polytechnic Institute, 6
Goddard Scientific Satellite Symposium, 119, 236
    Ness, Dr. Norman F., 119, 120, 177
Goddard Space Flight Center:
    Administration, 45, 56, 57
    Communication, 73, 304
    Construction, 35, 53–59, 209, 211, 213, 215, 218, 219, 223, 226, 227, 232, 233, 239, 240, 243, 309, 314

INDEX

Dedication, 31–35, 32 illus., 33 illus., 34 illus., 214, 310–311, 312
  Baumann, Robert C., 31, 166
  Dowgiallo, Father Victor J., 311
  Wyatt, Rev. Kenneth B., 311
Division assignments, 286–289, 305–306, 308–309, 312–313
Educational opportunities, 51–52
Establishment, 280, 281, 282, 290, 312
Fabrication, 61, 293, 303
Functions, 28, 292–296, 299–300, 301–307, 335–339
Funding, 43–44, 243–244, 280, 282, 283, 285
Organization, 59, 292–296, 297, 300, 301–307, 308, 312–313
Organizational relationships, 300, 304, 339–340
Personnel, 35, 49–52, 295, 297–298, 312–314
Physical plant, 35, 53–57, 314
Procurement, 61
Research, 43, 293, 303
Goddard Space Flight Center Colloquium, 238
  O'Keefe, John A., 238
Goddard Space Flight Center and satellite space probe projects, 155–179
Goddard Space Flight Center Symposium on the Physics of Solar Flares, 249
Goett, Dr. Harry, J., ix, 29, 31, 34 illus., 131, 133 illus., 211, 232, 235 illus., 253, 267–277, 312
Goldstone, Calif., 66, 72 illus., 89, 214, 214 illus.
Goonhilly, England, 114, 230
Grand Central Rocket Co., 19
Gray, Robert H., 288
Great radiation belt (*see also* Van Allen belt), 149
Greenbelt, Md., 39, 241, 249, 250, 280, 299, 308, 310–311, 314
Groetzinger, G., 157
Ground Communications Network, 72
Grumman Aircraft Corp., 47
Guam, 127
Guaymas, Mexico, 250
Guggenheim, Daniel, 8
Guggenheim, Harry, 8, 204
Gulf of Guinea, 247
Gulf of St. Lawrence, 95
Gummel, H. K., 171

Hagen, Dr. John P., 17, 18, 18 illus., 284, 288, 292, 297
Hagerty, James C., 282
Hagg, E. L., 169
Hallam, K., 165
Hanel, R., 161, 163, 165
Hartesbeesthoek, South Africa, 249

Hartz, T. R., 169
Harvard College Observatory, 234
  Menzel, Donald H., 234
Hawaii, 209
Helios. *See* AOSO
Heppner, Dr. J. P., 157, 160–161
Herring, Jackson, 60 illus.
Hess, Dr. Wilmot, 165, 170
Hickman, Dr. Clarence N., 7
Hirao, Dr. Kunio, 223 illus.
Hispaniola, 247
Hodgson, Alfred S., 292, 298
Holloman Air Force Base, 204
Holmdel, N.J., 89
Horowitz, R., 175
House Committee on Science and Astronautics, 218, 221, 234, 238
House joint resolution, 251
Houston, Tex., 41, 59
Hughes Aircraft Co., 47, 216, 224, 226, 240
Humphreys & Harding, Inc., 55
Hunter, C., 179
Huntsville, Ala., 208
Hutchison, Robert W., 222

Ibaraki Prefecture, Japan, 251, 252 illus.
IMP I (Interplanetary Monitoring Platform). *See Explorer XVIII*
Imperial College, London (P–21a), 107, 167
Indian Department of Atomic Energy, 231
Industrial Engineering Corp., 237
Infrared sensor (Tiros), 93, 110
Ingomish, Nova Scotia, 229
Injun III, 236
Institute of World Affairs, 211
Instrument Construction and Installation Laboratory, 55
Intercontinental ballistic missile (ICBM), 14, 18
Intermediate range ballistic missile (IRBM), 121
International Business Machines Corp., 47, 69, 72, 75
International Conference on the Ionosphere, 224
International Council of Scientific Unions, 208
International Geophysical Year (IGY) 15, 16, 18, 19, 22, 43, 65, 66, 77, 80, 121, 127, 147, 205, 206, 208, 211, 235
International Geophysics Bulletin, 232
International Ionosphere Satellite. *See* Ariel I
International Scientific Radio Union, 16
International Telecommunication Union, 209
International Telephone & Telegraph (ITT), 227
International Union of Geodesy and Geophysics, 16
International Year of the Quiet Sun. *See* IQSY

Interplanetary Monitoring Platform *(IMP I)*.
  See *Explorer XVIII*
Ion trap *(Explorer VIII)*, 93
IQSY (International Quiet Sun Year), 232, 234, 235, 241
Iris rocket, 192, 311, 315
Italian Space Commission, 226, 240, 254
Italy, 217 illus.

Jackson Building, 57
Jackson, H. G., 289
Jackson, John E., 164–165, 168–169, 241
Jaffe, Leonard, 231
Japan, 222, 223, 245, 247, 251
Jastrow, Dr. Robert, 60, 60 illus., 216 illus., 288, 292, 297-298, 313
JATO (jet-assist takeoff), 9, 204
Javelin *(Argo D-4)*, 127, 128 illus., 211, 215, 315
Jenkins, T. E., 287, 289, 292, 296, 297–298, 301
Jet Propulsion Laboratory (JPL), 14, 22, 89, 90, 155, 159, 204, 206, 208, 209, 304, 315
Jodrell Bank Tracking Station, England, 212
Johns Hopkins University, 242
Johannesburg, South Africa, 229
Johnson, Lyndon B., 208, 226, 230 illus., 237, 246
Journeyman *(Argo D-8)*, 127, 315
JPL. *See* Jet Propulsion Laboratory
Judge, D. L., 157, 159
Jungquist, N. L., 10 illus.
Juno II, 125, 156. 158, 162, 206, 208, 211
Juno V, 29, 295
Jupiter C, 206

Kagoshima, Japan, 240
Kalmia Construction Co., Inc., 243
Kennedy, Mrs. John F., 230 illus.
Kennedy, President John F., 43, 110, 114, 118, 215, 230 illus., 231, 235, 238, 239, illus, 241, 244, 247, 251
Kerr, Sen. Robert S., 31, 219, 220, 310
King, J. W., 169
Kiruna Geophysical Observatory, 227
Kisk, A., 10 illus.
Knecht, R. W., 169
Kollsman Instrument Division, 220
Kraushaar, W., 163
Kreplin, R. W., 157
Kronogard rocket range, 243, 244
Kupperian, Dr. J. E., Jr., 162, 288

Lacklen, R. 298
Lagos Harbor, Nigeria, 231, 244, 245, 246, 247
LaGow, H. E., 134 illus., 156–159, 288
Lakehurst, N.J., 244, 245, 246, 247, 252
Langley Medal, 212

Langley Research Center, 27, 29, 30, 160, 162, 215, 218, 234, 254, 281, 290, 295, 305, 309, 315
Langmuir probe, 93, 107, 150, 222, 241, 247
Launch Operations Center. *See* Cape Canaveral
Launch Phase Simulator (LPS), 251
  Northrop Electronics, 251
Leavy, William A., 228
Lewis Flight Propulsion Laboratory, 281
Lewis Research Center, 27, 29, 281, 290, 291, 295, 305, 308
Lick Observatory, 215
Lima, Peru, 65, 66
Lincoln Laboratory (MIT), 69, 222
Lindbergh, Charles A., 8
Lindsay, Dr. John C., 156, 158, 164, 228, 236, 288
Linfield Research Institute, 157
"Liquid Propellant Rocket Development," 204
Little, C., 157
Litton Industries, 57
Lockheed Missiles & Space Co., 231
Lockwood, G. E., 169
Loki rocket, 205
Longanecker, G., 163
Ludwig, G., 157, 177
Lyman-alpha, 85, 107, 315

McCracken, C., 159, 161, 165
McCulloch, A., 175
McDiarmid, L. B., 169
McDonald, Dr. F. B., 162–163, 170–171, 176–177
McElroy, Neil, 65
McIlwain, C., 171–173
Mackey, R. J., 158
Madley, Jesse M., 248
Magnetic coil *(Tiros II)*, 93
Magnetometer *(Explorer VII, X, XII, SLU-5)*, 85, 95, 100, 150, 209
Malraux, Madame André, 230 illus.
Malraux, Minister André, 230 illus.
Management Services and Supply Division, 240
Manned Flight Network, 67–78
Manned Spacecraft Center (MSC), 41, 51, 59
Manned Space Flight Program, 31, 308
Manned Space Flight Support Division, 59, 247
Manring, E., 159
Mansur, C., 10 illus.
Mansur, L., 10 illus.
Marcotte, Paul G., 170, 231
Mariner, 231, 232
Mariner II, 85
Mars, 6, 150
Marshall Space Flight Center (MSFC), 221, 315
Martin Co., 16, 18, 19, 85, 157
Massachusetts Institute of Technology (MIT), 69, 98, 155, 161, 163, 177, 220, 222
  Rossi, Dr. Bruno, 161, 220

INDEX

Matthews, N. Whitney, 288, 313
Mayer, Xopher W., 248
Mazur, D. G., 18 illus., 134 illus., 288, 313
Mediterranean Sea, 217 illus., 267
Medrow, K. R., 288
Mengel, John T., 18 illus., 29, 133 illus., 134 illus., 209, 288, 308, 313
Menzel, Donald H., 234
Mercury (planet), 151
Mercury Control Center, 69, 74 illus.
Meredith, Leslie H., 288, 312
Meyer, P., 157
Micrometeoroid detector *(Explorer VII, VIII)*, 85, 92, 311
Miller, Rep. George, 220
Miller, N. P., 289
Millstone Hill, Mass., 222
Miner, Marcia S., 240
Minitrack, 20, 29, 65–66, 91, 103, 218, 254, 311
   Radio interferometer, 66, 140 fn. 76
Minitrack Network, 34, 65, 75, 91, 209
   See also:
   Antigua, West Indies Federation
   Antofagasta, Chile
   Blossom Point, Md.
   East Grand Forks, Minn.
   Esselen Park, Union of South Africa
   Fairbanks, Alaska
   Fort Monmouth, N.J.
   Goldstone, Calif.
   Lima, Peru
   Quito, Ecuador
   Santiago, Chile
   St. John's, Newfoundland
   Winkfield, England
   Woomera, Australia
   See also:
   Bell Telephone Laboratories
   Bendix Corp.
   Burns & Roe, Inc.
   International Business Machines Corp., Inc.
   Lincoln Laboratory
   Western Electric Co.
Minneapolis, Minn., 240
Minneapolis-Honeywell Corp., 56
Moffett Field, Calif., 211
"Mona Lisa," 230 illus.
Motorola, Inc., Military Electronics Division, 219
Muldrew, D. B., 169
Muraoka, Toshio, 223 illus.
Mylar, 89, 95, 102, 211, 214

NACA. *See* National Advisory Committee for Aeronautics
Nagy, A. P., 289, 298.

NASA (National Aeronautics and Space Administration), 23, 24, 25, 27, 28, 39, 40, 65, 66, 67, 121, 280, 281, 282, 284–285, 286, 292–293, 302–307, 312, 315
   Founding, 207, 208, 282, 283
NASA Exceptional Scientific Achievement Award, 228
NASA Group Achievement Award, 228, 234
NASA Medal for Outstanding Leadership, 228
NASA Office of International Programs, 233
National Academy of Neurology, 240
National Academy of Sciences, 16, 18, 32, 147, 205, 206, 209, 214, 235, 245, 310
   Bronk, Dr. Detlev W., 32, 214, 310, 311
   Dossin, Dr. Francois V., 243 illus., 245
   Technical Panel for the Earth Satellite Program, 205
National Academy of Sciences Board, 147
National Advisory Committee for Aeronautics (NACA), 27, 28, 30, 203, 206, 281, 284, 356
   Langley Laboratory, 30
National Aeronautics and Space Act, 24, 208, 220, 281, 282, 283, 284
National Aeronautics and Space Council, 215, 232, 282, 317
National Broadcasting Co. (NBC), 237, 251
National Bureau of Standards, 157
National Capital Award, 233
National Conference of the American Society for Public Administration, 222
National Meteorological Center, 218
National Oceanographic Data Center, 247
National Operational Meteorological Satellite System, 218
National Research Council, 209, 235
National Research Council, Canada, 155, 169
National Rocket Club, 237, 240
National Satellite Weather Center, 238
National Science Foundation, 15, 16, 205, 206, 216
National Science Teachers Association, 248
National Weather Satellite Center, 241, 246
Naval Research Laboratory, 14, 16, 17, 18, 20, 23, 27, 29, 39, 40, 44, 65, 84, 89, 91, 121, 155, 157, 159, 204, 205, 206, 284–285, 286–287, 295, 305, 309, 314
   Newell, Homer E., Jr., 18, 18 illus., 125
   Rocket Sonde Branch, 29, 204
   See also Vanguard launch vehicles, SLV
Navy Bureau of Aeronautics, 14, 204
Nelms, G. L. B., 169
Neptune, 14
NERV. *See* Nuclear Emulsion Recovery Vehicle
Ness, Dr. Norman F., 119, 120, 177
Neupert, W., 165
New York University, 59

Newell, Dr. Homer E., Jr., 18, 18 illus., 125
Newton, G., 175
Nigeria, 231, 244, 245, 246, 247
Nike, 223
Nike-Apache, 127, 128 illus., 223, 227, 228–229, 238, 240–241, 242
Nike-Asp, 125, 211, 224, 233, 315
Nike-Cajun, 125, 126 illus., 127, 128 illus., 211, 216, 219, 221, 222–224, 227, 228, 229, 240, 241, 243, 244, 251, 315
Nike-Deacon, 125, 127
Nimbus, 47, 67, 131, 218, 221, 223, 229, 234, 246, 268, 270, 272, 315
    General Electric Co., 47, 246
Norair Engineering Corp., 55–57, 233, 240, 309
Nordberg, W., 161, 163, 165
Northrop Electronics, 251
Nova rocket, 29, 215, 295
NRL. *See* Naval Research Laboratory
NSF. *See* National Science Foundation
Nuclear Emulsion Recovery Vehicle (NERV), 316
Nutley, N.J., 114, 229–230

Office of Naval Research, 205
O'Brien, Brian J., 163, 171
O'Keefe, John A., 238
Orbiter, Project, 16, 205
Orbiting Astronomical Observatory (OAO), 47, 226, 231
    General Electric Co., 226
    Grumman Aircraft Corp., 47
OGO (Orbiting Geophysical Observatory), 47, 231, 232
    Thompson Ramo Wooldridge Space Laboratories, 47
OSO I (Orbiting Solar Observatory, S–16), 47, 104–105, 104 illus., 123, 124, 164–165, 221, 222, 223, 224, 232, 234, 236, 244, 273
    Ball Brothers, 47

P–21, 103, 105, 164–165, 219
P–21a, 105, 164–167, 221, 238
Palewski, Gaston, 230
Paris International Air Show, 241
Pasteur, Louis, 32
Patents, 227
Paull, Stephen, 228
Peake, Harold J., 227
Pennsylvania State University, 157
Peterson, L., 165
Petrie, L. E., 169
Philippine Islands, 245
Photodiode *(Explorer VII)*, 101
Piccioni, Attilio, 226
Pieper, Dr. George F., 134 illus.
Pioneer, 232

Pioneer III, 82
Pioneer IV, 81, 209
Pioneer V, 47, 85–88, 86 illus., 158–159, 211, 315
    Achievements, 86–88
    Thompson Ramo Wooldridge Space Laboratories, 47
Piracci Construction Co., 57
Plasma probe *(Explorer X)*, 96–97
    Rossi, Dr. Bruno, 161, 220
Pleasant Pond, Maine, 243, 245
Pleumeur-Bodou, France, 114
PMR (Pacific Missile Range), 60, 110, 209, 295, 316
POGO (Polar Orbiting Geophysical Observatory), 221, 223, 232
Point Mugu, Calif., 204, 246
Pomerantz, M., 157
Pope John XXIII, 241
Pope Paul VI, 241
Porter, Richard W., 205
President's Science Advisory Committee, 24
Pressly, Eleanor C., 238, 240
Princeton University, 6, 59
Procurement Division, 240, 245
Project Mercury, 29, 34, 68–72, 132, 208–209, 215, 219, 221, 224, 240, 246, 292, 295, 308, 311
Pronton analyzer *(Explorer XII)*, 101
Public Law 85–325 (NASA), 284
Public Law 85–568 (NASA), 24, 281, 282, 283, 284
Public Law 87–624 (Communications Satellite Act), 110, 317–318

Quilon, India, 234
Quito, Ecuador, 22, 65–66

Radar tracking, 68–70
Radio Corp. of America (RCA), 47, 76, 215, 216, 237, 246
Radio frequency (RF) impedance prober *(Explorer VIII)*, 92
Radio frequency resonance probe, 247
Radio interferometer, 66, 139 fn. 76
Radio Research Laboratory, 247, 248
Rados, Robert, 162, 164, 166, 168, 174, 178
RAND, 14–15, 204
Range and range rate tracking, 219
    Motorola, Inc., Military Electronics Division, 219
Ranger, 231
RCA Space Environment Center, 47, 215, 216, 230
Reber, C., 175
Rebound, Project, 214, 234
Redstone, 14, 16, 68, 205, 208
Reid, C., 157

INDEX

Relay I, 47, 114, 115 illus., 123, 132, 170–173, 214, 219, 229–230, 231, 233, 236, 237, 238–239, 241, 246, 247, 249, 251, 252–253, 268, 269
    National Academy of Neurology, 240
    See Radio Corp. of America
Relay II, 47
Republic Aviation Corp., 232, 249
Rio de Janeiro, Brazil, 114, 231, 239, 246
Robert H. Goddard Memorial Library, 11
Robert Hutchings Goddard Day, 232–233
Robinson, G., 165
Rocket and Satellite Research Panel, 205, 206
Rocket Development Branch, 204
Rocketdyne MB–3 engine. *See* Thor
Rocket Sonde Branch (NRL), 29, 204
Rockoons, 125
Rohr Aircraft Corp., 219
Roman, Dr. Nancy C., 240
Rose, R. C., 169
Rosen, Allen, 157
Rosen, M. W., 18 illus.
Ross, W., 157
Rossi, Dr. Bruno, 161, 220
Roswell, N. Mex., 8, 203, 204
Rover, Project, 215
Rumford, Maine, 114

S–66 Ionosphere Beacon Satellite, 228, 233
Saltonstall, Sen. Leverett, 232
San Marco Satellite Program, 226–234, 240, 244, 254
    Piccioni, Attilio, 226
San Nicolas Island, Calif., 99, 127
Santiago, Chile, 65–66, 99, 229
"Satellite Communications Corporation," 219, 220
Satellite Systems Building, 56
Satellite Tracking and Data Acquisition Network (STADAN), 249, 251
Saturn (rocket), 208
Savedoff, M., 165
Sayers, James, 167, 224
SCAMA (Switching, Conferencing, and Monitoring Arrangement), 72
Scandinavian Committee for Satellite Telecommunications, 246
Scearce, C. S., 161
Scherb, F., 161
Schindler, William, 233, 235 illus.
Schirra, Walter M., Jr., 109, 110, 227, 231, 232 illus., 246
Schjeldahl Co., 215
Schroeder, Clarence A., 313
Schwed, P., 157
Score, Project, 81, 208
Scout, 63, 102, 103, 124, 124 illus., 160, 162, 164, 212–213, 218, 219, 226, 229, 236

Scriven, Brig. Gen. George P., 203
Seamans, Dr. Robert C., Jr., 34 illus., 234
Secretan, Luc, 163, 243 illus.
Seddon, J. C., 288
Senate Committee on Aeronautical and Space Sciences, 221, 310
Senate Committee on Commerce, 222, 233
Serbu, G. P., 159, 167, 177
Sergeant-Delta rocket, 211
Shotput vericle, 240
Siepert, Albert P., 291, 298
Silver Spring, Md., 27, 31, 57, 243
Silverstein, Dr. Abe, 28, 29, 33, 290, 292, 297–298, 307
Simpson, J. A., 159, 177
Simpson, T. A., 157
Singer, Dr. S. Fred, 238
Siry, Dr. J. W., 18 illus., 214, 288, 313
Sisco, Bernard, 298
Skillman, T. L., 161
SLV. *See* Vanguard launch vehicles
Smith, C. P., Jr., 166, 174
Smith, Charles V. L., 313
Smith, E. J., 157
Smith, Sen. Margaret Chase, 240
Smithsonian Astrophysical Observatory, 20
Smithsonian Institution, 7–11, 33, 203–204, 209, 212, 311
Sonett, C. P., 157
Sounding rockets, 79, 125, 295, 313–315
    Launches (*see also* Wallops Island, White Sands), 181–202
South Point, Hawaii, 249
Soviet Academy of Sciences, 236
Space Communications Laboratory, 251, 252 illus.
Space Data Acquisition Division, 59
Space Environment Center (RCA), 47, 215, 216, 230
Space Environment Simulator, 62, 62 illus.
Space exploration:
    Astronomy, 152, 315
    Atmosphere exploration, 147–148, 314–315
    Biological science, 153
    Electric and magnetic fields, 150, 315
    Energetic particles, 149, 315
    Gravitational fields, 150–152
    Ionospheres, 148–149, 315
Space Operations Control Center, 73
Space Projects Integration Office, 59
Space Science and Satellite Applications, 59–60, 209, 305, 308, 312
Space Science Laboratory, 55, 211
Space Sciences Division, 29, 286–287, 288, 290, 292, 297, 305, 308, 312
Space Technology Laboratories (STL), 82–84, 155, 157, 159, 315

Space Technology Magazine, 253
Space Tracking and Data Acquisition Network (STADAN), 249, 251
Spacecraft Test Facility, 61–63
Spaid, G. H., 165
Spencer, N. W., 174–175
Sperry Rand Corp. Univac Division, 250
Spitsbergen Islands, 82, 209
Sputnik I, 15, 17, 21–23, 43, 206
Sputnik III, 92
STADAN. *See* Satellite Tracking and Data Acquisition Network
Stampfl, R. A., 160
Stanford Research Institute, 238
Stanford University, 82
State University of Iowa, 85, 157, 163, 171, 205, 217
Stever, H. Guyford, 206
Stewart, Homer J., 16
St. Hugh's Catholic Church, 311
St. John's, Newfoundland, 66
Stokes' Theorem, 151
Storey, Dr. Owen, 249
Stroud, William G., 134 illus., 158, 221
Stroup, R. W., 288
Stump Neck, Md., 89
Sun sensor *(Explorer X)*, 96
Sunderlin, Wendell, 170
Suomi, V., 99, 157, 163, 165, 175
Supplemental Appropriations Act., 1955, 205
Surveyor, 232
Swedish Committee for Space Research, 244
Swenson, G., 157
Swept Frequency Topside Sounder. *See* Alouette
Synchronous Meteorological Satellite (SMS), 232
    Republic Aviation Corp., 232
Synchronous orbit:
    Darwin Mobile Station, 236
    Hughes Aircraft Co., 47, 216, 224, 226, 240
    Syncom, 47, 132, 217, 219, 224, 226, 231, 233, 268, 269
Syncom I, 115, 116 illus., 123, 172–173, 233, 234
Syncom II, 117, 118, 123, 132, 174–177, 243, 244–245, 246–247, 249, 251

Takeuchi, Ryuji, 251
TAVE (Thor-Agena vibration experiment), 228
Technical Information Division, 240
Technical Panel for the Earth Satellite Program, 205
Technical services, 59
Tegea Knoptik lens (Tiros satellites), 104, 224, 227
Telespazio, 230
Telstar, 132, 219, 232, 233, 268
Telstar I, 109–110, 110 illus., 123, 166–169, 224, 226, 228
Telstar II, 116, 123, 174–175, 240
Tepper, Morris, 234
Test and Evaluation Division, 59
Theoretical Division, 60, 286–287, 288, 290, 292, 297, 305, 308, 313
Theory and Analysis Staff, 59
Thompson Ramo Wooldridge Space Laboratories, 47, 155, 157, 159
Thor, 121, 218, 226, 228
Thor-Aable rocket, 85, 121–122, 156, 158, 211
Thor-Agena, 63, 168, 221, 228
Thor-Agena A, 209
Thor-Agena B, 227
Thor-Delta, 29, 158, 160, 162, 213, 215, 216–217, 220, 221, 222, 224, 227, 233, 238, 240, 295
Thor-Hustler, 208
Thor-Vanguard. *See* Delta
Thule, Greenland, 87
Thumba, India, 231, 251
Tiros (Television Infrared Observation Weather Satellite), 25, 131, 209, 216, 219, 234, 238, 239, 246, 268, 270, 271, 311, 315
Tiros I, 88–89, 93, 124, 158–159, 211, 222, 223, 315
Tiros II, 93–95, 99, 123, 160–161, 213–214, 216, 218, 219, 222, 315
Tiros III, 94–95, 99, 104, 123, 162–163, 216, 218, 219, 222
Tiros IV, 103–104, 123, 164–165, 220, 222, 224
Tiros V, 107–109, 108 illus., 123, 166–167, 224, 227, 234, 246
    Elgeet-lens camera, 224
Tiros VI, 123, 168–169, 227, 229, 234, 245, 246, 249
Tiros VII, 117, 123, 174–175, 241, 245, 247
Tiros VIII, 120, 123, 178–179, 253
Titan, 18
Tokyo, Japan, 247, 248, 251
    Radio Research Laboratory, 247, 248
Topside Sounder Program, 110, 149, 315
    See also Alouette, *Alouette I*
TOSS (Tiros Operational Satellite System), 120, 131
Townsend, Dr. John W., 29, 133 illus., 134 illus., 170, 209, 228, 253, 288, 292, 297–298, 308, 312
Tracking, 20, 71, 219, 235
Tracking and data acquisition, 221, 235
Tracking and Data Systems, 59, 209, 246, 305, 308, 313
Tracking and Telemetry Laboratory, 57, 227
Tracking Systems Division, 59, 246, 305, 313
Triple coincidence telescope *(Explorer VI)*, 83

INDEX

Truax, Cmdr. Robert, 15
Truman, President Harry S. 204
Tyuratum Range, Kazakhstan, U.S.S.R., 21

Ulmer, R., 298
Unified S-band, 251
United Kingdom, 106, 311
United Nations, 208, 315
    Committee on Peaceful Uses of Outer Space, 234
    General Assembly, 247
United States Congress, 9, 24–25, 31, 43, 206–207, 218, 251, 280, 282, 310, 317–318
United States Engineering & Constructors, Inc., 56
United States Scientific Satellite Program, 282, 284
United States Weather Bureau, 120, 216, 218, 222, 223, 224, 229, 234, 246
University College, London (P–21a), 107, 167
University of Alaska, 67, 157, 223
University of Birmingham (P–21a, *Ariel I*), 107, 167, 224
University of California, 165, 171, 173, 177
University of Chicago, 82, 83, 157, 159, 177
University of Illinois, 157
University of Michigan, 216
University of Minnesota, 82, 83, 157, 159, 165
University of New Hampshire, 163, 171, 216, 217, 248
University of Rochester, 165
University of Rochester's Institute of Optics, 213
University of Stockholm Institute of Meteorology, 227, 244
University of Wisconsin, 85, 99, 157, 163, 165, 175
UPI (United Press International), 231
Upper Atmosphere Rocket Research Panel, 44, 205
USAF (U.S. Air Force) Cambridge Research Laboratories (AFCRL), 155, 157, 159, 227, 243
USIA (U.S. Information Agency), 245
USNS *Coastal Sentry*, 69 illus., 250
USNS *Kingsport*, 231, 244, 245, 246, 247
USNS *Rose Knot Victory*, 69 illus., 250

V–1, 204
V–2, 9, 13–14, 17, 127, 204
Vaccaro, Dr. Michael J., 29, 31, 133 illus., 134 illus., 209, 308, 312
Valley Forge, Pa., 220
Van Allen, Dr. James, 113, 157, 171, 205, 206, 228, 236
Van Allen radiation belt, 80, 84, 98–102, 120, 149, 209, 219, 272, 275, 315–316

Vandenberg Air Force Base, 221, 227
Vanguard I, 18 illus., 22, 23 illus., 80, 95, 206, 209, 232, 311
Vanguard II, 81, 131, 208
Vanguard III, 24 illus., 84, 125, 156–157, 211, 215, 315
Vanguard Division (Goddard Space Flight Center), 28, 286–287, 288, 290, 292–296, 297–298, 308
    See also:
    Baird, L. I.
    Canney, H. E., Jr.
    Hagen, J. P.
    Matthews, N. W.
    Mazur, D. G.
    Mengel, J. T.
    Siry, J. W.
    Stroud, R. W.
Vanguard launch vehicles, 20–22, 23, 156, 205, 206, 207, 208, 209, 211, 212 illus.
    SLV–1, 23, 207
    SLV–2, 23, 207
    SLV–3, 23, 208
    SLV–4, 25
    SLV–5, 209
    SLV–7, 25, 211
    TV–0, 20, 205
    TV–1, 20, 205
    TV–2, 20, 21, 23, 206
    TV–3, 21, 22, 206
    TV–4, 22, 211
    TV–5, 207
    See also:
    Aerojet-General Corp.
    Allegany Ballistics Laboratory
    Grand Central Rocket Co.
    Martin Co.
Vanguard, Project, 16–25, 27, 28, 29, 30, 44, 77, 80–81, 84, 121–124, 205–206, 207, 208, 218, 282, 283, 284, 314
Van Zandt, T. E., 169
Vega, 29, 295
Venus (planet), 150, 151
Viking, 14, 16–18, 20–21, 125, 205
Villard, O., Jr., 157
Voice of America, 231, 245, 247
Voorhees, Walker, Smith, Smith & Haines, 53

WAC Corporal rocket, 14, 125
Waddel, Dr. Ramond, 170–171
Walcott, Dr. Charles D., 203
Wallops Island, Va., 27, 68, 99, 103, 127, 209, 211, 212, 213, 214, 215, 217, 218, 219, 221–224, 226, 227, 228, 229, 231, 233, 236, 238, 239, 240, 241, 244, 246, 247, 249, 251, 295
Walsh, Dr. J. P., 18 illus.
War Department, 9
Warren, E. S., 169

Washington Academy of Sciences, 234
Wasielewski, Eugene W., 47, 133 illus., 134 illus., 312
Waterman, Dr. Alan T., 15
Webb, James E., v illus., 31, 133, 214, 216, 222, 231, 240, 245, 251, 310
Weilheim, West Germany, 114
Welsh, Dr. Edward, 232
Western Electric Co. (Mercury), 47, 68–69, 209
Western Union, 75, 76, 219
Westford, Mass., 222
West German Post Office, 219
Whale, H., 165
Whelpley, H., 157
Whipple, E., 159, 165
White, H., 165
White, William, 165, 236
White Sands Missile Range, N. Mex. (WSMR), 13, 19, 127, 227

Whitehead, J. D., 241
Wichita Falls, Tex., 211
Winckler, J. R., 157, 159, 165
Winkfield, England, 66
Winkler, Leopold, 29, 133 illus., 313
Wolfe, John, 179
Woomera, Australia, 65, 66, 209, 214, 229
Worcester, Mass., 5
Worcester Polytechnic Institute, 6
World Flight Memorial, 311
WSMR. *See* White Sands Missile Range, N. Mex.
WSPG (White Sands Proving Ground), 204, 205
Wyatt, DeMarquis, 292, 298
Wyatt, Rev. Kenneth B., 311

Yale University, 59
Youth Science Congress, 248

The Author

ALFRED Rosenthal has been the Historian of the Goddard Space Flight Center of the National Aeronautics and Space Administration since 1962. He is responsible for documentation and preparation of historical monographs covering over 50 major satellite programs as well as the other important activities at the Goddard Center with regard to space science, tracking, and advanced technology. Mr. Rosenthal is also Deputy Public Affairs Officer and published a series on space-science-oriented mathematics developed in cooperation with the U.S. Office of Education.

Before joining NASA, Mr. Rosenthal was with the U.S. Army Corps of Engineers, preparing studies on civil works programs and military projects, including a series on the development of U.S. water resources. He attended Charles University in Prague, Czechoslovakia. During World War II he served with the 88th Division in Italy.

NASA Historical Publications

HISTORIES
- * Robert L. Rosholt, *An Administrative History of NASA, 1958–1963,* NASA SP–4101, 1966; for sale by Supt. of Documents ($4). (Management History Series)
- * Loyd S. Swenson, James M. Grimwood, and Charles C. Alexander, *This New Ocean: A History of Project Mercury,* NASA SP–4201, 1966; for sale by Supt. of Documents ($5.50). (Program History Series)

HISTORICAL STUDIES
- * *History of Rocket Technology,* edited by Eugene M. Emme, special issue of *Technology and Culture* (Fall 1963); augmented and published by Society for the History of Technology (Detroit: Wayne State Univ., 1964).
- * *Space Medicine in Project Mercury,* by Mae Mills Link, NASA, SP–4003, 1965; for sale by Supt. of Documents ($1).

CHRONOLOGIES AND SPECIAL STUDIES
- * *Aeronautics and Astronautics: An American Chronology of Science and Technology in the Exploration of Space, 1915–1960,* compiled by Eugene M. Emme, Washington: NASA, 1961 (out of print).
- * *Aeronautical and Astronautical Events of 1961,* published by the House Committee on Science and Astronautics, 1962 (out of print).
- * *Astronautical and Aeronautical Events of 1962,* published by the House Committee on Science and Astronautics, 1963; for sale by Supt. of Documents ($1).
- * *Astronautics and Aeronautics, 1963,* NASA SP–4004, 1964; for sale by Supt. of Documents ($1.75).
- * *Astronautics and Aeronautics, 1964,* NASA SP–4005, 1965; for sale by Supt. of Documents ($1.75).
- * *Astronautics and Aeronautics, 1965,* NASA SP–4006, 1966; for sale by Supt. of Documents ($2.25).
- * *Astronautics and Aeronautics, 1966,* NASA SP–4007, 1967; for sale by Supt. of Documents ($1.50).
- * *Astronautics and Aeronautics, 1967,* NASA SP–4008 (1968).
- * *Project Mercury: A Chronology,* by James M. Grimwood, NASA SP–4001, 1963; for sale by Supt. of Documents ($1.50).
- * *Historical Sketch of NASA,* NASA EP–29, 1965; for sale by Supt. of Documents ($0.25).
- * *Project Gemini Technology and Operations: A Chronology,* by James M. Grimwood and Barton C. Hacker, with Peter J. Vorzimmer, NASA SP–4002 (1968).
- * *The Apollo Spacecraft: A Chronology,* Vol. I, through November 7, 1962, by Ivan D. Ertel and Mary L. Morse, NASA SP–4009 (1968).

www.ingramcontent.com/pod-product-compliance
Lightning Source LLC
Chambersburg PA
CBHW081717170526
45167CB00009B/3606